Springer Series in **Materials Science** 7

Edited by Morton B. Panish

Springer Series in **Materials Science**

Editors: U. Gonser · A. Mooradian · K. A. Müller · M. B. Panish · H. Sakaki
Managing Editor: H. K. V. Lotsch

Dr. Marian A. Herman

Institute of Physics, Polish Academy of Sciences, al. Lotnikow 46
PL-02-668 Warszawa, Poland

Dr. Helmut Sitter

Institut für Experimentalphysik, Johannes-Kepler-Universität
A-4040 Linz/Auhof, Austria

Series Editors:

Prof. Dr. *K. A. Müller*

IBM, Zürich Research Lab.
CH-8803 Rüschlikon, Switzerland

Prof. Dr. *U. Gonser*

Fachbereich 12/1
Werkstoffwissenschaften
Universität des Saarlandes
D-6600 Saarbrücken, FRG

M. B. Panish, Ph. D.

AT&T Bell Laboratories,
600 Mountain Avenue,
Murray Hill, NJ 07974, USA

Prof. *H. Sakaki*

Institute of Industrial Science,
University of Tokyo,
7-22-1 Roppongi Minato-ku,
Tokyo 106, Japan

A. Mooradian, Ph. D.

Leader of the Quantum Electronics Group, MIT,
Lincoln Laboratory, P.O. Box 73,
Lexington, MA 02173, USA

Managing Editor:

Dr. Helmut K. V. Lotsch

Springer-Verlag, Tiergartenstrasse 17
D-6900 Heidelberg, Fed. Rep. of Germany

ISBN-13: 978-3-642-97100-6   e-ISBN-13: 978-3-642-97098-6
DOI: 10.1007/978-3-642-97098-6

Library of Congress Cataloging-in-Publication Data. Herman, Marian A. Molecular beam epitaxy : fundamentals and current status / Marian A. Herman, Helmut Sitter. p. cm.–(Springer series in materials science ;
7). Bibliography: p. Includes index. 1. Molecular beam epitaxy. I. Sitter, Helmut, 1951-. II. Title.
III. Series.   QC611.6.M64H47 1989   530.4'1–dc 19   88-29492

© Springer-Verlag Berlin Heidelberg 1989
Softcover reprint of the hardcover 1st edition 1989

2154/3150-543210 – Printed on acid-free paper

M. A. Herman   H. Sitter

Molecular Beam Epitaxy

Fundamentals and Current Status

With 249 Figures

Springer-Verlag Berlin Heidelberg New York
London Paris Tokyo

Preface

Since a molecular-beam apparatus was first successfully used by Cho and Arthur in the late 1960s to crystallize and investigate GaAs epilayers, high vacuum epitaxial growth techniques using particle beams have developed rapidly. This development accelerated when different semiconductor devices with quantum-well structures were invented in the 1970s. The important implementation of these structures in devices like quantum-well lasers, high electron mobility transistors or superlattice avalanche photodiodes gave added impetus to research work and to increasing production aims. Coincident with this development, original research papers and reviews devoted to problems concerning these growth techniques have also rapidly grown in number, and in addition they have become very diversified. At present several hundred original papers on this subject appear in the literature each year. However, in contrast to this there is a lack of comprehensive monographs comprising the whole variety of problems related to epitaxial growth of semiconductor films from atomic and molecular beams.

This book, which presents a review of the state of the art of molecular beam epitaxy (MBE), as applied to the growth of semiconductor films and multilayer structures, may serve the reader as a convenient general guide to the topics related to this crystallization technique. However, special emphasis has been placed on three themes: the technological equipment, the physics of the crystallization processes of thin films and device structures, and the characterization methods which allow the physical properties of the grown film or structure to be connected with the technological parameters of the crystallization procedure.

In the conventional molecular-beam-epitaxial growth technique, thin films of different compounds are crystallized via reactions between thermal-energy molecular or atomic beams of the constituent elements and the substrate maintained at an elevated temperature in ultrahigh vacuum. During the past decade this early MBE technique has developed into a variety of different growth techniques: gas source MBE, and its derivative using metalorganic compounds; phase-locked epitaxy; atomic layer epitaxy; and migration-enhanced epitaxy. All currently known modifications of the MBE growth technique are discussed in the text, together with ultrahigh vacuum growth and processing procedures using MBE-growth and focused ion beam processing facilities.

In consideration of the continuing stream of new information concerning different growth and characterization techniques, which not only provides additional data but also often modifies the interpretation of a particular concept or experimental result, we have attempted to be sufficiently fundamental in our treatment of the subject that the conclusions drawn here will not be superseded. We hope therefore that future publications occurring in the field will be easily understood by readers referring to the principles presented here.

Linz, March 1988   *M.A. Herman and H. Sitter*

Acknowledgements

The authors have benefited from the kind cooperation of many scientists actively working in research on MBE who have supplied them with an up-to-date collection of their papers on this subject. We express hereby our gratitude to:

- Zh.I. Alferov and P.S. Kop'ev; A.F. Ioffe Physico-Technical Institute, USSR Academy of Sciences, Leningrad, USSR
- T.G. Andersson; Chalmers University of Technology, Gothenburg, Sweden
- H. Berger; Central Institute of Electron Physics, Academy of Sciences of the GDR, Berlin, GDR
- A.Y. Cho, H.J. Gossmann, R.C. Miller, M.B. Panish and W.T. Tsang; AT&T Bell Laboratories, Murray Hill and Holmdel, USA
- J.P. Faurie; University of Illinois, Chicago, USA
- B.A. Joyce and P.K. Larsen; Philips Research Labs., Redhill, UK, and Eindhoven, The Netherlands
- E. Kasper; AEG Research Institute, Ulm, FRG
- M. Kawabe; University of Tsukuba, Ibaraki, Japan
- A. Madhukar; University of Southern California, Los Angeles, USA
- A. Million; Laboratoire Infrarouge, Grenoble, France
- J. Nishizawa; Tohoku University, Sendai, Japan
- K. Ploog; Max Planck Institute, Stuttgart, FRG
- M. Pessa; Tampere University of Technology, Tampere, Finland
- T. Sakamoto and T. Yao; Electrotechnical Laboratory, Ibaraki, Japan
- H. Sakaki; University of Tokyo, Tokyo, Japan
- J. Singh; University of Michigan, Ann Arbor, USA
- S.I. Stenin; Institute of Semiconductors, USSR Academy of Sciences, Novosibirsk, USSR
- K. Takahashi; Tokyo Institute of Technology, Tokyo, Japan
- W.I. Wang; IBM T.J. Watson Research Center, Yorktown Heights, USA
- G. Weimann; Forschungsinstitut der Deutschen Bundespost, Darmstadt, FRG

The updated technical data concerning the MBE equipment and the color prints included in the book were provided by VG-Semicon and ISA-Riber. This illustrative material helped us to increase the perspicuity of the current status of technological equipment.

Our sincere thanks are expressed to Prof. Dr. Helmut Heinrich, head of the Department of Solid State Physics of the Institute of Experimental Physics at the Johannes Kepler University, Linz, Austria, for encouraging discussions and for carefully reading the text. The possibility of working on this book in the well-organized and friendly atmosphere of the Department of Solid State Physics provided the additional advantage of access to a broad range of experts on many diverse subjects. Detailed discussions with a guest of the Department, Prof. Dr. J.M. Langer, were helpful concerning the optical characterization methods. Our sincere appreciation is extended to Mr. Krzysztof Marek Herman, who drew all the illustrations, and to Mr. Rafal Pietrak for computer typesetting one of the versions of the manuscript. The continued interest in the preparation of this book of the whole scientific staff, as well as the technical assistance of co-workers of the Department of Solid State Physics of the Institute of Experimental Physics at the Johannes Kepler University is gratefully acknowledged.

Last but not least we wish to thank the team of experts at Springer-Verlag, Heidelberg, supervised by Dr. H.K.V. Lotsch. The careful linguistic corrections made by Miss D. Hollis are especially appreciated.

This work was partly supported by the "Fond für Förderung der wissenschaftlichen Forschung in Österreich".

Contents

Part V Conclusion

1. Introduction

Molecular beam epitaxy (MBE) is a versatile technique for growing thin epitaxial structures made of semiconductors, metals or insulators [1.1–8]. In MBE, thin films crystallize via reactions between thermal-energy molecular or atomic beams of the constituent elements and a substrate surface which is maintained at an elevated temperature in ultrahigh vacuum. The composition of the grown epilayer and its doping level depend on the relative arrival rates of the constituent elements and dopants, which in turn depend on the evaporation rates of the appropriate sources. The growth rate of typically 1 μm/h (1 monolayer/s) is low enough that surface migration of the impinging species on the growing surface is ensured. Consequently, the surface of the grown film is very smooth. Simple mechanical shutters in front of the beam sources are used to interrupt the beam fluxes, i.e., to start and to stop the deposition and doping. Changes in composition and doping can thus be abrupt on an atomic scale.

What distinguishes MBE from previous vacuum deposition techniques is its significantly more precise control of the beam fluxes and growth conditions. Because of vacuum deposition, MBE growth is carried out under conditions far from thermodynamic equilibrium and is governed mainly by the kinetics of the surface processes occurring when the impinging beams react with the outermost atomic layers of the substrate crystal. This is in contrast to other epitaxial growth techniques, such as liquid phase epitaxy or vapor phase epitaxy, which proceed at conditions near thermodynamic equilibrium and are most frequently controlled by diffusion processes occurring in the crystallizing phase surrounding the substrate crystal.

In comparison to all other epitaxial growth techniques, MBE has an unique advantage. Being realized in an ultrahigh vacuum environment, it may be controlled in situ by surface sensitive diagnostic methods such as reflection high-energy electron diffraction (RHEED) [1.9–15], Auger electron spectroscopy (AES) [1.16–19] or ellipsometry [1.20–24]. These powerful facilities for control and analysis eliminate much of the guesswork in MBE, and enable the fabrication of sophisticated device structures using this growth technique.

The remaining sections of this chapter provide a general outline of MBE and briefly describe its particular features. They survey phenomena typical of high vacuum epitaxial growth from molecular beams and indicate the variety of present-day MBE growth techniques. The main body of the text, beginning in Chap. 2, is organized into three parts, dealing with technological equipment

(Chaps. 2–3), characterization methods (Chaps. 4–5), and MBE growth processes (Chaps. 6–7). These chapters include:

— A description of operating principles and construction design of different sources of atomic and molecular beams
— A survey of commercially manufactured MBE equipment
— An introduction to in-growth and post-growth characterization methods as applied to MBE-grown epitaxial structures
— A consistent presentation of fundamentals of the MBE growth process
— A review of material-related peculiarities relevant to the MBE technique.

The book concludes with a chapter in which the MBE growth technique is compared with related epitaxial growth techniques used in the manufacture of semiconductor devices. Perspectives for MBE are reviewed and forecasts made concerning its future development.

A list of references can be found towards the end of the book. Due to the almost daily publication of papers on this subject, it would not have been possible to avoid omissions in attempting to produce a comprehensive and complete list of references. Therefore, rather than trying to include all papers, enough representative references are given to permit the interested reader to start a library search on a particular topic.

1.1 Thin Film Growth from Beams in a High Vacuum Environment

The various techniques of growing thin films in high vacuum can be divided roughly into two categories depending on whether the species are transported physically or chemically from the source to the substrate.

In the physical transport techniques [referred to as physical deposition techniques (PDT)], the compound to be grown or its constituent elements are vaporized from polycrystalline or amorphous sources at high local temperatures, and subsequently transported through the vacuum reactor toward the substrate in the form of streams of vapor or as thermal-energy beams without any chemical change.

In the chemical transport techniques [referred to as chemical deposition techniques (CDT)] volatile species containing the constituent elements of the film to be grown are produced first in, or outside, the vacuum reactor, and then transported as streams of vapor or as beams through the reactor toward the reaction zone near the substrate. These gaseous species subsequently undergo chemical reactions in the reaction zone, or dissociate thermally, to form the reactants which participate in the growth of the film on the surface of the substrate crystal.

Both the above-mentioned deposition modes are used in present-day MBE techniques.

1.1.1 Vacuum Conditions for MBE

It is convenient to distinguish between two categories of vacuum in which thin film growth may be realized. For total residual gas pressures in the reactor in the interval $1.33 \times 10^{-1}\,\text{Pa} \geq p \geq 1.33 \times 10^{-7}\,\text{Pa}$ (10^{-3}–10^{-9} Torr) the term high vacuum (HV) will be used hereafter, while for pressures $p \leq 1.33 \times 10^{-7}\,\text{Pa}$ (10^{-9} Torr) the vacuum will be called an ultrahigh vacuum (UHV) [1.25].

The characteristic feature of all MBE techniques is the beam nature of the mass flow toward the substrate. From this point of view it is important to consider the indispensable vacuum conditions, i.e., the admissible value of the total pressure of the residual gas in the vacuum reactor, which have to be ensured in order to preserve the beam nature of the mass transport in the reactor.

Two parameters, closely related to pressure, are important for the characterization of the vacuum. The first is the mean free path of the gas molecules penetrating the vacuum, while the second is the concentration of the gas molecules (the number of molecules per unit volume).

The mean free path (denoted hereafter by L) is defined as the average distance (averaged over all of the molecules of the gas considered) traversed by the molecules between successive collisions. Using the ordinary assumptions of the kinetic theory of an ideal gas, namely, that (i) the gas molecules are identical point masses, (ii) interacting forces between molecules are neglected, (iii) the velocity distribution of the molecules is Maxwellian, and (iv) the gas is isotropic, one may derive the following relation for the mean free path [1.25–27]:

$$L = \frac{1}{\sqrt{2}\pi n d^2} \quad , \tag{1.1}$$

where d is the molecular diameter, and n is the concentration of the gas molecules in the vacuum. The last quantity is related to pressure p and temperature T by the expression

$$n = \frac{p}{k_B T} \quad . \tag{1.2}$$

Here k_B denotes the Boltzmann constant ($k_B = 1.381 \times 10^{-23}\,\text{JK}^{-1}$). Substituting this expression into (1.1) one obtains for the free path of the gas molecules

$$L = 3.11 \times 10^{-24} \frac{T}{p d^2} \quad , \tag{1.3}$$

where all quantities are expressed in SI units.

In standard MBE reaction chambers the molecular beams, generated in effusion cells or introduced into the vacuum environment through special gas inlets, traverse a distance approximately equal to 0.2 m between the outlet orifices of the beam sources and the substrate crystal surface. On the way to the substrate the beam molecules may encounter molecules of the residual gas if the pressure in the chamber is not low enough. Thus, scattering processes degrading the beam nature of the mass flow may occur.

The highest admissible value of the residual gas pressure may be estimated from the condition that the mean free path L_b of the molecules of the reactant beam penetrating the environment of the residual gas has to be larger than the distance from the outlet orifice of the beam source to the substrate crystal surface ($L_b > 0.2$ m). Assuming, for simplicity, that the beam and the residual gas create together a mixture of two gases, one may use the following formula [1.26] to calculate the admissible concentration of the molecules for the residual gas:

$$L_b^{-1} = \sqrt{2}\pi n_b d_b^2 + \pi n_g d_{bg}^2 \sqrt{1 + v_g^2/v_b^2} \quad \text{with} \tag{1.4}$$

$$d_{bg} = \frac{d_b + d_g}{2} \quad ,$$

where n_b, d_b and v_b are the concentration, diameter and average velocity of the molecules in the molecular beam, respectively, while n_g, d_g and v_g are the concentration, diameter and average velocity of the molecules of the residual gas in the reaction chamber. Usually the average velocities of the residual gas molecules are much smaller than the velocities of the beam molecules ($v_g \ll v_b$). Therefore, (1.4) may be simplified by omitting the term containing the ratio of the molecular velocities. Rearranging the simplified expression, one gets for the concentration of the residual gas molecules

$$n_g = \frac{L_b^{-1} - \sqrt{2}\pi n_b d_b^2}{\pi d_{bg}^2} \quad . \tag{1.5}$$

After substituting (1.2) into (1.5), the partial pressure of the residual gas may be expressed as

$$p_g = k_B T \frac{L_b^{-1} - \sqrt{2}\pi n_b d_b^2}{\pi d_{bg}^2} \quad . \tag{1.6}$$

The following numerical data are typical for conventional MBE growth of GaAs:

$$T = 300\,\text{K} \quad , \quad k_B = 1.381 \times 10^{-23}\,\text{JK}^{-1} \quad , \quad L_b = 0.2\,\text{m} \quad ,$$
$$d_b = 2.70 \times 10^{-10}\,\text{m} \quad , \quad d_g = 3.74 \times 10^{-10}\,\text{m} \quad , \quad \text{and}$$
$$n_b = 2.6 \times 10^{16}\,\text{m}^{-3} \quad .$$

The value for d_b is the Ga atom diameter [1.28]; d_g has been taken from viscosity measurements at 273 K for air [1.25, 26]; n_b has been calculated assuming a Ga beam flux of 10^{19} atoms m^{-2}s^{-1} and a Ga source temperature of 1250 K [1.29]. Using these data in (1.6) one obtains the maximum value of the residual gas pressure $p_{g,max} = 7.7 \times 10^{-2}$ Pa. It is evident that the beam nature of the mass transport is preserved even in HV conditions. However, the low growth rates typical for conventional MBE techniques (about 1 μm/h, or 1 monolayer/s) coupled with the obvious requirement of negligible unintentional impurity levels in the crystallized epilayer lead to much more rigorous limitations for the total pressure of the residual gas in the MBE reactor.

The condition for growing a sufficiently clean epilayer can be expressed by the relation between the 1 monolayer deposition times of the beam $t_1(b)$ and the background vapor $t_1(v)$:

$$t_1(b) = 10^{-5} t_1(v) \quad . \tag{1.7}$$

That means that the time $t_1(v)$ during which one monolayer of contaminants is deposited on the substrate surface from the residual gas contained in the vacuum reactor is one hundred thousand times longer than the time $t_1(b)$ necessary for the deposition of a film of one monolayer thickness from the molecular beams. For a beam flux of 10^{19} atoms m^{-2}s^{-1} reaching the substrate surface (this is a value typical for the growth of GaAs by MBE [1.29]) and an equilibrium surface density of the substrate crystal atoms equal to 10^{19} atoms m^{-2} one gets for $t_1(b)$ the value of 1 s. Consequently, the above condition will be fulfilled if the time $t_1(v)$ is not shorter than 10^5 s $\approx$ 28 h.

The number of molecules of the residual gas species i which strike the unit area of the substrate surface in unit time may be related to the partial pressure (p_i) of the species i, the molecular weight (M_i) of this species, and the residual gas temperature (T) in the vacuum reactor by the following formula [1.26, 27]:

$$w_i = p_i \sqrt{\frac{\mathcal{N}_A}{2\pi k_B M_i T}} \quad ; \tag{1.8}$$

$\mathcal{N}_A$ and k_B are the Avogadro and Boltzmann constants, respectively. Equation (1.8) takes the simplified form

$$w_i = 8.33 \times 10^{22} \frac{p_i}{\sqrt{M_i T}} \quad \text{molecules m}^{-2}\text{s}^{-1} \tag{1.9}$$

when all the quantities in (1.8) are expressed in SI units. Assuming now that the residual gas in the vacuum reactor consists only of nitrogen molecules ($M_i = 28.02 \times 10^{-3}$ kg/mol), and that the temperature of the residual gas is equal to 300 K, one gets the relation $w_i = 2.87 \times 10^{22} p_i$. Taking into consideration that each nitrogen molecule consists of two atoms, one has to multiply the numerical factor in this relation by 2, if the deposition rate is to be expressed in atoms m^{-2}s^{-1}. Consequently, the calculated time $t_1(v) = 10^{19}(5.74 \times 10^{22} p_i)^{-1}$. This value must be larger than or equal to 10^5 s. From this requirement the following condition results for the pressure of the residual gas in the vacuum reactor: $p_{i,\text{max}} < 1.7 \times 10^{-9}$ Pa. This numerical result indicates that MBE growth at low rates (1 μm/h) should be carried out in an UHV environment.

In general, the concentrations of unintentional contaminants incorporated into a growing film depend also on their sticking coefficients (s_i) on the substrate at the growth temperature. Consequently, the total contamination incorporated in the growing film from the residual gas in the vacuum reactor may be written as [1.30]

$$n = \sum_{i=1}^{n} n_i = \sum_{i=1}^{n} s_i w_i r_g^{-1} \quad , \tag{1.10}$$

where n_i is the impurity concentration of the gas molecules of species i, and r_g is the growth rate of the epitaxial film.

The residual gas pressure routinely achieved in MBE vacuum systems is in the range of low 10^{-8} Pa to mid 10^{-9} Pa [1.31]. These values increase somewhat during operation due to the increased heat load from the effusion cell furnace and the substrate. It has been found that the substrate rotation may cause high levels of elemental carbon, methane, carbon oxide and carbon dioxide to be released, which may be incorporated into the MBE layer [1.32]. The given values are larger than the calculated value for $p_{i,max}$. However, despite this, very clean thin films may be grown by MBE. This fact means that the present success of MBE technology stems more from the small sticking coefficient of the residual gas species on the heated substrate than from the attainable low residual gas pressure.

1.1.2 Basic Physical Processes in the MBE Vacuum Chamber

The essential elements of a MBE system are shown schematically in Fig. 1.1. It is apparent from this illustration that each MBE arrangement may be divided into three zones where different physical phenomena take place [1.33]. The first zone is the generation zone of the molecular beams. Next is the zone where the beams from different sources intersect each other and the vaporized elements mix, together creating a very special gas phase contacting the substrate area. This area, where the crystallization processes take place, can be regarded as the third zone of the MBE physical system.

The molecular beams are generated in the first zone under UHV conditions from sources of the Knudsen-effusion-cell type [1.33], whose temperatures are

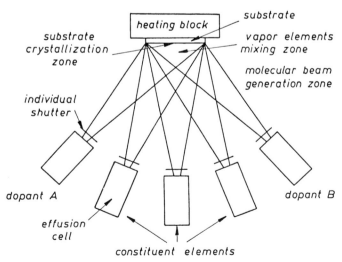

Fig. 1.1. Schematic illustration of the essential parts of a MBE growth system. Three zones where the basic processes of MBE take place are indicated [1.33]

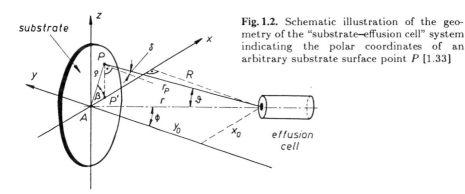

Fig. 1.2. Schematic illustration of the geometry of the "substrate–effusion cell" system indicating the polar coordinates of an arbitrary substrate surface point P [1.33]

accurately controlled (see Sect. 2.3 for details). Conventional temperature control, based on high performance proportional-integral-derivative (PID) controllers and thermocouple feedback, enables a flux stability of better than $\pm 1 \%$ [1.31]. By choosing appropriate cell and substrate temperatures, epitaxial films of the desired chemical composition can be obtained. Accurately selected and subsequently precisely controlled temperatures of the substrate and of each of the sources of the constituent beams have thus a direct effect upon the growth process.

The uniformity in thickness as well as in the composition of the films grown by MBE depends on the uniformities of the molecular beam fluxes and also on the geometrical relationship between the configurations of the sources and the substrate. Let us consider the substrate–source geometry shown in Fig. 1.2. The number of molecules or atoms impinging on the substrate surface per unit area and per second at the surface point P is equal to [1.33] (for details see Sect. 2.1.2)

$$N_P = I_A \left(\frac{r}{r_P} \right)^2 \cos \vartheta \, \cos \, (\vartheta + \varPhi) \cos \delta \quad , \tag{1.11}$$

where δ is the condensation angle and I_A is the beam flux reaching the substrate center A when the substrate surface is perpendicular to the axis of the effusion cell (Fig. 2.2a). The position of the point P on the substrate surface is given by a pair of polar coordinates ϱ and β. It is clear that the distance r_P and the angles ϑ and δ depend on the definite values of these two coordinates, whereas the distances r, R and x_0 are constant for a given source configuration. So, N_P at a given source–substrate geometry can be expressed as a function of ϱ and β. The mass of species i deposited on the substrate per second and per unit area is given by

$$m_i(\varrho, \beta) = m_{1i} N(\varrho, \beta) = M_i \frac{N(\varrho, \beta)}{\mathcal{N}_A} \quad . \tag{1.12}$$

Here m_{1i} is the mass of one molecule of the species i, $\mathcal{N}_A$ is the Avogadro constant and M_i is the molecular mass of the species i. We have assumed, for simplicity, that the sticking coefficient s_i for the molecules of the species i impinging on the substrate equals 1.

The growth rate of the epilayer for the given species i can be calculated from the formula

$$r_{gi}(\varrho, \beta) = \frac{m_i(\varrho, \beta)}{d_i} \quad , \tag{1.13}$$

where d_i is the density of this species material. Assuming that the beam flux is not dependent on time, the thickness of the layer grown from the molecules of species i is equal to

$$l_i(\varrho, \beta) = t r_{gi}(\varrho, \beta) \quad , \tag{1.14}$$

where t is the growth time.

In the case when molecular beams of the same species i are generated by n sources, the resulting thickness of the epilayer at the point $P(\varrho, \beta)$ is described by the sum

$$l_i(\varrho, \beta) = \sum_{j=1}^{n} l_j(\varrho, \beta) \quad , \tag{1.15}$$

where j denotes now a definite source of the species i. The value of l_i depends not only on ϱ and β but also on the angle Φ between the axis of the jth effusing source and the normal to the substrate surface.

The inhomogeneity in the layer thickness can now be established by defining the thickness homogeneity coefficient

$$\gamma = \frac{l_i(\mathrm{max}) - l_i(\mathrm{min})}{l_i(\mathrm{max})} \times 100\,\% \quad , \tag{1.16}$$

where $l_i(\mathrm{max})$ and $l_i(\mathrm{min})$ are the maximum and the minimum thicknesses of the epilayer, respectively.

In the case of k different species, each effused from one source, a composition homogeneity coefficient η can be defined. If the concentration Q_i of the species i in the substrate surface at point $P(\varrho, \beta)$ is equal to

$$Q_i(\varrho, \beta) = N_i(\varrho, \beta) \left(\sum_{j=1}^{k} N_j(\varrho, \beta) \right)^{-1} \tag{1.17}$$

then the composition homogeneity is described by

$$\eta = \frac{Q_i(\mathrm{max}) - Q_i(\mathrm{min})}{Q_i(\mathrm{max})} \times 100\,\% \quad . \tag{1.18}$$

Estimating now the values of γ and η for different values of x_0 one can choose the proper geometry of the arrangement for substrate and system of sources. This is, however, very often an extremely difficult task [1.34, 35]. Therefore, if it is possible, the substrate is rotated during the growth with a constant angular velocity around the axis perpendicular to its surface. The substrate rotation causes a considerable enhancement in thickness and composition homogeneity of the grown epilayer [1.36] but, unfortunately, excludes the possibility of growing patterned structures, which is often required in the technology of present-day semiconductor devices.

The second zone in the MBE vacuum reactor is the mixing zone, where the molecular beams intersect each other. Little is known at present about the physical phenomena occurring in this zone. This results from the fact that usually the mean free path of the molecules belonging to the intersecting beams is so long that no collisions and no other interactions between the molecules of different species occur there.

Epitaxial growth in MBE is realized in the third zone, i.e. on the substrate surface. A series of surface processes are involved in MBE growth, however, the following are the most important [1.37]:

i)   adsorption of the constituent atoms or molecules impinging on the substrate surface,
ii)   surface migration and dissociation of the adsorbed molecules,
iii)   incorporation of the constituent atoms into the crystal lattice of the substrate or the epilayer already grown,
iv)   thermal desorption of the species not incorporated into the crystal lattice.

These processes are schematically illustrated in Fig. 1.3 [1.38]. The substrate crystal surface is divided there into so-called crystal sites with which the impinging molecules or atoms may interact. Each crystal site is a small part of the crystal surface characterized by its individual chemical activity. A site may be created by a dangling bond, vacancy, step edge, etc. [1.39].

The surface processes occurring during MBE growth are characterized by a set of relevant kinetic parameters that describe them quantitatively.

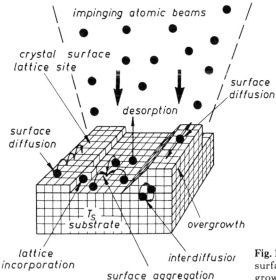

Fig. 1.3. Schematic illustration of the surface processes occurring during film growth by MBE [1.38]

9

The arrival rate is described by the flux of the arriving species, which gives the number of atoms impinging on the unit area of the surface per second. For MBE growth of compound semiconductors, the required fluxes are typically between 10^{18} and 10^{20} atoms m^{-2} s^{-1} [1.29]. The atoms arriving at the substrate surface will generally have an energy distribution appropriate to the temperature of their place of origin. In the MBE case being considered this will most frequently be the effusion cell temperature T_i. After arrival at the substrate surface, which has a temperature T_s, usually lower than T_i, the atom may reevaporate immediately, carrying with it an energy corresponding to temperature T_e. The impinging atoms may, however, also exchange energy with the atoms of the substrate until they are in thermodynamic equilibrium at T_s. The quantitative description of this process is possible by defining the thermal accommodation coefficient as

$$a = \frac{T_i - T_e}{T_i - T_s} \quad .$$
(1.19)

Clearly, when T_e is equal to T_s the accommodation coefficient is unity. Thus, it emerges as a measure of the extent to which the arriving atoms reach thermal equilibrium with the substrate [1.39]. It is important to differentiate between the accommodation coefficient defined above and the sticking, or condensation, coefficient s. The latter is defined as the ratio of the number of atoms adhering to the substrate surface to the number of atoms arriving there:

$$s = \frac{N_{adh}}{N_{tot}} \quad .$$
(1.20)

In many cases the sticking coefficient is less than unity and it may be a very small fraction in cases when the adsorption energy of atoms on the substrate is low, or the substrate temperature is high. Assuming that $a = 1$, all the arriving atoms are accommodated on the substrate surface and achieve thermodynamic equilibrium. This does not mean, however, that they will remain there permanently. They still have a finite probability related to the substrate temperature of acquiring sufficient energy to overcome the attractive forces and leave the substrate. If condensation, i.e., aggregation of adatoms, does not occur, all adatoms will eventually reevaporate. Thus, the sticking coefficient can be almost zero even though the accommodation coefficient is unity.

There are two types of adsorption. The first is physical adsorption, often called physisorption, which refers to the case where there is no electron transfer between the adsorbate and the adsorbent, and the attractive forces are van der Waals type. The second is chemisorption, which refers to the case when electron transfer, i.e., chemical reaction, takes place between the adsorbate and the adsorbent. The forces are then of the type occurring in the appropriate chemical bond. In general, adsorption energies for physical adsorption are smaller than for chemical adsorption.

It is obvious that for these two kinds of adsorption different sticking coefficients can be defined. According to the definition of the chemisorption process the sticking coefficient s_c for the chemisorbed phase may be dependent on the crystallographic orientation of the substrate surface, as well as on the nature and spatial distribution of atoms already adsorbed on this surface. Physisorption, on the other hand, generally shows little or no dependence on surface site arrangement. Consequently, the sticking coefficient s_p for the physisorbed phase may be taken to be independent of the local environment, i.e., of the orientation and coverage of the substrate surface.

It is experimentally well documented [1.29] that crystal growth in MBE, in many cases in which molecular species are involved, proceeds through a two-step condensation process in which the molecular species reach the chemisorbed state via a precursor physisorbed phase [1.40]. The simplest version of such a stepwise condensation process may proceed according to the scheme [1.41]

$$A_g \underset{k_c}{\overset{k_d}{\longleftarrow}} A_p \overset{k_a g(\theta)}{\longrightarrow} A_c \quad ,\tag{1.21}$$

where A_g, A_p and A_c represent the adsorbate in the gaseous, precursor and chemisorbed states, respectively. The rate constants k_c and k_d apply to condensation and desorption of the precursor state, and $k_a g(\theta)$ is the rate constant for conversion from the precursor state to the chemisorbed state. The conversion rate is assumed to be proportional to $g(\theta)$, the probability of having suitable vacant sites in the chemisorbed state, where θ is the surface coverage counted in monolayers.

Another possible version of the condensation process via the precursor state may be sketched as [1.42]

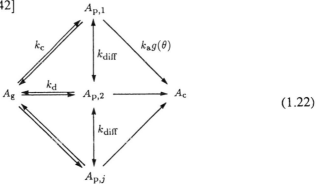

$$\tag{1.22}$$

This model is similar to that implied by the scheme (1.21) except that if the site visited by the molecule in the precursor state is occupied by a chemisorbed species, the molecule is allowed to diffuse over the surface with a rate constant k_{diff} to another site.

The interaction potentials due to the surface seen by an incoming molecule perpendicular to the surface, for chemisorbed and precursor states, are shown schematically in Fig. 1.4 [1.38]. It is evident from that figure that the molecule adsorbed in the precursor state has to overcome a lower barrier when it is subsequently chemisorbed at the surface, than in the case when it re-evaporates into the vacuum, because $E_a < E_{d_p}$. The illustrated situation is a special example; a variety of other potential configurations can also exist.

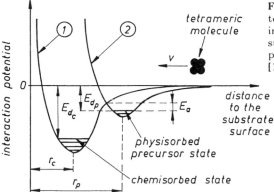

Fig. 1.4. The interaction potential due to the surface, as seen by a molecule impinging perpendicularly to the surface for chemisorbed *(curve 1)* and physisorbed precursor *(curve 2)* states [1.38]

It is generally accepted that three possible modes of crystal growth on surfaces may be distinguished. These modes are illustrated schematically in Fig. 1.5 [1.43, 44]. In the island, or Volmer–Weber mode, small clusters are nucleated directly on the substrate surface and then grow into islands of the condensed phase. This happens when the atoms, or molecules, of the deposit are more strongly bound to each other than to the substrate. This mode is displayed by many systems of metals growing on insulators, including many metals on alkali halides, graphite and other layer compounds such as mica.

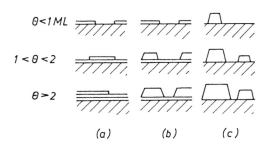

Fig. 1.5a–c. Schematic representation of the three crystal growth modes **(a)** Layer-by-layer or Frank–van der Merwe; **(b)** layer plus island or Stranski-Krastanov; **(c)** island or Volmer-Weber mode. Θ represents the coverage in monolayers [1.44]

The layer-by-layer, or Frank–van der Merwe mode, displays the opposite characteristics. Because the atoms are more strongly bound to the substrate than to each other, the first atoms to condense form a complete monolayer on the surface, which becomes covered with a somewhat less tightly bound second layer. Provided the decrease in binding is monotonic toward the value for a bulk crystal of the deposit, the layer growth mode is obtained. This growth mode is observed in the case of adsorbed gases, such as several rare gases on graphite and on several metals, in some metal-metal systems, and in semiconductor growth on semiconductors.

The layer plus island, or Stranski–Krastanov, growth mode is an intermediate case. After forming the first monolayer, or a few monolayers, subsequent layer growth is unfavorable and islands are formed on top of this "intermediate" layer. There are many possible reasons for this mode to occur and almost any factor which disturbs the monotonic decrease in binding energy characteristic for layer-by-layer growth may be the cause [1.44].

1.2 Evolution of the MBE Technique

Many materials have been crystallized as epitaxial thin films by MBE [1.45–47], but semiconductors of the III–V group in general [1.29, 48] and GaAs and $Al_x Ga_{1-x} As$ in particular have received most attention as prototype materials [1.46]. This is because of the superior high-frequency properties and the unique optical properties of the III–V semiconductors as compared to Si. Furthermore, the MBE growth mechanism of III–V compounds is more fully understood from the early studies at Bell Laboratories in the USA [1.49] and at Philips Research Laboratories in England [1.50]. Despite this apparent imbalance, MBE growth of II–VI [1.51], and IV–VI [1.52] compounds as well as of metals [1.53], insulators [1.54] and Si [1.55–58] is attracting more and more interest. In principle, MBE is applicable to the growth of epitaxial layers of a wide variety of materials. In addition, because of the extreme dimensional control and low growth temperatures with MBE, one can in fact tailor deposition on an atomic layer-by-layer basis [1.59], and it is now possible to grow artificial crystals with periodicities not available in nature [1.38, 47].

1.2.1 The Early Stages of MBE

The development of MBE to a practical thin film growth technique evolved from two complementary approaches [1.46]; first, the "three-temperature method" developed by *Günther* [1.60] in 1958 and, second, the surface kinetic studies of the interaction of Ga and As_2 beams with GaAs substrates [1.61] performed in the middle 1960s.

Most of the III–V compound semiconductors consist of constituents of greatly differing vapor pressures, and they exhibit considerable decomposition at their evaporation temperatures. Günther was the first to develop the basic ideas for one of the key requirements of MBE, viz., the deposition of stoichiometric films of III–V semiconductors from molecular beams of the constituent elements impinging onto a heated substrate, and he employed them in his three-temperature method. He used the group V element source oven, kept at temperature T_i to maintain a steady pressure in his static vacuum chamber. The source oven with the group III element, kept at a much higher temperature T_j, provided a flux of atoms incident on the substrate that was critical for the condensation rate. The choice of the substrate temperature was most crucial. This intermediate temperature T_s had to be increased to a value ensuring that the excess group V component which had not reacted was reevaporated from the substrate surface.

In the first experiments, Günther was successful in growing stoichiometric films of InAs and InSb, which were, however, deposited on glass substrates and therefore polycrystalline [1.60]. It was not until ten years later that epitaxy for monocrystalline GaAs was achieved on clean single-crystal substrates under improved vacuum conditions by *Davey* and *Pankey* [1.62] using Günther's method.

In the last two decades the three-temperature method has been developed in two directions: first, the so-called hot-wall technique [1.63–67], where the material is deposited at conditions close to thermodynamic equilibrium under high vacuum, yielding high growth rates, and second, the more sophisticated process of MBE [1.68–70], a cold-wall technique working far from thermodynamic equilibrium with low growth rates. Both methods result in layered semiconductor crystals with accurately tailored electronic properties [1.38].

Starting in 1970, the surface chemical processes involved in the MBE growth of binary and ternary III–V compounds were studied extensively by *Arthur* [1.61,71] and *Foxon* et al. [1.72–75], using a combination of the modulated molecular beam (MMB) technique and reflection high-energy electron diffraction (RHEED). Both groups made measurements of thermal accommodation coefficients, surface lifetimes, sticking coefficients, desorption energies, and reaction order. From these results rather detailed models became available for the evaporation of GaAs under equilibrium and nonequilibrium conditions [1.75] as well as for the growth of GaAs from beams of Ga and As_2 [1.74] and Ga and As_4 [1.73].

The established growth models turned out not to be unique to GaAs; they are also valid for other binary III–V compounds such as AlAs [1.49] and InP [1.76] and, with minor modifications, for ternary III–III–V compounds such as $Al_xGa_{1-x}As$ [1.77]. In practical terms, good compositional control of the growing alloy film can be achieved by supplying excess group V species (as in the case of binary III–V compounds) and adjusting the flux densities of the impinging group III beams. The principal limitation to the growth of ternary III–III–V alloys by MBE is the thermal stability of the less stable of the two III–V compounds of which the alloy is composed. At higher temperatures preferential desorption of the more volatile group III element occurs. Thus, the surface composition of the alloy reflects the relative flux ratio of the group III elements only, if growth is carried out at temperatures below which GaAs (in the case of $Al_xGa_{1-x}As$), InAs (in the case of $Ga_xIn_{1-x}As$) or InP (in the case of $Ga_xIn_{1-x}P$) are thermally stable [1.29].

The growth of ternary III–V–V alloy films by MBE, however, was found to be much more complex, since the relative amounts of the group V elements incorporated into the growing film are not simply proportional to their arrival rates. *Foxon* et al. [1.78] worked out that the Ga(In)-to-As_4 flux ratio is the critical parameter in the control of the overall composition for the growth of GaP_yAs_{1-y} and InP_yAs_{1-y} from Ga(In) plus P_4 and As_4. Due to the greater surface lifetime of As_4 on GaAs or InAs than that of P_4 on GaP or InP, a much higher incorporation probability (by a factor of up to 50) was found for arsenic than for phosphorus.

The incorporation of controlled amounts of electrically active impurities (dopants) in growing III–V semiconductor films was also studied to some extent using the MMB technique [1.79]. Most elements that have been used successfully as n-type, amphoteric, or semi-insulating dopants for GaAs grown by MBE exhibit unity sticking coefficients over a wide range of accessible growth conditions. This behavior is indicated by linear behavior of carrier concentration vs reciprocal effusion cell temperature, which has been established for the most suitable dopant elements.

As yet, detailed investigations of the dopant incorporation behavior are available only for GaAs and other III–V compounds [1.80–88]; there is thus an increasing need for further studies applied to II–VI, IV–VI and elemental semiconductors produced by MBE [1.46]. This is particularly true since growth takes place far from equilibrium, and kinetic limitations to incorporation (e.g., low sticking coefficient, accumulation at the growth surface) can occur [1.29].

The majority of the UHV equipment used for MBE growth until 1977 was "home" designed, i.e., it was built by commercial UHV manufacturers according to the instructions of the particular user. Although basically consisting of a single deposition–analysis chamber design, the early growth systems differed widely in conception, complexity, and automation.

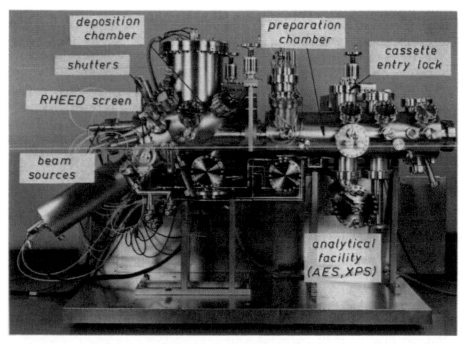

Fig. 1.6. III–V MBE system delivered commercially by VG Semicon. The components characteristic of most elements of the system are indicated [1.31]

15

Since 1978, more-standardized MBE systems consisting of several basic UHV "building blocks", such as growth (reaction) chamber, substrate loading and preparation chamber, and surface analysis chamber, have become commercially available. This more sophisticated, modular design allows the whole instrumentation to be matched favorably to most individual requirements [1.31]. An example of such a commercially available MBE growth system is shown in Fig. 1.6.

1.2.2 MBE in the 1980s

Four milestones may be distinguished along the path of evolution of the MBE technique in the 1980s. The first is connected with the introduction of gas sources into the arsenal of tools available for MBE, to make possible the well-controlled use of P and P plus As for the growth of compound semiconductors [1.89, 90]. The second results from the observation of oscillations in intensity of various features in the RHEED pattern occurring during MBE growth, which has been reported by *Neave* et al. [1.91, 92] and *van Hove* et al. [1.93]. This has been used for precise control of the growth of very thin films and superlattices [1.94–96], as well as for studying the growth mechanisms in MBE [1.50]. The next milestone concerns the pulsed mode of supplying the reactant species to the growing epilayer surface during the MBE process [1.59], frequently assisted by external radiation like ultraviolet light [1.97] or ionized hydrogen [1.98]. The main advantage of the pulsed beam growth technique is a considerable lowering of the growth temperature of III–V and II–VI compounds, and consequently no diffusion during growth. This is advantageous for obtaining abrupt heterojunctions and very sharp doping profiles for both n- and p-type impurities, as well as for developing new processing technologies, including the growth of epilayers on finely structured substrates. The last milestone is connected with coupling of all the MBE-related technological and characterization operations with the UHV processing steps in one preparation plant, designed as a multichamber system. For example, focused ion beam implantation and etching operation facilities have been coupled together with MBE growth and characterization equipment [1.99, 100]. Thus, the production of complicated device structures realized entirely in the UHV environment became possible, with a reasonable output [1.101].

1.3 Modifications of the MBE Technique

The evolution of the MBE technique in the 1980s resulted in a whole variety of modifications of this technique. Unfortunately, these modifications have been introduced under new names, which very often do not relate to the original technique, i.e., the MBE technique. The most important modifications will be discussed in more detail, and will then be listed in a classification scheme in Sect. 1.3.5.

1.3.1 Gas Source MBE

Since the initial work on gas source MBE (GS MBE) by *Panish* et al. [1.102–104], *Tsang* [1.105, 106] and *Tokumitsu* et al. [1.107–109], it has become clear that the use of gas sources for the cracking of AsH_3 and PH_3 does in fact permit a degree of control that makes possible the epitaxy of ternary and even quaternary compounds on InP substrates, with dimensional precision similar to that achieved by conventional MBE for $Al_x Ga_{1-x} As$/GaAs heterostructures. Modifications of the basic GS MBE process [1.102], mostly involving the use of metalorganic gas sources [1.105–111], further enhance the usefulness of the gas source method.

For the modification of GS MBE that uses metalorganic gas sources, the names chemical beam epitaxy (CBE) [1.105, 106] and metalorganic MBE (MO MBE) [1.107–109] have been introduced. Following the practice of many scientists actively involved in research into and the practice of MBE [1.112], we will use exclusively the name MO MBE hereafter. MO MBE [1.107] combines many important advantages of MBE and metalorganic chemical vapor deposition (MO CVD) [1.113] and advances epitaxial technology beyond both these techniques. In CVD, the gas pressure in the reactor is between 10^5 and 1 Pa. As a result, the transport of the reactant gases is by viscous flow. If the pressure is further reduced down to 10^{-2} Pa, the gas transport becomes a molecular beam. Hence, the process evolves from vapor deposition to beam deposition. If the thin film deposited is an epitaxial layer, we call this process metalorganic molecular beam epitaxy, which can use exclusively organometallic gas sources or a combination of organo- and inorganometallic gas sources. In MO MBE, unlike MBE, which employs beams (e.g., Al, Ga, and In) evaporated at high temperature from elemental sources, the sources are gaseous or they are charged with reactants (solid or liquid) that have sufficiently high vapor pressure that they produce gaseous beams, so that they need not be heated significantly above room temperature. Al, Ga and In are derived by the pyrolysis of their organometallic compounds, e.g., trimethylaluminum, triethylgallium and trimethylindium, at the heated substrate surface. As_2 and P_2 are obtained by the thermal decomposition of their hydrides passing through a heated baffled cell. The use of hydrides was first introduced into high vacuum film growth in 1974 by *Morris* and *Fukui* [1.114]. They grew polycrystalline films of GaAs and GaP by cracking the relevant hydrides in their apparatus to group V elements and hydrogen. In MBE, the hydrides were first used, in 1980, by *Panish* [1.102] and then by *Calawa* [1.115] for the growth of GaAs and GaInAsP. Unlike MO CVD, in which the chemicals reach the substrate surface by diffusing through a stagnant carrier gas boundary layer above the substrate, the chemicals in MO MBE are admitted into the high vacuum growth chamber in the form of a beam. Therefore, comparing with conventional MBE, the main advantages include [1.105]:

1. The use of group III organometallic sources that are solids or liquids which have sufficiently high vapor pressure that they need not be heated significantly above room temperature, what simplifies multiwafer scale-up

2. Semi-infinite source supply and precision electronic flow control with instant flux response (which is suitable for the production environment)
3. A single group III beam that guarantees material composition uniformity
4. No oval defects even at high growth rates (important for integrated circuit applications)

Comparing with MO CVD, the advantages include:

1. No flow pattern problem encountered in multiwafer scale-up
2. Beam nature produces very abrupt heterointerfaces and ultrathin layers conveniently
3. Clean growth environment
4. Easy implementation of in situ diagnostic instruments, e.g., RHEED
5. Compatibility with other high vacuum thin-film processing methods, e.g., metal evaporation, ion beam milling, and ion implantation

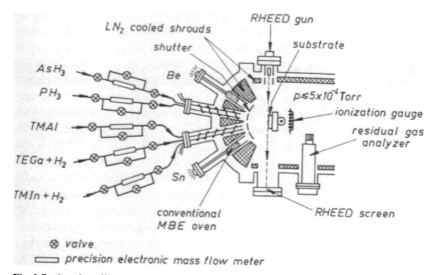

Fig. 1.7. Gas handling system and growth chamber with in situ surface diagnostic capabilities incorporated in a MO MBE system [1.106]

A gas handling system with precision electronic mass flow controllers is used in MO MBE for controlling the flow rates of the various gases admitted into the growth chamber as shown in Fig. 1.7. Hydrogen is used here as the carrier gas for transporting the low vapor pressure group III organometallic compounds. Separate gas inlets are used for group III organometallics and group V hydrides. A low-pressure arsine (AsH_3) and phosphine (PH_3) cracker with a reduced input pressure of 3×10^4 Pa is maintained on the high pressure side of the electronic mass flow controller. To avoid material condensation, the manifold is warmed up to $\sim 40°$ C. The cracking temperature is $\sim 920°$ C. Decomposition of arsine and phosphine into arsenic, phosphorus, and hydrogen may be routinely achieved as observed with an in situ residue gas analyzer [1.106]. In earlier studies on MO MBE of III–V compounds [1.105, 116] group V alkyls were thermally decomposed after mixing with H_2. Though these alkyls are safer than hydrides, their purity at present is rather poor. Hence, hydrides are preferred at present.

Triethylgallium (TEGa) maintained at 30°C, trimethylindium (TMIn) at 25°C and trimethylaluminum (TMAl) at 25°C are used. The TEGa, TMIn and TMAl flows are combined to form a single emerging beam impinging line-of-sight onto the heated substrate surface. This automatically guarantees composition uniformity [1.116]. The typical growth rates are 2–3 μm/h for GaAs, 4–6 μm/h for AlGaAs, 3.65 μm/h for GaInAs, and 1.5–2.5 μm/h for InP. Such growth rates are higher than those typically used in conventional MBE.

1.3.2 Phase-Locked Epitaxy

In 1983 damped oscillations in the intensity of the specular and the diffracted beams of the RHEED pattern were observed for the first time during MBE growth [1.91–93]. A typical example of the specular beam in the RHEED pattern from MBE-grown GaAs (001) surface is shown in Fig. 1.8 [1.91]. The period of oscillation corresponds exactly to the growth of a single monolayer, i.e., a complete layer of Ga and As atoms (($a_0/2$) in the [001] direction) and is consequently independent of the azimuth of the incident beam and of the particular diffraction features being measured. The amplitude, however, is strongly dependent on both of these parameters. For evaluation of growth dynamics most of the information is contained in the specular beam. The observation of oscillatory effects in thin film growth studies [1.117–119] is usually associated with a layer-by-layer growth process (i.e., two-dimensional growth), which is in qualitative agreement with the theoretical treatment of *Weeks* and *Gilmer* [1.120]. They considered transients in crystal growth and predicted oscillatory rate behavior under conditions where a significant proportion of incident atoms desorb immediately after growth commences. Although for MBE growth of GaAs under normal conditions there is no significant desorption of the growth-rate-controlling group III elements, there is nevertheless a close analogy with the Gilmer-Weeks theory.

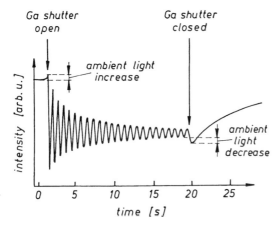

Ga shutter open

Ga shutter closed

ambient light increase

ambient light decrease

intensity [arb. u.]

time [s]

0 5 10 15 20 25

Fig. 1.8. Intensity oscillations of the specular beam in the RHEED pattern from a GaAs (001) 2 × 4 reconstructed surface [110] azimuth. The period exactly corresponds to the growth rate of a single Ga+As layer, and the amplitude gradually decreases. Note that the marked inflections at the beginning and end of growth result from ambient light changes as the shutters are opened and closed [1.91]

In their model, the equilibrium surface is assumed to be smooth, and at the start of growth random clusters are generated, with a stable cluster requiring some minimum number of atoms. The layer is complete when all the clusters have coalesced and all the holes are filled. According to the Gilmer-Weeks model, at the beginning of the growth process most atoms desorb before they are incorporated in growing clusters, but as stable clusters nucleate and spread, the growth rate increases, reaching a maximum at some fraction of a complete layer. It then decreases as nucleation of the next layer commences. This sequence repeats, but with gradual damping as growth becomes distributed over several layers. Impingement and nucleation are both random processes, so local surface regions experience different deposition rates. The complete surface therefore becomes distributed over many levels and maximum growth rates in different regions are not synchronised. Consequently there is an overall time-independent asymptotic rate once the surface is "rough", and the amplitude of the oscillations is therefore a measure of the localization (or degree) of roughening.

If we now equate changes in intensity of the specular beam in the RHEED pattern with changes in surface roughness, the equilibrium surface is smooth, corresponding to high reflectivity. Any roughening of this surface during growth can be related to local growth rate differences, although the overall growth rate does not vary and is determined only by the Ga flux. On application of the driving force (i.e., at commencement of growth) clusters are formed at random positions on the crystal surface, leading to a decrease in the reflectivity. This decrease can be predicted for purely optical reasons, since the de Broglie wavelength of the typically used electrons is ≈ 0.12 Å, while the bilayer step height is ≈ 2.8 Å, i.e., the wavelength is at least an order of magnitude less than the size of the scatterer, so diffuse reflectivity will result. Growth is not restricted to a single layer, but can recommence in a new layer before the preceding layer is complete. In the early stages, however, one layer is likely to be almost complete before the next layer starts, so the reflectivity will increase as the surface again becomes smooth on the atomic scale, with subsequent roughening as the next layer develops. This repetitive process will cause the oscillations in reflectivity to be gradually damped as the surface becomes statistically distributed over several incomplete atomic layers.

In Fig. 1.9 a real space representation of the formation of the first two complete layers is shown [1.121]. This illustrates how the oscillations in the intensity of the specular beam occur. There is a maximum reflectivity for the initial and final smooth surfaces and a minimum (or maximum in diffuse scattering) for the intermediate stage when the growing layer is approximately half complete. It has, however, also been established that the amplitude of the intensity oscillations is dependent on the direction of the primary electron beam, being greater for the beam incident along [110] than [$\bar{1}$10] azimuths. This would suggest, if we take the optical model one stage further, that most of the steps which develop on the surface are along the [$\bar{1}$10] direction, as shown in Fig. 1.9, i.e., they cause maximum diffuse scattering when their longer edges are normal to the

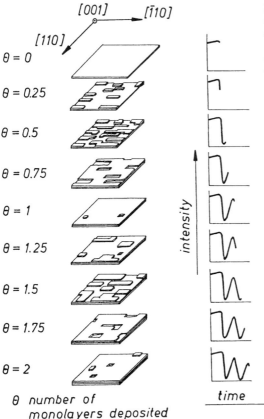

incident beam. One may conclude, therefore, that growth occurs principally by a two-dimensional layer-by-layer process, but new layers are able to be started before preceding ones have been completed. The oscillation period provides a continuous and absolute growth rate monitor with atomic layer precision.

The phenomenon of RHEED pattern intensity oscillation has been routinely used by many MBE research groups to study the initial phase of the MBE growth mechanism [1.92, 96, 122–124] of different materials. It has, however, also been applied for a considerable improvement of the growth procedure of very thin films and superlattice structures [1.95]. This has been achieved with the so-called phase-locked epitaxy (PLE) technique. A schematic diagram of the PLE system is shown in Fig. 1.10.

The idea of the method is as follows. Long continuing intensity oscillations of the RHEED pattern in the [100] azimuth on a (001) oriented substrate are recorded during MBE growth of GaAs and $Al_xGa_{1-x}As$ (Fig. 1.11a). Using these oscillations growth rates of these compounds and the Al mole fraction x of the ternary compound are accurately monitored during the growth. The phase of the RHEED oscillations is analyzed by computer and molecular beam shutters are operated at a particular phase of the oscillations.

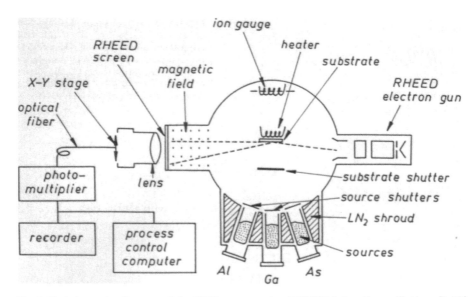

Fig. 1.10. Schematic diagram of the PLE system using RHEED intensity oscillations [1.95]

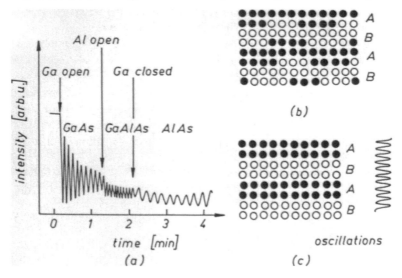

Fig. 1.11. (a) RHEED intensity oscillations of the specular beam observed in the [100] azimuth of (001)GaAs substrate during the continuous growth of GaAs, $Al_xGa_{1-x}As$ and AlAs. **(b, c)** Simplified models of the cross sectional view of superlattice structures grown by two methods: the conventional non-PLE method **(b)**, and the PLE method **(c)** [1.95]

This computer-controlled PLE was used to grow precisely defined (GaAs)$_2$ (AlAs)$_2$ bilayer superlattices. Raman scattering spectra of the bilayer superlattice showed split lines characteristic of superlattices [1.94, 95]. From TEM observation of a GaAs-AlAs multilayered structure, it was verified that one cycle of oscillations corresponds to one monolayer growth of GaAs and AlAs. The PLE has a great advantage over the conventional MBE growth method for the precise control of the growth of very thin films and superlattices because it is invulnerable to fluctuations of molecular beam flux intensity. Simplified models of the cross-sectional view of superlattice structures grown by these two methods are shown in Fig. 1.11. In the case of the non-PLE method (Fig. 1.11b) molecular beam shutters are operated at a random phase of the surface molecular coverage, and as a result the heterostructure interfaces between A and B have steps. On the other hand, in the case of the PLE method, the interfaces between A and B become very flat, as shown in Fig. 1.11c, since the molecular beam shutters are operated at the maximum amplitude of the oscillations when the monolayer is completed. A $(GaAs)_2(AlAs)_2$ bilayer superlattice with 699 periods, i.e., with 2796 monolayers, was grown using this computer-controlled PLE method, with excellent structural quality [1.94, 95].

1.3.3 Atomic Layer Epitaxy

The next milestone concerning the evolution of MBE in the 1980s actually originated in the mid 1970s, when Suntola and Antson patented the so-called atomic layer epitaxy (ALE) thin film growth technique [1.59, 125]. The basic idea behind ALE is to ensure a surface-controlled growth instead of the source-controlled growth used in conventional thin film techniques. In ALE it is a question of material growth instead of material transfer.

The ALE process is carried out stepwise by separate surface reactions, each forming one atomic layer only. A particularly simple way to describe the ALE approach is to say that it makes use of the difference between chemisorption with bonds typical of the compound to be grown, and chemisorption with much weaker bonds, typical of the pure constituent element, or weak physisorption, when molecular species are involved in the growth. When the first layer of atoms or molecules of a reactive species reaches a solid surface there is usually a strong interaction (chemisorption); subsequent layers tend to interact much less strongly (chemisorption with weaker bonds of the pure element atoms or physisorption). If the initial substrate is heated sufficiently one can achieve a condition such that only the first chemisorbed layer remains attached. In the earliest and perhaps simplest example of ALE, viz., the growth of ZnS by evaporation in a vacuum [1.126], Zn vapor was allowed to impinge on the heated glass. Evaporation was then stopped and any weakly adsorbed Zn present because of the impinging flux reevaporated. The process was then repeated with sulfur, the first layer of which chemisorbed on the initial Zn layer; any subsequent weakly adsorbed sulfur gradually came away from the heated substrate when the sulfur flux was cut off, leaving one (double) layer of ZnS. This complete cycle could be repeated indefinitely, the number of layers grown being determined solely by the number of cycles (Fig. 1.12, upper panel).

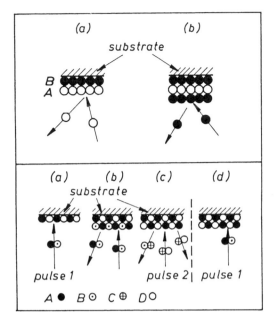

Fig. 1.12. Schematic illustration of film growth by ALE. Upper panel shows the growth from elemental source materials [variant (1) of the ALE process]. Lower panel shows the growth from compound source materials via surface exchange reactions [variant (2) of the ALE process] [1.59]

It has been shown by straightforward measurements that a chemical vapor deposition (CVD) of ZnS from $ZnCl_2$ and H_2S could also be operated in an ALE mode, the cycle being somewhat more complex (Fig. 1.12, lower panel) in that a chemisorbed $ZnCl_2$ monolayer loses its chlorine to the hydrogen of the later-arriving H_2S molecules, forming HCl, again resulting in a (double) layer of ZnS [1.127]. The importance of this CVD variant is that it enables ALE to be applied to compounds with one or more involatile constituents for which the evaporative approach would clearly not be viable.

One can thus arrive at a definition of ALE: it is based on chemical reactions at the solid surface of a substrate, to which the reactants are transported alternately as pulses of neutral molecules or atoms, either as chopped beams in high vacuum, or as switched streams of vapor possibly on an inert carrier gas. The incident pulse reacts directly and chemically only with the outermost atomic layer of the substrate. The film therefore grows stepwise – a single monolayer per pulse – provided that at least one complete monolayer coverage of a constituent element, or of a chemical compound containing it, is formed before the next pulse is allowed to react with the surface. Given that, any excess incident molecules or atoms impinging on the film do not stick if the substrate temperature T_s is properly chosen, and one therefore obtains precisely monolayer coverage in each cycle. The formation of a "layer per cycle" is the specific feature that conceptually distinguishes the ALE mode from other modes of vapor phase deposition; the latter all give a growth rate, ALE gives a growth per cycle.

As already indicated, ALE in its basic form operates by growing complete atomic layers on top of each other, there being two basic variants based on (i) evaporative deposition (with, as one extreme, MBE), relying on heated elemental source materials, or (ii) CVD, relying on sequential surface exchange reactions between compound reactants. Therefore, ALE is not so much an entirely new method of crystal growth as a special mode of these well-established growth techniques.

Both ALE variants, illustrated in Fig. 1.12, assume that at the substrate temperature the vapor pressures of the source materials as solid phases are much larger, perhaps by several orders of magnitude, than those of the compound films eventually formed from them. This requirement is usually met in the case of II–VI materials and can hold reasonably well in many other cases.

The precaution must be taken of turning off the source beam for a sufficiently long time (of the order of 1 s) after each deposition pulse. In this way the surface is allowed to approach thermodynamic equilibrium at the end of each reaction step, a growth condition that is not usually met in other techniques of deposition by evaporation. It is then clear that the gas dynamics is entirely eliminated from the growth problem. While this original ALE procedure was invented for growing II–VI semiconductor compounds, it was later extended to many other materials, including also GaAs and related III–V compounds [1.59].

The idea of the pulsed mode of supplying reactant species of the growing film surface has been introduced to the MBE technology of III–V compounds by *Kawabe* et al. [1.128, 129] as a new method of composition control of the $Al_xGa_{1-x}As$ compound. This idea has also been applied to GaAs in the growth techniques named molecular layer epitaxy (MLE) [1.97] and migration enhanced epitaxy (MEE) [1.130], respectively.

In principle, MLE is equal to the second variant of ALE, illustrated in Fig. 1.12 (lower panel). Its main advantage in comparison to conventional MBE is a considerable lowering of the growth temperature of GaAs. When using TMGa and AsH_3, MLE growth of GaAs at $500° C$ has been achieved, while when using TEGa and AsH_3, the growth temperature was as low as $300° C$ [1.131].

Growth temperature in MBE of GaAs and related compounds may also be considerably lowered with the MEE technique. This technique is characterized by using an alternate supply of Ga or Al and As to the growing surface and is thus similar to the first variant of ALE (Fig. 1.12, upper panel). MEE is based on the rapid surface migration effect which is characteristic of Ga and Al atoms in an As-free atmosphere over the GaAs surface, i.e., the migration is enhanced in the absence of As with respect to As-rich surfaces. Epilayers of GaAs of high quality have been grown at $200° C$, and GaAs-AlAs quantum well structures of high quality have been grown at $300° C$, with this method [1.130].

1.3.4 FIBI-MBE Processing Technology

Vacuum-based techniques currently cover the entire spectrum of semiconductor processing from simple metal evaporation and thermal vacuum annealing to more complex processes such as ion implantation, plasma and reactive ion etching, and plasma deposition. Under ever-increasing pressure to achieve greater reproducibility and finer control, with ever-shrinking device dimensions, such techniques are turning to the advantages of a cleaner, UHV environment.

MBE has proven to be a particularly versatile technique for growing semiconductor epitaxial layers, the basic starting point for all such processing steps, and it is at its heart a UHV technique. As MBE growth and UHV processing techniques mature, a welding of the two technologies into a combined semiconductor growth and processing scheme completely in UHV may eventually make feasible semiconductor device fabrication entirely in vacuum. Manufacturers of MBE equipment are already producing machines capable of performing some extended processing steps in auxiliary chambers which are all UHV vacuum interlocked with the main semiconductor growth chamber [1.31, 99]. This trend in the development of technological equipment is characteristic for the fourth milestone of MBE evolution in the 1980s.

An interesting microfabrication process performed in UHV, implemented with a focused ion beam implanter (FIBI)-MBE crystal growth system has been described by *Miyauchi* et al. [1.99, 132]. Using this system, the practical feasibility of the FIBI-MBE processing technology has been demonstrated by growing $Al_xGa_{1-x}As/GaAs$ heterojunction laser structures with high quality layer-to-layer interfaces. One of the most significant contributions of the focused ion beam (FIB) technology to device fabrication processing may be direct ion implantation of doping impurity atoms without masks or resists. Direct implantation offers an ultraclean process without contamination from resist or other organic and inorganic materials. Implantation by scanning a high-resolution ion beam opens up the possibility not only of submicron doping geometries, but also of controlling lateral doping density. Liquid alloy metal ion sources can emit various types of ion species simultaneously, providing arbitrary and instant selection of dopant atoms even during the process. Pattern design of the microfabrication process through software alone enables rapid device modification [1.100].

Figure 1.13 schematically represents the FIBI-MBE growth system [1.99], which consists of FIBI and MBE growth chambers connected via a sample transfer vacuum chamber, and its application process. The oil-free vacuum of the complete system (MBE growth chamber, FIBI chamber, and transfer corridor) is maintained at better than 10^{-8} Pa. These chambers can be isolated from each other by gate valves.

A molybdenum sample holder (2 in.) with GaAs substrate is put into the preparation chamber to bake out water vapor and then brought into the transfer chamber and set on a sample carriage trolley which can convey four holders at one time. This trolley, which is pulled along rails by a chain manipulated from outside, can convey the sample from the MBE chamber to the FIBI chamber, and vice versa. During the sample transfer the vacuum level deteriorates to about 10^{-7} Pa level because of outgas from mechanical friction. However, the vacuum level recovers instantly after the trolley stops. This short degradation of vacuum level will not seriously contaminate the sample surface.

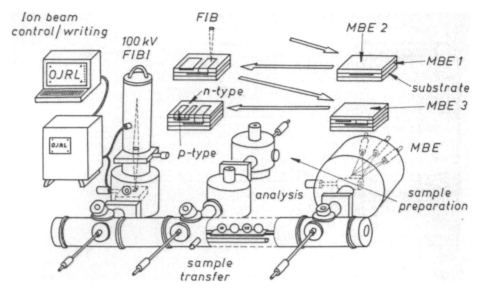

Fig. 1.13. FIBI-MBE crystal growth and processing system. The pattern-doping multilayer growth process which can be performed with this system is shown in the upper part of the picture [1.99]

As in the process drawn in the upper part of Fig. 1.13, the purpose of this system is to arbitrarily fabricate p- and n-type pattern doping geometries controlled in depth and lateral directions with fine resolution over GaAs and AlGaAs multilayers. To realize this structure, actual ion beam writing in an MBE-FIBI process is performed.

Considering the experimental results published so far [1.99, 100], one has to conclude that the FIBI-MBE processing technology offers an exciting prospect of freedom in device design and remarkable performance improvements that cannot be provided by any conventional, separated processing technology.

1.3.5 A Classification Scheme for the MBE Techniques

Completing this section it is worthwhile constructing a classification scheme for all current modifications of the MBE technique. Such a scheme is shown in Fig. 1.14, according to the state of the art at the end of 1987.

A comprehensive bibliography of over 1600 references spanning the first 25 years of activities in MBE of III–V compounds was compiled by *Ploog* and *Graf* in 1984 [1.133]. The publications concerning MBE in general, including papers on II–VI and IV–VI compounds, as well as silicon and silicon-related compounds, which appeared before the end of 1986 may be found in [1.1–8, 58, 134], while recent publications, up-dated to the end of 1987 but including also some papers and books which will appear in 1988, are cited in the reference list at the end of this book.

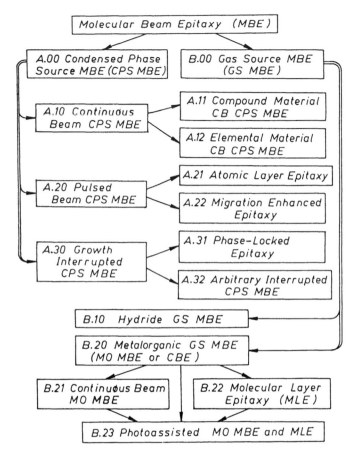

Fig. 1.14. A classification scheme of current modifications of the MBE technique

2. Sources of Atomic and Molecular Beams

The application of MBE to the growth of compound semiconductors for devices and monolithic circuits requires excellent film uniformity and reproducibility of growth conditions [2.1–7]. The uniformity in thickness, as well as in composition, of films grown by MBE depends on the uniformity of the molecular beams across the substrate. As already discussed (Sect. 1.1.2), the uniformity of the molecular beam patterns upon the substrate depends on the geometry of the "sources-substrate" system, and on the angular flux distribution of the individual sources in the system. The best uniformity of beam patterns is obtained with a sufficiently large source-to-substrate spacing, and with flux distributions at the source orifices which are isotropic in the solid angle subtended by the substrate [2.8]. The reproducibility of the growth process depends, on the other hand, on the long term stability of the beam fluxes, as well as on the flux transients resulting from the shutter operations, e.g., from cooling of the surface of the charge contained in the source upon shutter opening [2.9].

There is one other factor that is source-related and that influences considerably the MBE process. This is the structure of the molecular species used for the MBE growth. It has been shown by many experiments [2.10–13] that the properties of compound films grown by MBE depend on the kind of molecules, dimeric or tetrameric, involved in the growth. This results from the differences in the surface chemistry of the growth process relevant to different molecular species of the same element. For example, the dissociation mechanisms of the two arsenic species As_2 and As_4 are quite different during the MBE growth of GaAs and related III–V compounds [2.11]. This may be generalized for the case of phosphorus (P_2 and P_4) and antimony (Sb_2 and Sb_4) when compounds with these elements as constituents are grown [2.10]. Replacing molecular species by corresponding atoms leads to further changes in the growth mechanism of compound films [2.14].

All source-related parameters that influence the MBE process will be discussed in this chapter. A presentation of different constructions of the beam sources used in present-day MBE systems will also be given here. This includes beam sources applied in CPS MBE systems (conventional and dissociation effusion cells, electron beam and laser radiation heated sources), as well as gas sources used in GS MBE (arsine and phospine cracker sources, MO MBE gas sources).

2.1 The Effusion Process and the Ideal Effusion Cell

The effusion process belongs to the class of evaporation phenomena, and is thus described by evaporation theory [Ref. 2.15; Sect. 1.3]. Early attempts to express quantitatively the rates at which condensed materials enter the gaseous state are connected mainly with the names of Hertz, Knudsen and Langmuir. The observation of deviations from the originally postulated ideal behavior led to refinements of the transition mechanisms, which became possible as the understanding of molecular and crystalline structures increased. As a result, the evaporation theory includes concepts of reaction kinetics, thermodynamics and solid-state theory. The questions pertaining to the directionality of evaporating molecules are primarily answered by statistical considerations derived from gas kinetics and sorption theory. The first systematic investigation of evaporation rates in a vacuum was conducted by *Hertz* in 1882 [2.16]. He distilled mercury at reduced air pressure and observed the evaporation losses while simultaneously measuring the hydrostatic pressure exerted on the evaporating surface by the surrounding gas. From these observations, he drew the important and fundamental conclusion that a liquid has a specific ability to evaporate and cannot exceed a certain maximum evaporation rate at a given temperature, even if the supply of heat is unlimited. Furthermore, the theoretical maximum evaporation rates are obtained only if as many evaporant molecules leave the surface as would be required to exert the equilibrium pressure p_{eq} on the same surface and none of them return. The latter condition means that a hydrostatic pressure of $p = 0$ must be maintained. Based on these considerations, the number of molecules dN_e evaporating from a surface area A_e during the time dt is equal to the impingement rate given by (1.8) with the equilibrium pressure p_{eq} inserted, minus a return flux corresponding to the hydrostatic pressure p of the evaporant in the gas phase:

$$\frac{dN_e}{A_e dt} = (p_{eq} - p)\sqrt{\frac{\mathcal{N}_A}{2\pi M k_B T}} \ [\mathrm{m^{-2}s^{-1}}] \quad , \tag{2.1}$$

where M is the molecular weight of the evaporating species, and k_B and $\mathcal{N}_A$ are the Boltzmann and Avogadro constants, respectively.

The evaporation rates originally measured by Hertz were only about one-tenth as high as the theoretical maximum rates. The latter were actually obtained by *Knudsen* [2.17] in 1915. Knudsen argued that molecules impinging on the evaporating surface may be reflected back into the gas rather than incorporated into the liquid. Consequently, there is a certain fraction $(1 - a_v)$ of vapor molecules which contribute to the evaporant pressure but not to the net molecular flux from the condensed phase into the vapor phase. To account for this situation, he introduced the evaporation coefficient a_v, defined as the ratio of the observed evaporation rate in vacuum to the value theoretically possible according to (2.1). The most general form of the evaporation rate equation is then

$$\frac{dN_e}{A_e dt} = a_v (p_{eq} - p) \sqrt{\frac{\mathcal{N}_A}{2\pi M k_B T}} \; [\mathrm{m^{-2} s^{-1}}] \quad , \tag{2.2}$$

which is commonly referred to as the Hertz-Knudsen equation.

Knudsen found the evaporation coefficient a_v to be strongly dependent on the condition of the mercury surface. In his earlier experiments, where evaporation took place from the surface of a small quantity of mercury, he obtained values of a_v as low as 5×10^{-4}. Concluding that the low rates were attributable to surface contamination, he allowed carefully purified mercury to evaporate from a series of droplets which were falling from a pipette and thus continually generated fresh, clean surfaces. This experiment yielded the maximum evaporation rate

$$\frac{dN_e}{A_e dt} = p_{eq} \sqrt{\frac{\mathcal{N}_A}{2\pi M k_B T}} \; [\mathrm{m^{-2} s^{-1}}] \quad . \tag{2.3}$$

2.1.1 Langmuir and Knudsen Modes of Evaporation

It was first shown by *Langmuir* in 1913 [2.18] that the Hertz-Knudsen equation also applies to evaporation from free solid surfaces. He investigated the evaporation of tungsten from filaments in evacuated glass bulbs and assumed that the evaporation rate of a material at pressures below 1 Torr is the same as if the surface were in equilibrium with its vapor. Since recondensation of evaporated species was thereby excluded, he derived the maximum rate as stated by (2.3). Phase transitions of this type, which constitute evaporation from free surfaces, are commonly referred to as Langmuir or free evaporation.

An alternative evaporation technique was established by *Knudsen* [2.19] and is associated with his name. In Knudsen's technique, evaporation occurs as effusion from an isothermal enclosure with a small orifice (Knudsen cell). The evaporating surface within the enclosure is large compared with the orifice and maintains the equilibrium pressure p_{eq} inside. The diameter of the orifice must be about one-tenth or less of the mean free path of the gas molecules at the equilibrium pressure, and the wall around the orifice must be vanishingly thin so that gas particles leaving the enclosure are not scattered or adsorbed and desorbed by the orifice wall. Under these conditions, the orifice constitutes an evaporating surface with the evaporant pressure p_{eq} but without the ability to reflect vapor molecules; hence $a_v = 1$. If A_e is the orifice area, the total number of molecules effusing from the Knudsen cell into the vacuum per unit time, which will be called hereafter the total effusion rate Γ_e, is given by

$$\Gamma_e \equiv \frac{dN_e}{dt} = A_e (p_{eq} - p_v) \sqrt{\frac{\mathcal{N}_A}{2\pi M k_B T}} \; [\mathrm{molecules \; s^{-1}}] \quad , \tag{2.4}$$

where p_v is the pressure in the vacuum reservoir to which the molecules effuse from the cell orifice. This is the Knudsen effusion equation. It may be simplified by setting $p_v = 0$, which is reasonable for effusion into an UHV environment.

The Knudsen equation is often written in the form

$$\Gamma_e \equiv 3.51 \times 10^{22} \frac{p A_e}{\sqrt{MT}} \ [\text{molecules s}^{-1}] \quad , \tag{2.5}$$

where p is the pressure in the effusion cell in Torrs, and all other quantities are in cgs units (see [Ref. 2.20; Chap. 6] or [Ref 2.21; Chap. 1]). Expressing these quantities in SI units, one has to replace the numerical factor in (2.5) by 8.33×10^{22}, see (1.9).

Langmuir's and Knudsen's modes of evaporation have been employed in many experimental methods of determining the vapor pressure of materials and heats of vaporization [2.22, 23]. A critical examination of both techniques with their limitations has been published by *Rutner* in [2.24]. Langmuir's method suffers from the uncertainty of whether or not an observed rate of weight loss truly reflects the equilibrium rate of evaporation. It is often used, however, to determine a_v by comparing its results with independently known vapor-pressure data or with evaporation-rate measurements from Knudsen cells. The principal problem with Knudsen's technique is that an ideal cell with an infinitely thin-walled orifice yielding free molecular flow can only be approximated. In practice, orifices of finite thickness must be used, which necessitates the application of corrective terms in the effusion equation.

2.1.2 The Cosine Law of Effusion

Let us consider an ideal Knudsen effusion cell. It contains a condensed phase and vapor of the charge material that are in thermodynamical equilibrium with each other. The effusion aperture of this cell is a small orifice in an infinitesimally thin cell lid (the diameter of the orifice is much smaller than the mean free path of the gas particles in the cell, and there is no orifice wall). The gas reservoirs outside and inside the cell are large enough that the molecules encounter each other more frequently than the walls. The gas pressure inside the cell is much higher than the pressure outside ($p_{eq} \gg p_v$). As is shown in Fig. 2.1, the molecules or atoms of the charge vapor escape from the cell orifice into the ultrahigh vacuum environment in all directions encompassed by a hemisphere with its center coincident with the orifice center. An individual molecule m moving along a straight line in the direction ϑ toward the cell orifice center passes it and escapes into the vacuum in the same direction, because the thickness of the orifice wall is assumed to be infinitesimally thin ($L_0 \approx 0$). The differential angular effusion rate $d\Gamma_\vartheta$ from the orifice area A_e into the vacuum in the directions contained between ϑ and $\vartheta + d\vartheta$ is porportional to:

1. The surface of the orifice seen by the molecule

$$d\Gamma_\vartheta \sim A_e \cos \vartheta \quad .$$

2. The number of molecules entering unit area of the orifice in unit time

$$d\Gamma_\vartheta \sim \frac{\Gamma_e}{A_e} = 3.51 \times 10^{22} \frac{p}{\sqrt{MT}} \quad .$$

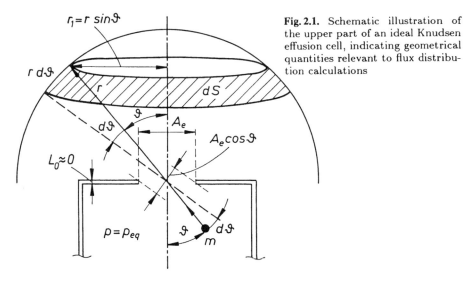

Fig. 2.1. Schematic illustration of the upper part of an ideal Knudsen effusion cell, indicating geometrical quantities relevant to flux distribution calculations

3.   The probability P that a molecule m enters the orifice in a direction between ϑ and $\vartheta + d\vartheta$. This probability is proportional to the solid angle $d\omega$ at dS subtended by these directions. The solid angle element $d\omega$ is measured by the area on a sphere of radius r (Fig. 2.1) of the circular zone between the angle ϑ and $\vartheta + d\vartheta$, and is equal to

$$d\omega = \frac{dS}{r^2} = 2\pi \, \sin \vartheta \, d\vartheta \quad . \tag{2.6}$$

The total solid angle into which the molecules escape from the cell orifice is equal to 2π, so the probability that ϑ lies between ϑ and $\vartheta + d\vartheta$ is

$$P = \frac{d\omega}{2\pi} = \sin \vartheta \, d\vartheta \quad . \tag{2.7}$$

Thus $d\Gamma_\vartheta \sim \sin \vartheta \, d\vartheta$.

The differential angular effusion rate $d\Gamma_\vartheta$ of molecules escaping from the cell orifice may be expressed by the formula

$$d\Gamma_\vartheta = (C_0 A_e \, \cos \vartheta) \left[\left(\frac{\Gamma_e}{A_e} \right) \sin \vartheta \, d\vartheta \right] \quad , \tag{2.8}$$

where C_0 is a proportionality constant. This constant can be estimated, taking into consideration the obvious fact that $0 \leq \vartheta \leq \pi/2$ and that the total effusion rate of molecules escaping from the orifice must be equal to Γ_e given by (2.5). Thus

$$\Gamma_e = \int_0^{\pi/2} d\Gamma_\vartheta = \int_0^{\pi/2} C_0 \Gamma_e \, \cos \vartheta \, \sin \vartheta \, d\vartheta = \frac{C_0 \Gamma_e}{2} \quad , \tag{2.9}$$

which means that the proportionality constant $C_0 = 2$.

33

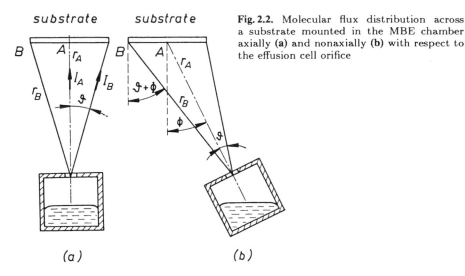

Fig. 2.2. Molecular flux distribution across a substrate mounted in the MBE chamber axially **(a)** and nonaxially **(b)** with respect to the effusion cell orifice

(a)   *(b)*

The formula (2.8) may now be written in the form

$$dI_\vartheta = \frac{I_e}{\pi} \cos \vartheta \, d\omega \quad . \tag{2.10}$$

Equation (2.10) represents the so-called cosine law of emission, which is equivalent to Lambert's law in optics [Ref. 2.15; Sect. 1.3]. We will call it hereafter the cosine law of effusion.

Using the Knudsen effusion equation in the form (2.5) and the cosine law (2.10), one may easily calculate the impingement rate I_A (also called flux in [2.25]) of the molecular beam at the central point A of the substrate axially mounted in front of the effusion cell (Fig. 2.2a). This impingement rate is given by the angular effusion rate for $\vartheta = 0$ per unit substrate area $dS = d\omega r_A^2$ around point A :

$$I_A \equiv \left. \frac{dI_\vartheta}{dS} \right|_{\vartheta=0} = \frac{I_e}{\pi r_A^2}$$

$$= 1.118 \times 10^{22} \frac{pA_e}{r_A^2 \sqrt{MT}} \text{ [molecules cm}^{-2}\text{s}^{-1}] \quad , \tag{2.11}$$

where p is expressed in Torrs, and the other quantities in cgs units [Ref. 2.20; Chap. 6 or Ref. 2.21; Chap. 1]. If SI units are used, the numerical factor must be changed to 2.653×10^{22}.

The flux at the edge point B is defined by $I_B = dI_\vartheta/dS(\vartheta)$, where $dS(\vartheta)$ is the surface element around point B given by $dS(\vartheta) = [(d\omega r_A^2)/\cos \vartheta] \cdot (r_B^2/r_A^2)$. Thus I_B can be written in the form

$$I_B = \frac{I_e \cos^2 \vartheta}{\pi r_A^2} \cdot \frac{r_A^2}{r_B^2} = I_A \cos^4 \vartheta \quad . \tag{2.12}$$

Proceeding in an analogous way the impingement rate from a source which is tilted by an angle ϕ from the perpendicular substrate axis can be calculated (Fig. 2.2b):

$$I'_A = I_A \cos \phi \quad , \tag{2.13}$$

$$I'_B = I_A \frac{r^2_A}{r^2_B} \cos \vartheta \cos (\vartheta + \phi) \quad . \tag{2.14}$$

2.2 Effusion from Real Effusion Cells

The effusion cells used in MBE systems are usually either cylindrical crucibles with a single circular collimating orifice in the crucible lid [Ref. 2.25; Sect. 2.2.2], or single-channel-type crucibles cylindrical or conical in shape [2.9, 26, 27].

Let us first consider a cylindrical, near-ideal (approximately true Knudsen type) effusion cell, i.e., a cell in which the vapor pressure exhibits a near-equilibrium value at the exit orifice. Such cells are used in practical MBE systems as dopant or as calibration cells.

2.2.1 The Near-Ideal Cylindrical Effusion Cell

The theoretical analysis of near-ideal cylindrical cells was given by *Clausing* [2.28, 29], who assumed a random return of molecules to the gas reservoir from the orifice walls according to a cosine law of return.

According to Clausing's theory the effusion equation (2.4) and consequently also (2.11) should be multiplied by a correction factor W_a called the orifice transmission factor. This factor has the meaning of the probability that a molecule which enters the orifice in the effusion cell lid at the side of gas reservoir 1 goes directly to the second reservoir (in which $p_2 \approx 0$) without having been back to the first one. Clausing's factor W_a depends only on the geometry of the orifice. As an example, in Table 2.1 some values are given of W_a for a straight tubular orifice of length L_0 and diameter d_0.

Table 2.1. Values of Clausing's transmission factor W_a for straight tubular orifices with various orifice dimensions [2.29]

L_0/d_0	W_a	L_0/d_0	W_a
0	1.0000	0.75	0.5810
0.1	0.9092	1.00	0.5136
0.2	0.8341	1.50	0.4205
0.4	0.7177	2.00	0.3589
0.5	0.6720	4.00	0.2316

A modification to Clausing's theory has been proposed by *Winterbottom* and *Hirth* [2.30]. They have discussed the likelihood that molecules, rather then being randomly reflected, are temporarily adsorbed when they strike an orifice wall, diffuse along the surface, and eventually re-evaporate into the vapor phase. This leads to a correction factor for (2.4) larger than Clausing's factor W_a. The numerical data for this factor can be found in [2.31].

Further modifications to the effusion equation have been introduced by *Motzfeldt* [2.32], who has taken into account the influence of the main body of the cell.

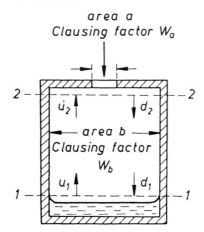

Fig. 2.3. Diagram of a cylindrical effusion cell after *Motzfeldt* [2.32]

Let us consider the cell shown in Fig. 2.3, for which we will discuss the pressure in plane 1 just above the evaporation surface, and in plane 2 just below the lid of the cell. At steady state and with no hole, the pressure will be evenly distributed throughout the cell. With a finite hole size, this will no longer be the case. The upward pressure u will be different from the downward pressure d for each plane, and the pressures will be different in the different planes. Motzfeldt gave the following relations between these pressures:

$$\alpha p_{eq} = \alpha d_1 + f u_2 \quad , \tag{2.15}$$

$$u_1 = \alpha p_{eq} + (1 - \alpha)d_1 \quad , \tag{2.16}$$

$$d_2 = (1 - f)u_2 \quad , \tag{2.17}$$

$$u_2 = W_b u_1 + (1 - W_b)d_2 \quad . \tag{2.18}$$

The equations express:

(2.15)   The rate of evaporation, which is proportional to αp_{eq} must be equal to the rate of recondensation plus the rate of escape through the hole.

(2.16)   The upward pressure u_1 above the surface is made up of the evaporation from this surface plus the rebounding of molecules coming from above.

(2.17)   The downward pressure just below the lid is due to reflection of the upward molecules minus those that escape through the hole.

(2.18)   The upward pressure u_2 at the top of the cell is due to the fraction W_b of the molecules that are heading upward from the bottom of the cell plus the fraction $(1 - W_b)$ of the molecules heading downward from the top of the cell, because this fraction is reflected upwards again by the walls of the cell.

The pressure u_2 can be expressed using (2.4) with Clausing's transmission factor W_a for the cell orifice and the condition $p_v = 0$:

$$u_2 = \frac{\Gamma_e}{A_e W_a} \sqrt{\frac{2\pi M k_B T}{N_A}} . \tag{2.19}$$

There are now as many equations as unknowns (u_1, d_1, d_2 and p_{eq}), so p_{eq} can be expressed in terms of u_2, α, f and W_b. The verified expression for the total effusion rate for the cell shown in Fig. 2.3 is given as

$$\Gamma_e^{Mot} = \frac{A_e p_{eq} W_a}{1 + f(\alpha^{-1} + W_b^{-1} - 2)} \sqrt{\frac{N_A}{2\pi M k_B T}} , \tag{2.20}$$

where p_{eq} is the equilibrium pressure of the gas at the surface of the evaporant condensed phase at temperature T, α is the gas condensation coefficient, W_a is the Clausing transmission factor for the orifice [2.28], W_b is the Clausing transmission factor for the main body of the cell, and f is the escape coefficient for the orifice of the cell [$f = W_a(a/b)$]. This is the verified effusion equation applying to a near-ideal cylindrical cell.

For real effusion cells, the cosine law (2.10) is also not valid exactly, and consequently has to be multiplied by relevant correction factors. An example of how such a correction factor may be determined is presented below on the basis of Clausing's theory [2.28, 29], according to Dayton's calculations for a cylindrical near-ideal effusion cell [2.33, 34].

Figure 2.4 shows a cylindrical tube located between a large chamber in which gas is everywhere at rest at a constant pressure p_1, except near the entrance to the tube, and a large chamber in which the pressure is approximately zero except near the exit of the tube where gas molecules emerge after passing through the tube under conditions of "molecular flow". The term "molecular flow" has been introduced to denote cases in which the mean free path of the gas molecules is large with respect to the orifice dimensions.

The total number of molecules entering the tube in unit time from all directions is given by the effusion rate of reservoir 1:

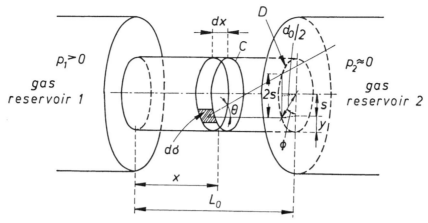

Fig. 2.4. Schematic diagram of a gas particle escape line through circular exit from a cylindrical tube wall [2.34]

$$\Gamma = w_1 \pi \frac{d_0^2}{4} \quad , \tag{2.21}$$

where w_1 is the number of molecules per unit time and unit area crossing any plane from one side only within the first chamber as given by the effusion law for a gas at rest at pressure p_1, see (1.8) and (2.5). Some of the molecules which enter the tube will then pass in a straight line from entrance to exit without striking the wall or colliding with other gas molecules. The number of these molecules per unit time can be calculated making use of the sketch shown in Fig. 2.5. Referring to this sketch it is evident that only the overlapping (shaded) area of projected entrance and exit will allow molecules to pass in the direction θ. This area is

$$S_s = \frac{d_0^2}{2} \arccos\left(\frac{2s}{d_0}\right) - 2s\sqrt{\frac{d_0^2}{4} - s^2} \quad , \tag{2.22}$$

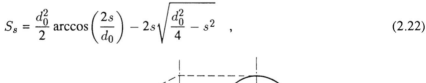

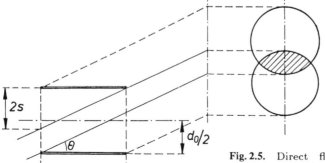

Fig. 2.5. Direct flight through both the entrance and exit areas of a cylindrical tube of diameter d_0 and length L_0 [2.34]

where $s = (L_0/2) \tan \theta \leq d_0/2$. It can be shown (Sect. 2.1.2) that the angular effusion rate from the first chamber in a unit solid angle whose axis is perpendicular to the entrance plane is w_1/π. The total effusion rate directed within the solid angle $d\omega$ having the form of an infinitesimally narrow cone whose axis makes an angle θ with the tube axis, as shown in Fig. 2.5, such that the molecules can escape directly through the exit is therefore

$$d\Gamma_\theta = \frac{w_1}{\pi} \cos \theta \, d\omega \, S_s \quad , \tag{2.23}$$

where S_s is given by (2.22). This fraction of the molecules which enter the tube will emerge from the exit without any interaction with the wall of the tube or with other gas molecules. The rest of the molecules entering the tube will strike the wall for the first time at various distances from the entrance.

A molecule striking the wall of the tube has equal probability of leaving the wall in any given direction within 90° to the normal to the wall, when molecular roughness of the tube wall is assumed. This will result in a cosine law distribution of molecules coming from any element of the wall area $d\sigma$ small enough to be treated as a plane. Some of the molecules leaving the wall will pass directly through the exit, some will return directly through the entrance, and the rest will again strike the wall at another place. This model of diffuse scattering and the assumption that molecules that will escape through the exit never come back to the tube are supplemented in Clausing's and Dayton's theories.

Let us now return to Fig. 2.4. The total number of molecules colliding with the wall per square centimeter per second at x will be given by some function of x, d_0 and L_0, which will be designated $g(x)$. *Clausing* [2.29] has derived an exact integral equation for this function, but for a practical solution of the problem of gas flow through short tubes he presented the following approximate formula giving good results when $L_0 \leq 2d_0$:

$$g(x) = w_1 \left(\alpha + (1 - 2\alpha) \frac{L_0 - x}{L_0} \right) \quad \text{with} \tag{2.24}$$

$$\alpha = \frac{\sqrt{L_0^2 + d_0^2} - L_0}{d_0 + d_0^2/\sqrt{L_0^2 + d_0^2}} \quad . \tag{2.25}$$

Some of the molecules which have arrived at the segment $\pi d_0 dx$ will leave the wall and pass directly through both the plane C and exit D without striking the wall again. Consider only those molecules leaving an element of area $d\sigma = (d_0/2) d\phi \, dx$ whose direction of flight lies within a given infinitesimal solid angle $d\omega$ which we shall assume to have the form of a cone whose axis is parallel to a ray drawn in the x, y plane at an angle θ to the tube axis. Then to pass through both C and D, these molecules must come from those elements $d\sigma$ of the segment $\pi d_0 \, dx$ for which $y \leq (d_0/2) - s$, where

$$s = \frac{L_0 - x}{2} \tan \theta \quad . \tag{2.26}$$

The projection of the area $d\sigma$ in the direction corresponding to θ will be

$$dA = \cos\phi \sin\theta \, d\sigma = \frac{d_0}{2}\cos\phi \sin\theta \, d\phi \, dx \quad , \tag{2.27}$$

where ϕ is the angle between the radius of the tube directed to $d\sigma$, and the x, y plane. Since $g(x)/\pi$ is the number of molecules leaving $d\sigma$ per unit time per unit area per unit solid angle in a direction perpendicular to $d\sigma$, the corresponding effusion rate $d\Gamma(d\sigma)$ of $d\sigma$ within the solid angle $d\omega$ in the direction given by θ will be

$$d\Gamma(d\sigma) = \frac{g(x)}{\pi}dA \, d\omega \quad . \tag{2.28}$$

The total effusion rate of those elements $d\sigma$ for which $y \leq (d_0/2) - s$ in the direction of θ is obtained after substituting (2.27) into (2.28) and by integrating this equation with respect to ϕ between the limits $-\phi_s$ and ϕ_s, where

$$\phi_s = \arccos\left(\frac{2s}{d_0}\right)$$

corresponding to the value of ϕ for which $y = (d_0/2) - s$. This effusion rate will be

$$\begin{aligned}
d\Gamma_x &= \frac{g(x)}{\pi}d_0 d\omega \, \sin\theta \, dx \int_0^{\phi_s} \cos\phi \, d\phi \\
&= \frac{g(x)}{\pi}d_0 d\omega \, \sin\theta \, dx \, \sin\phi_s \quad .
\end{aligned} \tag{2.29}$$

This equation becomes, after substituting

$$\sin\phi_s = \frac{2\sqrt{d_0^2/4 - s^2}}{d_0} \quad \text{and} \quad dx = \frac{2ds}{\tan\theta} \quad ,$$

$$d\Gamma_x = -\left(\frac{4g(x)}{\pi}\right)d\omega \, \cos\theta\sqrt{\frac{d_0^2}{4} - s^2}\right)ds \quad . \tag{2.30}$$

Substituting now (2.24) and (2.26) in (2.30) and integrating from $s = (L_0/2)\tan\theta$ to $s = 0$, corresponding to the limits $x = 0$ and $x = L_0$, gives the effusion rate $d\Gamma_W$ of the molecules leaving the wall of the tube and escaping directly through the exit in the direction defined by $d\omega$ and the angle θ, when $\tan\theta \leq d_0/L_0$:

$$d\Gamma_W = \frac{w_1}{4}\cos\theta \, d\omega \, d_0^2 B(p \leq 1) \quad , \quad \text{where} \tag{2.31}$$

$$B(p \leq 1) = \frac{2\alpha}{\pi}(\arcsin p + p\sqrt{1 - p^2})$$

$$+ \frac{4(1 - 2\alpha)}{3\pi}\left(\frac{1 - (\sqrt{1 - p^2})^3}{p}\right) \quad \text{and} \tag{2.32}$$

$$p = \frac{L_0}{d_0}\tan\theta \leq 1 \quad . \tag{2.33}$$

Equation (2.31) gives the effusion rate of the molecules leaving the wall of the tube and escaping directly through the exit in the given direction for $p \leq 1$ (this inequality is equivalent to $\tan \theta \leq d_0/L_0$).

The total effusion rate $d\Gamma_C$ from the exit in the direction defined by $d\omega$ and the angle θ, when $\tan \theta \leq d_0/L_0$, is

$$d\Gamma_C(p \leq 1) = d\Gamma_\theta + d\Gamma_W \quad . \tag{2.34}$$

This equation can be written in the form

$$d\Gamma_C(p \leq 1) = \frac{w_1}{4} \cos \theta \, d\omega \, d_0^2 C_0 \quad , \quad \text{where} \tag{2.35}$$

$$C_0(p \leq 1) = 1 - \frac{2}{\pi}(1 - \alpha)\left(\arcsin p + p\sqrt{1 - p^2}\right)$$
$$+ \frac{4}{3\pi p}(1 - 2\alpha)\left[1 - \left(\sqrt{1 - p^2}\right)^3\right] \quad . \tag{2.36}$$

When $\tan \theta \geq d_0/L_0$, we can integrate only from $s = d_0/2$ to $s = 0$, corresponding to the limits $x = L_0 - (d_0/\tan \theta)$ and $x = L_0$. The result has the same form as (2.35), however, the correction factor is now given by

$$C_0(p \geq 1) = \alpha + \frac{4(1 - 2\alpha)}{3\pi p} \quad . \tag{2.37}$$

Equation (2.35) presents the modified version of (2.10), the cosine law of effusion, for the near-ideal cylindrical effusion cell.

The function C_0 depends only on the angle θ and the ratio $2L_0/d_0$. The quantity α occurring in the expressions defining C_0 for both ranges of p is given by (2.25) when $L_0 \leq 2d_0$. Some illustrative data taken from Clausing's work are shown in Fig. 2.6 and Table 2.2 for the case of a symmetrical effusion cell orifice ($L_0 = d_0$). It is seen that the plot of $d\Gamma_C$ in polar coordinates is given in this case as an egg-shaped curve, while the plot of the same quantity for an ideal orifice (according to Knudsen theory) has the shape of a circle [2.28, 35]. Moreover, for $\theta = 0$, $C_0 \cos \theta = 1$ if $d\omega$ is chosen small enough. Even at $\theta = 10°$, $C_0 \cos \theta = 0.88$, so that an angular beam of considerable width through a short tube gives nearly the same flow in the forward direction as a hole of the same area. This point is of considerable interest because, by using an effusion cell with one orifice of this sort, the distribution of molecules in the forward direction (i.e., in the direction needed for the beam) is nearly equal to that for an ideal orifice while the total gas escaping is reduced to one-half (see Table 2.1, $W_a = 0.5136$ for $L_0 = d_0$). This makes it possible to increase the outflow from the effusion cell orifice in the direction of the substrate wafer without spilling too many molecules into the vacuum environment outside the substrate.

Clausing's theory has been re-examined later by different authors. Using the model of Winterbottom and Hirth of surface adsorbed and diffusing molecules [2.30], *Ruth* and *Hirth* [2.31] have shown that the beams effused from a near-

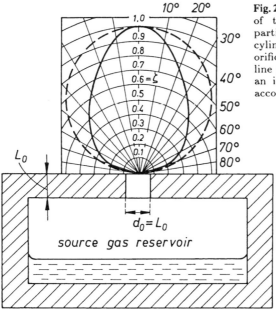

Fig. 2.6. A plot in polar coordinates of the angular distribution of gas particles effused from a near-ideal cylindrical cell with a symmetrical orifice ($L_0 = d_0$) [2.28]. The dashed line gives the relevant distribution for an ideal effusion cell ($L_0 = 0$), i.e., according to the cosine law (2.10)

Table 2.2. Values of the correction factor C_0 for the case of symmetrical orifices $L_0 = d_0$ for various evaporation angles θ [2.28]

θ	C_0	θ	C_0
0°	1.0000	50°	0.4259
10°	0.8882	60°	0.3687
20°	0.7721	70°	0.3221
30°	0.6483	80°	0.2811
40°	0.5183	90°	0.2426

ideal effusion cell are less focused than predicted by the cosine law corrections of Clausing.

Bröhl and *Hartmann* [2.36] analysed Clausing's theory with regard to long cylindrical tubes ($L_0 > 2d_0$). These authors have considered the transmission probabilities of gas particles colliding with the tube wall at sites symmetric with respect to the middle of the tube. In conclusion they have formulated two theorems. The first states that if a constant molecular gas flow is considered in a long straight tube ($L_0 = 10d_0$) with diffuse particle-wall scattering, then equal numbers of the particles colliding with the unit surface areas at x and at $(L_0 - x)$ in unit time will fly directly or indirectly into the reservoir 2 (without having been back into reservoir 1 intermediately). The second theorem states that if the same conditions are fulfilled, then equal numbers of the particles colliding with the tube left from $L_0/2$ and right from $L_0/2$ will fly into reservoir 2 directly or indirectly.

2.2.2 The Cylindrical Channel Effusion Cell

It is evident that a significant collimation of the molecular beam generated in an effusion cell can be reached by enlarging the value of the ratio L_0/d_0 for the orifice of the cell. This problem has been analysed by *Dayton* [2.33], who proposed the following, more general, formula for α:

$$\alpha = \frac{u\sqrt{u^2+1} - v\sqrt{v^2+1} + v^2 - u^2}{[u(2v^2+1) - v]/\sqrt{v^2+1} - [v(2u^2+1) - u]/\sqrt{u^2+1}} \tag{2.38}$$

with $u = L_0/d_0 - v$ and $v = L_0\sqrt{7}/(3L_0 + d_0\sqrt{7})$. This formula gives good results when $L_0 > 2d_0$, i.e., in the case of a channel-type orifice, while Clausing's formula (2.25), gives a good approximation for thin circular orifices, i.e., when $L_0 \leq 2d_0$.

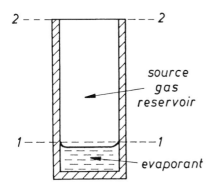

source
gas
reservoir

evaporant

Fig. 2.7. Effusion cell with the shape of a straight cylindrical crucible without a lid

Let us now consider a typical single-channel effusion cell with the shape of a straight cylindrical crucible without a lid, like the one shown in Fig. 2.7. Here the two planes perpendicular to the crucible axis marked as 1-1 and 2-2 correspond to the entrance and exit of a circular tube (Fig. 2.4), which is the subject of Clausing's and Dayton's theories. So, all results concerning the angular number distribution of molecules flowing through a cylindrical tube connecting two vacuum chambers can be directly exploited in the case of single channel effusion cell without a lid, assuming that the evaporation process from the surface of the crucible contents gives gas molecules in the infinitesimally thin gas layer contacting the evaporation surface, for which the assumptions concerning the Knudsen effusion equation (2.4) are fulfilled. In practical effusion procedures used in MBE this is just the case [2.35]. So, we can use the results of Dayton's calculations to sketch in polar coordinates the angular patterns of the gas jets effused from cylindrical crucibles (Fig. 2.8 and Table 2.3) [2.37,38].

The problem of the angular distribution of the flux emerging at the exit surfaces of cylindrical tubes has been recalculated and verified by a Monte Carlo simulation technique by *Krasuski* [2.39]. He has considered the so-called near field flux distribution, while in papers published before, the emphasis had been

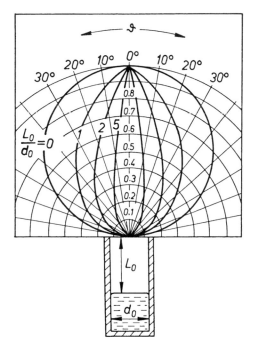

Fig. 2.8. The angular term $C_0 \cos \theta$ of the flux angular distribution for several positions of the charge surface in a single channel effusion cell of diameter d_0 [2.33–35]

Table 2.3. Values of the correction factor C_0 for one-channel effusion cells with different dimensions for various evaporation angles θ [2.33]. The values of C_0 for cells with $L_0/d_0 = 1$ are identical to the values given in Table 2.2

θ	L_0/d_0: 0.5	2	5
0°	1.0000	1.0000	1.0000
10°	0.9439	0.7785	0.4875
20°	0.8844	0.5660	0.2949
30°	0.8184	0.4108	0.2154
40°	0.7397	0.3335	0.1784
50°	0.6399	0.2831	0.1543
60°	0.5139	0.2457	0.1364
70°	0.4337	0.2151	0.1218
80°	0.3783	0.1883	0.1090
90°	0.3263	0.1631	0.0969

rather on the far-field distribution of the flux. The results presented in [2.39] show a considerable variation of angular distribution of the flux for different zones in the exit surface. This means that the shapes of the distribution curves are no longer the same when originating at different places on this surface. The data concerning the near field pattern of the flux distribution are especially important for the case when the exit of the effusion cell is partly closed by the shutter or by another diaphragm.

2.2.3 Hot-Wall Beam Cylindrical Source

Due to the large distance between source and substrate in conventional MBE systems (Sect. 1.1.1) the far-field distribution of the flux is the main interest from the practical point of view, and therefore much effort has been expended on theoretical calculations. However, in the case of hot-wall beam epitaxy (HWBE) (see Sect. 3.2.1 for a detailed description) the substrate is mounted very close to the source orifice. As shown in Fig. 2.9 the hot-wall beam cylindrical source consists of a source assembly part and a separately heated cylindrical tube. Because of the high vapor pressure existing inside the source, nearly all particles effusing from the source collide with the hot wall. Consequently this section of the tube acts as an equilibrating unit for the particle beam.

To calculate the near field distribution of the flux, Monte Carlo simulations were performed by *Humenberger* and *Sitter* [2.40]. A simplified model of the

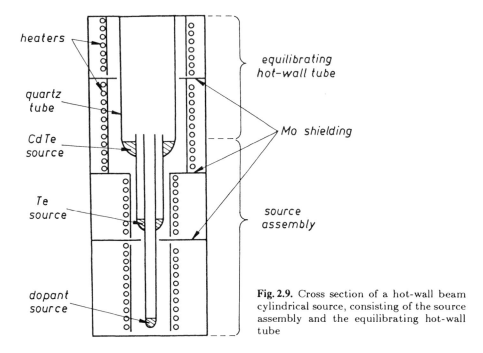

equilibrating
hot-wall tube

Mo shielding

source
assembly

Fig. 2.9. Cross section of a hot-wall beam cylindrical source, consisting of the source assembly and the equilibrating hot-wall tube

hot-wall beam cylindrical source was used for the computer simulation as depicted in Fig. 2.10. According to a possible path of the molecules through the tube as indicated in Fig. 2.10, the calculations are done in the following steps: Calculations begin by generating a molecule which is randomly selected according to the stoichiometry of the source (A). The particle starts moving by means of a velocity which is generated following the rules of kinetic theory of gases. The intersection point of the trajectory with the tube wall is calculated (B). The density of particles between points A and B in Fig. 2.10 determines the value of the

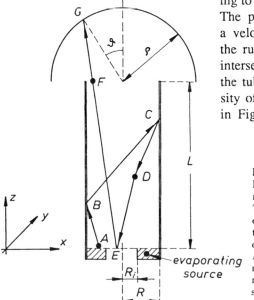

evaporating
source

Fig. 2.10. Geometry of the schematic hot-wall evaporation tube used as a model for the Monte Carlo simulation. The indicated path of a particle (A–G) contains all events which can happen to an evaporated molecule (B, C: wall collisions; D : intermolecular collision; E: collision with ground plane or recondensation). (L: tube length; R: radius of source; R_i: inner radius of ring-shaped source) [2.40]

mean free path $\overline{\lambda}$, which the particle under consideration can expect. Depending on $\overline{\lambda}$ and the distance between A and B the molecule may hit the wall at point B or undergo a collision on the path between A and B. In each case the event and the location of occurrence is recorded. For the interaction of the particle with the wall complete thermal accommodation is assumed and re-evaporation occurs according to a cosine law. No condensation of particles on the walls was allowed to occur. If a particle reaches a point (F) in the exit plane of the tube, this point is recorded. The point G, which is the intersection point of the particle path with a hemisphere of radius ϱ situated above the tube exit, is calculated to determine the angle distribution of effusing particles. The hemisphere with radius ϱ was divided into area elements of equal size, each corresponding to an angle element $\Delta\vartheta$. Since the angle distribution was calculated by simply counting particles hitting the area elements of the hemisphere at an angle $\vartheta \pm \Delta\vartheta$, ϱ was chosen to be $2R$ in order to get a significant statistical value for each element. In the case of a ring-shaped source $(R_i > 0)$ a particle hitting point E is re-evaporated; for a disk-shaped source $(R_i = 0)$ the point E is on the surface of the "source" and the particle is assumed to recondense. Each particle effusing or recondensing is followed by a new particle generated in the source.

For the calculations in the free and near-free molecule regime an iteration method was elaborated which is based on the test-particle method [2.41]. In this method, test particles moving among target particles can hit walls, collide with target particles, be ejected through the mouth of the tube or recondense in the source. By following the path of the test particles, they by themselves set up a density distribution, which gives the target distribution for the next iteration. This is an iterative method which starts with a plausible starting distribution of the target particles and ends when the density distribution does not change significantly with additional iterations.

The most important information that could be drawn from the calculations is given in Figs. 2.11 and 2.12. Figure 2.11a shows the relative density of particles crossing the exit plane of the tube for different source temperatures, i.e., different equilibrium vapor pressures p_{eq} over the source material. It is evident from these results that the vapor pressure over the source has little influence on the distribution of impingement rate in the exit plane and also at a distance $d = \varrho$ from the exit. In Fig. 2.11b the impingement rate onto a plane a distance ϱ above the tube opening is drawn. The central dip in the impingement rate in Fig. 2.11b was only observed for a tube with $L/R = 3$ and a ring-shaped source. For example, the simulations for a larger L/R ratio and a ring-shaped source did not reveal this feature for either low or high values of p_{eq}. The dip very clearly reflects geometrical effects of the cell, namely L/R value and ring source, which can be influenced by a proper design of the hot-wall tube. Geometrical effects can obviously only be observed when the flow is almost collisionless, at least in the upper part of the tube. In fact, for higher vapor pressure and whenever L/R is higher than 3, the dip decreases. This fact underlines again the important equilibrating effect of the hot wall tube.

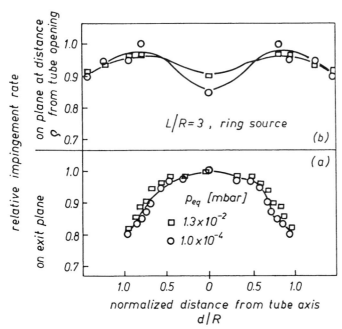

Fig. 2.11. Density of particles (a) crossing the exit of the hot-wall tube, (b) impinging on a plane situated at a distance ϱ above the tube opening [2.40]

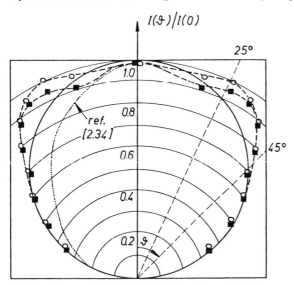

Fig. 2.12. Angular distribution of effusing particles for a tube of $L/R = 4$ with a disk-shaped source [2.40]. (o) $PQ_1 = 1 \times 10^{-4}$ mbar; (■) $PQ_1 = 2 \times 10^{-2}$ mbar). The dotted line represents data from [2.34]

Figure 2.12 shows the angular distribution of effusing particles from a tube with $L/R = 4$ and a disk-shaped source. This type of polar diagram was obtained for all calculations except for $L/R \leq 3$ and a ring-shaped source. For angles greater than 45° the distribution is cosine-like, for smaller angles the intensity is higher than due to a cosine distribution. Furthermore, the calculations show that up to a value of about 25° the impingement rate remains equal to the intensity at $\vartheta = 0°$. Thus, almost no collimation is observed, especially for higher L/R ratios. For comparison, calculated data for effusion characteristic of a cylindrical tube with $L/R = 4$ taken from [2.34] for far-field distributions are indicated in the figure, where a pronounced collimating effect of the tube is seen.

In conclusion, calculations show that the use of the near-field distribution in comparison to the far-field distribution of Knudsen cells results in uniform impingement rates. That means that no substrate rotation is necessary in the case of hot-wall beam epitaxy to achieve homogeneous layers. These considerations are based on relative geometrical ratios and show no theoretical limit for absolute dimensions of the substrate diameter.

2.2.4 The Conical Effusion Cell

Single-channel effusion cells, conical in shape, are at present frequently used as sources of the main molecular or atomic beams in MBE systems, i.e., for beams of the constituent elements of the film to be grown [2.26, 27]. They are, however, also used in special cases for doping beams [2.42]. They are nonequilibrium cells, which means that the vapor and the condensed phase of the contents are not in thermodynamic equilibrium with each other. Therefore, the effusion equation (2.4) as well as the cosine law (2.10) cannot be directly applied in analysing the effusion pattern of conical cells. Both of these equations should somehow be corrected for this case. One way of doing this is based on the procedure, proposed by *Shen* [2.43], treating the effusing surface of the cell as a combination of many small point sources, and subsequently superimposing their fluxes.

As an example of such a procedure we will follow the analysis given by *Curless* [2.26]. In this analysis the following model of the effusion process has been used. The source material is assumed to be at a single temperature (the evaporation rate for the material is independent of position). It is assumed that the effusion cell consists of a truncated cone shown in Fig. 2.13, or its limiting case of a cylinder, and is positioned with respect to the substrate as shown in Fig. 2.14. The rate of evaporation at a point on the wall of the crucible is taken to be equal to the flux incident on the wall at that point, unless that point is defined as being wetted by the source material, in which case the rate of evaporation is given by the rate of evaporation for the source material. The bottom of the crucible is assumed to be covered by the evaporating source material with its surface normal to the axis of symmetry of the crucible. All fluxes leaving surfaces are assumed to obey a cosine distribution. The effect of surface diffusion is ignored, as is any interaction between molecules.

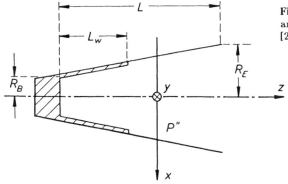

Fig. 2.13. Cross-sectional view of an idealized conical effusion cell [2.26]

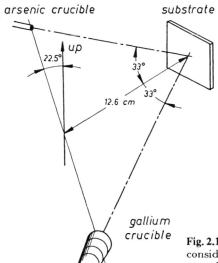

Fig. 2.14. Geometry of the "sources-substrate" system considered by *Curless* [2.26] for simulation of MBE growth of GaAs

In order to study the flux distribution upon the substrate during MBE growth, a computer program for the considered effusion process has been developed [2.26]. The first section of this program allows the operator to enter the geometry of the orifice portion of the crucible. The variables included are the radius of the crucible exit (R_E), the radius of the source material's surface (R_B), the distance from the source material in the bottom of the crucible to the crucible exit (L), the length of the crucible axis through that portion of the crucible which is wetted with source material (L_W), and the areal evaporation rate for the source material (E_S), (Fig. 2.13). The next section of the program calculates the areal effusion rate from the crucible wall.

Then the operator is allowed to define the positions of the crucibles with respect to the substrate (Fig. 2.14). There are some restrictions on the placement of the crucibles. The substrate is considered to be mounted on a block which is free to pivot about an axis which is parallel to and behind the substrate. Also, the crucible axes must pass through a single point on the arc through which the substrate's center point can swing.

The fluxes incident on the substrate surface are then calculated. The crucible wall and the surface of the source material in the bottom of the crucible are each divided into 720 regions of area. The flux at a point on the substrate surface is calculated by the summation of the fluxes due to each of the regions, checking of course that the line connecting the center of the region and the point on the substrate passes through the crucible's exit. The flux (dF) from one region (with center point P') to a point P on the substrate is approximately given by

$$dF = E \frac{(\boldsymbol{n}' \cdot \boldsymbol{v})(-\boldsymbol{n} \cdot \boldsymbol{v})}{\pi (\boldsymbol{v} \cdot \boldsymbol{v})^2} da' \quad , \tag{2.39}$$

where E is the areal effusion rate at P', $\boldsymbol{n}'$ is the unit vector normal to the surface at P', $\boldsymbol{v}$ is the position vector of P with respect to P', $\boldsymbol{n}$ is the unit vector normal to the substrate at P, and da' is the area of the region containing P'.

This equation is the vector form of the cosine law (2.10), with E standing for Γ_e, dF for $d\Gamma_\vartheta$ and da'/v^2 for $d\omega$. The terms $(\boldsymbol{n}' \cdot \boldsymbol{v})/v$ and $-(\boldsymbol{n} \cdot \boldsymbol{v})/v$ replace $\cos \vartheta$ in (2.10) and $\cos \phi$ in (2.13), respectively. Let us now proceed with the calculation by which the flux leaving the crucible walls may be determined. Because of the symmetry of the crucible, E is a function of the coordinate z only. It is thus only necessary to compute the values of E for the points where the x, z plane (with x positive) intersects the crucible wall. If the wall is defined as wetted by the source material, then E_S is used for E. In the case of a dry wall, E is equal to the flux incident upon the wall. The flux incident upon the wall can be evaluated in the same manner as the flux incident upon the substrate. A point on the crucible wall, P'', is used in place of P. It is convenient to choose the origin of the coordinates so as to make the point P'' have the coordinates $(r, 0, 0)$ and to split the incident flux into two parts: that from the wall and that from the bottom of the crucible. For the contribution from the bottom of the crucible the following values are used:

$$\boldsymbol{n} = \tilde{m}[-1, 0, m] \quad ,$$

$$\boldsymbol{n}' = [0, 0, 1] \quad ,$$

$$\boldsymbol{v} = [r - r' \cos \phi', -r' \sin \phi', -z'] \quad ,$$

$$da' = r' dr' d\phi' \quad ,$$

where $m = (R_E - R_B)/L$, $\tilde{m} = 1/\sqrt{m^2 + 1}$, r is the crucible's radius at P'', r' is the distance from the crucible's axis to the point P', z' is the z coordinate of P', and ϕ' is the angle from the x, z plane to the ray perpendicular to the z axis and ending at P'. Integrating dF over the bottom of the crucible one obtains for the flux arriving directly from the material in the bottom of the crucible (F_B) :

$$F_B = \frac{-z'\tilde{m}E_S R_B}{(2r)^2} \int_0^1 A f_1(A, B) dA \quad , \quad \text{where} \tag{2.40}$$

$$f_1(A, B) = \frac{1}{\pi} \int_0^\pi \frac{1 - A\cos\phi'}{[B - A\cos\phi']^2} d\phi' \quad \text{with} \tag{2.41}$$

$$A = \frac{r'}{R_B} \quad , \quad \text{and} \quad B = \frac{r^2 + z'^2 + r'^2}{2R_B r} \quad . \tag{2.42}$$

For the contribution from the wall of the crucible the following values are used:

$$n = \tilde{m}[-1, 0, m] \quad ,$$

$$n' = \tilde{m}[-\cos\phi', -\sin\phi', m] \quad ,$$

$$v = [r - r'\cos\phi', -r'\sin\phi', -z'] \quad ,$$

$$da' = \frac{r'}{\tilde{m}} dz' d\phi' \quad .$$

Integrating dF over the crucible wall results in the flux arriving at point P'' from the crucible wall (F_W) :

$$F_W = \frac{\tilde{m}}{2r} \int_{\text{bottom}}^{\text{exit}} E f_2(C) dz' \quad , \quad \text{where} \tag{2.43}$$

$$f_2(C) = \frac{1}{\pi} \int_0^\pi \frac{1 - \cos\phi'}{(C + 1 - \cos\phi')^2} d\phi' \quad \text{with} \tag{2.44}$$

$$C = \frac{z'^2}{2rr'\tilde{m}^2} \quad . \tag{2.45}$$

The flux incident at point P'' is calculated by first determining the value for all F_B. Then F_W is obtained by an iterative method. First F_W is set to zero and an approximate value for F_W is obtained by using $F_B + F_W$ for E for that portion of the wall which is dry and E_S for that portion of the wall which is wetted. Then F_W is recalculated with the new value of $F_B + F_W$ and the process repeated until the change in F_W becomes smaller than 1 % of the calculated F_W.

The final section of the program utilizes a contour plotting subprogram to depict the profiles of the fluxes and their ratios incident on the substrate. Using the described calculation procedure, a computer simulation of growth of GaAs was performed, for the position of the Ga and As beam sources as shown in Fig. 2.14 [2.26]. The Ga source was defined by the parameters $R_E = 1.05$ cm, $R_B = 0.38$ cm, $L = 6.72$ cm, and $L_W = 0$ cm, and the As source by $R_E = 0.22$ cm, $R_B = 0.22$ cm, $L = 0.1$ cm, and $L_W = 0$ cm. A 4 cm × 4 cm square substrate area was chosen, but no substrate rotation was taken into account. The results showed that the Ga flux varied by 30 % and the Ga to As ratio varied by a factor

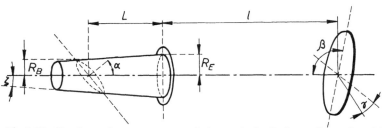

Fig. 2.15. Configuration parameters of the conical effusion cells analyzed by *Yamashita* et al. [2.27]. The following numerical data were used for the Ga (Al) cell: $L = 6.5(5.5)$ cm, $R_B = 0.78(0.61)$ cm, $R_E = 0.96(0.66)$ cm, $\alpha = 58(58)°$, $l = 12(12)$ cm, $\beta = 95(95)°$ and $\gamma = 11(-11)°$

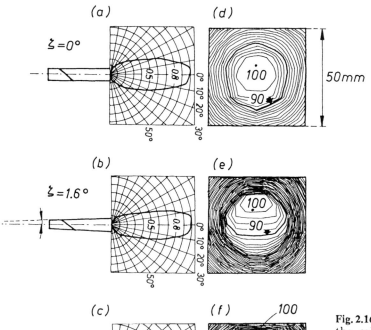

Fig. 2.16a–f. Distributions of the calculated beam fluxes (on the 5 cm × 5 cm substrate square) generated by conical liquid charge effusion cells with different tapering angles ζ. (**a, b, c**) Angular distributions; (**d, e, f**) surface distribution [2.27]

of 2 over the substrate. These results have been compared with experimental data gained by growing GaAs in a Varian GEN II MBE system, with the geometry of the sources-substrate system similar to that shown in Fig. 2.14. Reasonable agreement of the computer simulation with the experimental data has been found, which confirms that the described calculation procedure is correct.

Using the procedure of *Curless* [2.26], *Yamashita* et al. [2.27] have analyzed the beam fluxes from conical crucibles arranged slantingly with respect to the substrate and containing liquid Ga or Al. They have analyzed the influence of the shape of the conical crucible on the distribution of the beam flux upon the substrate surface, as well as the influence of the arrangement angle α of the beam source on the flux distribution. The configuration parameters of the beam source are shown in Fig. 2.15, while the results of the calculations are shown in Fig. 2.16. It is evident from these calculations that, in the case of a liquid source, a cylindrical crucible is more effective in improving the uniformity of the flux distribution (compare HWB sources in Sect. 2.2.3) than a conical crucible with a large tapering angle ζ, since the effusion cell is slantingly arranged. Since the direct flux from the source becomes dominant as the tapering angle of a conical crucible increases, the distribution of the beam flux is then mostly affected by the source surface determined by the arrangement angle.

2.3 Effusion Cells Used in CPS MBE Systems

Effusion cells are the basis of nearly all beam sources used in condensed phase source MBE (CPS MBE), although electron beam and laser radiation heated sources have also been successfully applied for growing certain materials by this technique [2.44–47]. Dissociation, or cracker, cells are used in CPS MBE, too. These cells are modified versions of conventional effusion cells in which the effusion beams are directed from a conventional Knudsen-type crucible via a higher-temperature (cracker) region onto the substrate [2.44]. The cracker cells are mainly used for the group V materials As and P, which produce the tetramers As_4 and P_4 when they are directly evaporated from elemental charges at lower temperatures. The cracking region provides an elevated temperature multiple collision path for the beams and thus enables effective dissociation of the tetrameric molecules into dimers [2.48–50].

2.3.1 Conventional Effusion Cells

In effusion cells used in CPS MBE, the solid or liquid source material is held in an inert crucible which is heated by radiation from a resistance-heated source. A thermocouple is used to provide temperature control.

Conventionally the heater is a refractory metal wire wound noninductively either spirally around the crucible or from end to end and is supported on insulators or inside insulating tubing. Care is taken to place the thermocouple in a position to give a realistic measurement of the cell temperature. This is either as a band around a midposition on the crucible or spring loaded to the base of the crucible. Experience has shown that Ta is the best refractory metal both for the heater and for the radiation shields, principally because it is relatively easy to outgas thoroughly, is not fragile after heat cycling, can be welded, and has a reasonable resistivity. Where temperatures exceed 150° C refractory materials should be used, as metallic impurities (Mn, Fe, Cr, Mg) have commonly been observed in layers grown in systems where stainless steel is heated [2.51]. It is also important that the refractory metal is of high purity ($> 99.9\%$) and has a low oxygen content. The insulator material has proven to be even more critical. Sintered alumina (Al_2O_3) proved unsuitable for this task; degraded optical and transport properties in grown layers have been associated with the use of this material [2.52]. In a classical series of modulated beam mass spectrometry experiments it was clearly shown that at temperatures as low as 850° C, Al_2O_3 was reduced when in contact with a refractory metal, and at temperatures above 1100° C significant dissociation of the ceramic occurred. In addition, metallic impurities were often found in layers when this material was present in the source, presumably associated with a volatile impurity species.

The preferred insulating material now used for sources is pyrolytic boron nitride (PBN), which can be obtained with impurity levels < 10 ppm. Although dissociation of this material does occur above 1400° C, the nitrogen produced has not yet been shown to have a deleterious effect on the grown layers.

The preferred crucible material is also PBN, although other materials have been successfully employed (notably ultrapure graphite [2.44] or sapphire [2.53] for high-temperature evaporation and quartz for temperatures below 500° C), and there is now some interest in the use of nonporous vitreous carbon, which is available with < 1 ppm total impurity level. Graphite, however, is difficult to outgas thoroughly and is by nature porous, with a large surface area, thus making it susceptible to gas adsorption when exposed to the atmosphere. Also, Al reacts with graphite at the usual Al evaporation temperature ($> 1100° C$) and so PBN is normally used as the crucible for this metal, but not without some problems [2.44]. Aluminum tends to "wet" the crucible internal surface and also moves by capillary action through the growth lamellae of the PBN crucible. When the Al solidifies on cooling the subsequent contraction can often crack the PBN crucible. To overcome this limitation in crucible lifetime it has been found that if only small quantities of Al are loaded into the crucible (viz., one-tenth the total crucible capacity) wetting is incomplete and the crucible does not break. A more practical solution is to position a second Al crucible inside the main one and replace this crucible when it is damaged or before significant leakage has occurred. It is also possible to fabricate a PBN crucible with a double skin of differing expansion coefficients specifically for Al evaporation, which then operates in a similar manner to the two separate crucibles but with obviously improved thermal properties.

The standard thermocouple material employed for the sources is W-Re (5 % and 26 % Re). These refractory alloys are suitable for operation at elevated temperatures and are inert to the reactive environment present. Multiple radiation shields surround each furnace to improve both the temperature stability and the thermal efficiency. Radiation heat transfer, however, is still not negligible at high temperatures and the shields themselves can become secondary centers for the generation of extraneous gases [2.54]. To this end, gas-tight radiation heat shielding is also used around the cell orifice to restrict the range of angles over which these gases can reach the substrate by a straight-line path.

Effusion cells designed as specified above have an operating temperature range of up to 1400° C, which makes possible short outgassing sequences at 1600° C. Although these temperatures are more than adequate for the common III–V materials, in practice most cells are limited to operating temperatures ≤ 1200° C, which is only just within the range of that required for Ga, Al, and Si (as a dopant source) evaporation. This is because of outgassing problems caused principally by the large temperature difference between the heater and the actual temperature of the charge in the cell (up to 300° C at 1200° C). In order to reduce this effect recent cell designs employing a much larger radiation surface heater area have been developed. Figure 2.17 shows a schematic illustration of an effusion cell used in VG Semicon MBE systems. An important feature of this design is the application of high efficiency, self-supporting tantalum radi-

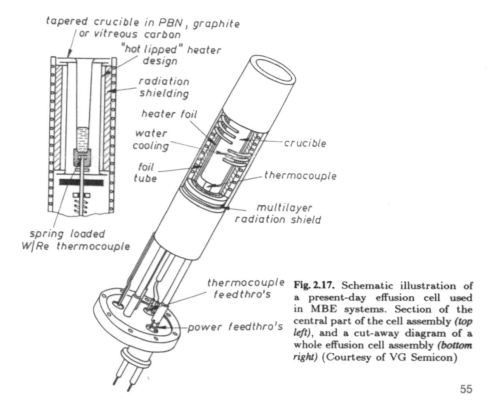

Fig. 2.17. Schematic illustration of a present-day effusion cell used in MBE systems. Section of the central part of the cell assembly *(top left)*, and a cut-away diagram of a whole effusion cell assembly *(bottom right)* (Courtesy of VG Semicon)

55

ant heater elements. Together, these elements make a large area radiation heater assembly capable of high temperature operation. Thus, the crucible runs at a temperature close to that of the foils. The need of ceramic insulators is minimized, because the heater foils are almost entirely self-supporting. This means less outgassing and reduced contamination during the MBE growth. The low thermal mass results in quick response to cell temperature adjustments. A spring-loaded W-Re thermocouple provides temperature monitoring and control. The whole cell has an integral water-cooled surround, effectively thermally isolating it from the surroundings.

Water cooling is considerably more efficient at dissipating heat than liquid nitrogen, which boils if heat is directly radiated onto the cryopanel surface. Intermittent hot spots can occur where the nitrogen gas forms an insulating barrier. Avoiding these hot areas of stainless steel of which the cryopanel is made prevents irregular outgassing, which can cause contamination of the films to be grown. For crucible materials high-purity graphite, vitreous carbon or PBN have been used in the VG Semicon cell (Fig. 2.17). The exit orifices of the crucibles are large, so that the cell is a nonequilibrium effusion source.

One consequence of the large exit orifices now employed in the source cells used in CPS MBE is that the radiation heat loss can be significant when the beam shutter is opened. This can result in a temperature drop at or near the orifice. In the worst situation the source material condenses at the end of the cell and reduces the exit orifice dimensions, thus changing the flux intensity.

Flux transients in MBE growth limit film reproducibility and are an inconvenience to the MBE user. These transients can be reduced by the use of partially filled crucibles so that the influence of the charges on the radiative shielding provided by the shutter are reduced. This approach is, however, impractical as the reduced charge limits machine operation between cell recharging. An interesting crucible arrangement with little or no flux transient has been propsed by *Maki* et al. [2.8, 9]. It involves a large volume crucible ($40 \, cm^3$) with a conical insert. Figure 2.18 shows the proposed crucible assembly and a geometrical construction for constant melt area as seen from the substrate position. Within the cone angle a nearly constant melt area is obtained, while outside this angle poor beam uniformity results from a shadowing by the crucible lip. The small flux transient observed from this cell reflects a more stable melt temperature. This is attained both as a result of the melt location deep in the furnace and through the additional radiation shielding provided by the crucible insert.

The described effusion cells have been used mainly for growing III–V compounds. When materials with high vapor pressure have to be grown the cell construction is usually changed. For growing CdTe and related compounds, a fused quartz cell with a special collimating tube has been designed [2.55], while for growing Cd and Zn II–VI compounds a high-rate nozzle effusion cell was proposed [2.56]. The most difficult among the II–VI compounds, from the MBE point of view, are, however, the Hg-based compounds (HgTe, $Hg_{1-x}Cd_xTe$ or $Hg_{1-x}Zn_xTe$) [2.57, 58], which are very important for infrared techniques [2.59].

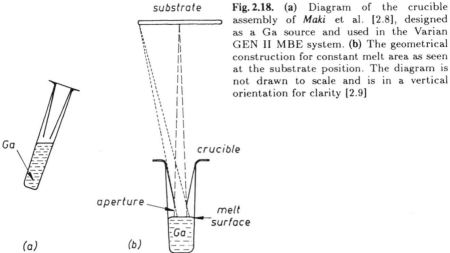

substrate

Fig. 2.18. (a) Diagram of the crucible assembly of *Maki* et al. [2.8], designed as a Ga source and used in the Varian GEN II MBE system. (b) The geometrical construction for constant melt area as seen at the substrate position. The diagram is not drawn to scale and is in a vertical orientation for clarity [2.9]

Ga

crucible

aperture

melt surface

Ga

(a)   (b)

This is based on the fact that the evaporation of these compounds is highly non-congruent [2.60]. The Hg effusion cell needs a special design, because a very high Hg atom flux is necessary to maintain the growth conditions for these compounds.

The equilibrium vapor pressure of Hg is 2×10^{-3} Torr at 300 K [2.61], so the Hg cell cannot be left in the growth chamber either during bakeout or prior to the growth, and consequently a transferable Hg effusion cell which can be loaded into the growth chamber through a vacuum interlock should be used [2.62, 63]. Figure 2.19 shows an example of a Hg cell used in CPS MBE. This is the cell of *Harris* et al. [2.63], which provides an extremely stable flux of Hg during film growth over long periods of use. This source is refillable without disturbing the UHV ambient of the growth chamber, and is capable of generating vapor flux densities as large as 10^{-3} Torr (this is the so-called beam equivalent pressure, BEP) at the substrate during film growth. The source consists of two Hg reservoirs and a heated tube separated by two UHV valves. The Hg source is attached to a 4.5 in. diameter UHV Conflat flange which mounts to the MBE growth chamber source flange. All internal surfaces of the Hg source are constructed of stainless steel. These surfaces are surrounded by solid aluminum heat sink material to provide good temperature stability. The temperature of the heated tube and the internal Hg reservoir are measured with platinum resistance thermometers and controlled with Eurotherm controllers. Platinum resistance thermometers are used rather than thermocouples because of their increased sensitivity over the temperature range employed ($T = 100°-200°$ C). To fill (or refill) the Hg source, the internal Hg reservoir is first evacuated to UHV and then the valve between the reservoir and MBE chamber is closed. Next, Hg is introduced into the larger external reservoir through the fill port. The valve between the two reservoirs is then momentarily opened so that Hg fills the internal reservoir. In this way, Hg is

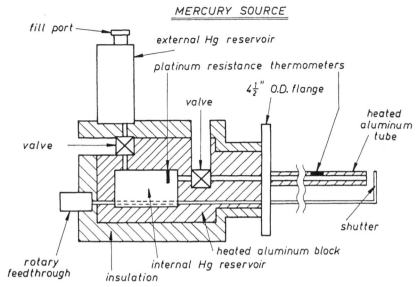

Fig. 2.19. Mercury MBE vapor source [2.63]

transferred to the internal reservoir with minimum oxide contamination. As can be seen in Fig. 2.19, the Hg source is a modified one-channel circular effusion cell.

2.3.2 Dissociation (Cracker) Effusion Cells

Cracker cells are used in CPS MBE mainly for producing dimeric arsenic molecules (As_2) from elemental charges. The phosphorus-containing compounds are at present most frequently grown by gas source MBE [2.13], where cracking processes are performed in gas leak sources [2.64] (Sect. 2.4).

The temperatures for efficient cracking of the As_4 or P_4 beams to dimers have been calibrated by modulated beam mass spectroscopy experiments [2.65] and are typically within the range $800°–1000°$ C for 100% dissociation of both As_4 and P_4 [2.44]. When designing the cracker region of the dissociation cell, similar rigorous precautions concerning the choice of refractory materials and heater construction should be taken as with conventional effusion cells. Any significant heat transfer between the hot zone (the cracking zone) and the effusion cell region should also be avoided, otherwise uncontrolled evaporation could occur.

Two examples of arsenic cracking sources are shown in Figs. 2.20 and 2.21. The first, a two-stage source, incorporates an integral getter pump which reduces arsenic oxides in the beam to approximately 10^{-10} Torr [2.49]. The two stages of this source are radiatively isolated by means of tantalum baffles and conductively isolated by the length of the tantalum tubing (Fig. 2.20a). The temperatures of the

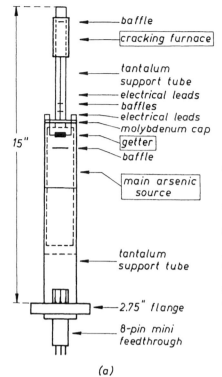

baffle

cracking furnace

tantalum
support tube
electrical leads
baffles
electrical leads
molybdenum cap

getter

baffle

main arsenic
source

tantalum
support tube

2.75" flange

8-pin mini
feedthrough

15"

(a)

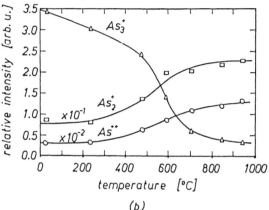

(b)

Fig. 2.20. (a) Schematic design of a two-stage MBE arsenic source incorporating an integral getter [2.49]. (b) Experimental data concerning the peak intensities of arsenic fraction as a function of the temperature of the cracking zone in the cell shown in (a). The lower-order species' intensities are reduced by the factors indicated

cracking stage tube furnace that surrounds the tantalum exit tube and the main arsenic source can be independently controlled and the temperatures measured using W-Re thermocouples. The entire assembly is mounted on a 2.75 in. Conflat flange with a mini-Conflat feedthrough depicted providing eight connections. In use, the source is surrounded by a liquid nitrogen cooled cryoshroud which reduces outgassing of As-O from the walls of the chamber.

The getter was incorporated into the source because of the observation that residual As-O background originated from the source itself. It removes effectively the oxygen from the arsenic stream generated by the source. The getter is incorporated in the main arsenic reservoir and baffles are used to increase the probability that all molecules leaving the source hit the getter surface. The getter is composed of Zr-C and has an active surface even at room temperature. The getter capacity, but not the pumping speed, is increased at elevated temperatures. The getter, if saturated, can be reactivated at $900°-1000°$ C. The As_4 to As_2 cracking efficiency of the described source has been evaluated in a MBE chamber with surface analysis equipment and a mass spectrometer. Representative curves showing the results of monitoring the arsenic fractions as a function of the temperature of the cracking zone are shown in Fig. 2.20b. The low temperature arsenic furnace was held at a constant temperature for each run and the getter was operated at a low enough temperature that no significant cracking was observed from its use.

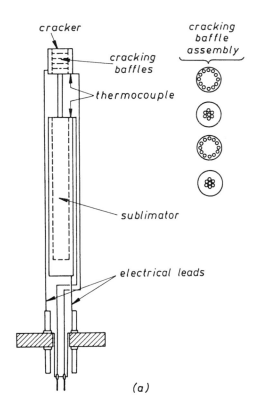

cracker

cracking
baffles

thermocouple

sublimator

electrical leads

cracking
baffle
assembly

(a)

Fig. 2.21. (a) Schematic sketch of the arsenic cracking cell of *Lee* et al. [2.48]. **(b)** Measured relative abundance of As_2 and As_4 molecular species in the beams generated by the cell shown in **(a)**

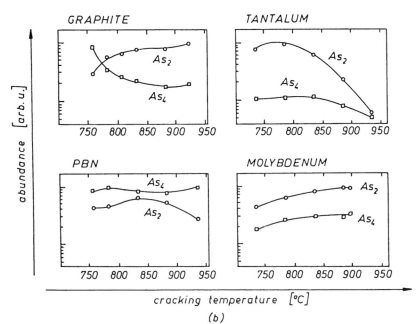

(b)

The second cracking source, shown in Fig. 2.21a, has been designed for testing the cracking patterns obtained with four different cracking baffle materials, namely graphite, tantalum, molybdenum and PBN. In this source an afterburner with these baffle materials is placed on the arsenic oven. The cracking patterns have been compared (Fig. 2.21b) in order to determine how the baffle materials react with arsenic at high temperatures and which material results in the highest cracking efficiency [2.48].

Graphite [2.65–67], tantalum [2.49], and molybdenum [2.68] have been used as the material for cracking furnaces. Graphite is easy to machine but may outgas at high temperatures. Tantalum and molybdenum give much freedom in design and fabrication, but may react with arsenic, as tantalum does with phosphine [2.12]. A nonrefractory material (PBN) was interesting because it had been suspected that cracking of arsenic was not solely due to simple thermal decomposition, but required a catalyst, as is the case for cracking of gaseous sources [2.69, 70].

The composition of the beam flux from the cracking furnace was measured using a UTI-100C residual gas analyzer in the beam path. The measurements were corrected for ionization efficiency, multiplier gain and quadrupole transmission for each species. Figure 2.21b shows the measured relative dimer and tetramer abundance as a function of cracking temperature. The graphite baffles give the highest dimer flux at high temperatures. Except for tantalum, each of the materials yields its own asymptotic dimer/tetramer composition ratio at high temperatures. These asymptotic values are below the equilibrium limit, which implies that the cracking reaction is kinetically limited. It is noteworthy that the results presented here are from a cracker which is not optimized for dimer content yet and can be improved by tuning the geometry of the cracker-sublimator assembly.

The dimer ratio for the tantalum baffles reaches a maximum as the temperature increases and it decreases with further increase in temperature. This tantalum baffle effect is similar to that of the phosphine cracking furnace [2.12]. It has been indicated that tantalum baffles may react with phosphorus at elevated temperatures and the cracking product composition is baffle-area sensitive. The data suggest that the tantalum baffle cracking is coupled with tantalum arsenide formation. The different plateau values of dimers indicate that the baffle materials crack the tetrameric arsenic molecules catalytically.

The cracking efficiency and the catalytic activity of several materials have also been studied by *Garcia* et al. [2.71] by modulated beam mass spectrometry [2.64]. They used an all-PBN double-oven arsenic dimer source in which tetramer molecules produced from crystalline arsenic placed in a conical PBN crucible were cracked in a cracker stage containing a series of baffles which ensured numerous molecule-wall collisions. The catalyst under investigation was introduced into the cracker cell's upper zone in the form of a wire or a baffle. It was found that the catalytic activity of the tested materials decreases as follows: Pt, Pt-Rh(alloy), Re, Ta, Mo, W-Re(alloy), graphite, PBN. In this work it was confirmed that PBN is catalytically inactive in the cracking of As_4 molecules and therefore cannot be used below 1200° C without adding a catalyst. Platinum and platinum-rhodium alloys exhibit the highest catalytic activity, being, however, unusable because of their reaction with arsenic. Platinum reacts with arsenic above 500° C, yielding definite compounds such as $PtAs_2$. It is evident that in the experiments of [2.71] the best results have been achieved with rhenium. This catalyst allows one to obtain a 95% cracking efficiency for arsenic tetramer-dimer molecule conversion at a temperature of about 700° C. That is more than 300° C lower than the temperature required with a conventional graphite cracker stage for an equivalent efficiency. Such a decrease of the operating temperature of the cracker cell is of great practical interest because of the reduction of the arsenic flux contamination by outgassing impurities.

In conclusion, it is worth mentioning that the cracking efficiency is usually proportional to the number of wall collisions. Therefore, general statements about the usefulness of different materials may be misleading unless large and comparable surface areas were used in the investigations.

2.3.3 Electron Beam and Laser Radiation Heated Sources

Instead of supplying energy by resistance heating, as in the case of the effusion cells described so far, the gas particle beams for MBE may also be generated with sources heated by electron bombardment or by laser irradiation. This is favorable because the interactions between the evaporant and the source walls may be greatly reduced in this way. The heating energy transported by electrons or photons can be concentrated on just the evaporating surface so that other portions of the evaporant and the source walls can be maintained at a lower temperature, causing the chemical interactions between the evaporant and the walls to be negligible. This mode of evaporation has another unique property: the fast response time of the evaporant to the heating source, and thus the ease in modulating the evaporant beam intensity [2.46, 47].

In the case of electron bombardment, a stream of electrons is accelerated through a field of typically 5–10 kV and focused onto the evaporant surface. Upon impingement, most of the kinetic energy of the electrons is converted into heat, and temperatures exceeding 3000° C may be obtained. Experimentally, the method can be implemented in a number of ways [Ref. 2.15; Sect. 1.4]. In MBE systems however, bent-beam electron guns are used most frequently [2.72–74]. In these guns the electrons are forced into a curved path through a transverse magnetic field provided by an electromagnet, which permits focusing of the electron beam during operation. This allows the gun structure to be effectively separated from the vapor source without resorting to long distances. As an example of an electron bent-beam heated MBE source, the silicon source used by *Ota* [2.74, 75] for silicon MBE will be described.

The silicon source was built on a deep crucible electron-gun unit with a 270° electron beam deflection. In a 270° deflection system, the tungsten filament is hidden from the Si substrate, which avoids tungsten contamination of the epitaxial

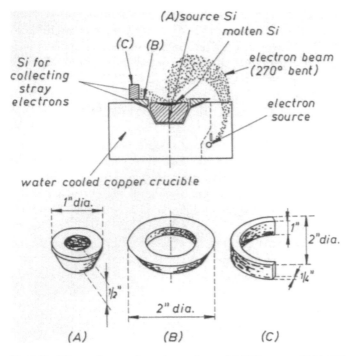

Fig. 2.22. Silicon electron bent-beam heated MBE source [2.74, 75]

film. Since the crucible was made of copper, the area adjacent to the source silicon was protected by high purity silicon to avoid Cu contamination.

The silicon source consisted of three sections. They were made of eight high purity floating zone (FZ) single crystal p-type, 5 k cm silicon or high purity polysilicon (impurity < 0.4 ppb). A cross-sectional view of the source assembly is shown in Fig. 2.22. In the center of the source section was a truncated cone of shaped silicon (part A). During film growth, the electron-beam impinged only on the center of this section. For an electron-beam current of 140 mA, the molten area was less than 13 mm in diameter. The doughnut-shaped silicon (part B) protected the copper crucible surface from bombardment by stray electrons. Since the main beam did not strike outside of part A and a small gap existed between parts A and B, the region B remained at a low temperature. With this arrangement, Cu contamination was avoided. Another possible source of contamination was sputtering of the stainless steel parts surrounding the electron-gun system by the electrons reflected from the silicon surface, mainly from part A. This source of contamination was eliminated by placing a semicircular silicon wall (part C) around the side opposite the incoming direction of the electron-beam. The focusing capability of the electron-beam was improved by the addition of magnetic field poles. They are not shown in Fig. 2.22. With this source design and modifications, stacking fault-free films were grown at a growth rate of 1.4 μm/h using a beam current of 250 mA at 4.5 kV.

Electron beam evaporators have found universal application in silicon MBE for the generation of the main matrix element [2.74], as well as for coevaporation of metal silicides [2.76, 77]. However, they are finding increasing applications, especially for the sequential evaporation of metal contact layers, particularly in the case of low-vapor-pressure and refractory metals [2.40, 78, 79].

The first use of electron beam heated sources in MBE growth of III–V compounds was reported by *Malik* [2.47]. The reported system was designed and constructed in collaboration with VG Semicon. The system contains three electron beam evaporators with $40\,cm^3$ capacity copper hearths. The electron beam evaporators are used to generate group III (Ga, Al, In) and/or Si molecular beams by thermal heating due to high energy electron bombardment. There are ports for fitting conventional effusion cells with high temperature cracker zones which are used to produce group V (As_2, Sb_2) dimer molecular beams. There are also three horizontally mounted ports for fitting dopant sources. A Si strip heater is used for n-type doping and a small, $2\,cm^3$ Be effusion cell is used for p-type doping. All the molecular-beam sources have separate shutters which are computer controlled. The MBE system also includes a reflection high energy electron diffraction system and a quadrupole mass spectrometer. The ultrahigh vacuum is obtained through a combination of ion pump, cryopump, Ti sublimation pump, and extensive liquid nitrogen cryopanels, giving the system a base pressure of $< 10^{-10}$ Torr.

Figure 2.23 shows the geometry of the electron beam sources in the MBE system. Each source has its own individual beam flux sensor, which is used to measure and control the evaporation rates by feedback. The sensor consists of a normal Vacuum Generators VIG-17 ion gauge, which is modified by a cutout in the grid and by offsetting the collector filament from the center of the gauge. This prevents the beam flux, which is collimated by a series of apertures, from coating the ion gauge, which would result in a change in sensitivity of the gauge. An electron shield at -250 V is used to prevent secondary electrons from entering the gauge. A chopper blade is placed in front of the gauge, which enables differential

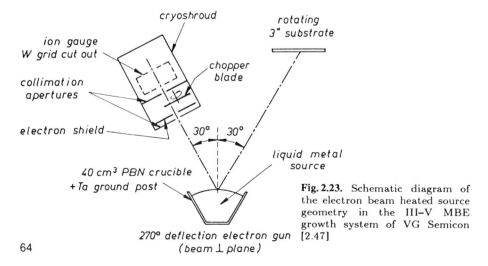

Fig. 2.23. Schematic diagram of the electron beam heated source geometry in the III–V MBE growth system of VG Semicon [2.47]

measurement of the beam flux through subtraction of the background pressure in the gauge. The sensor assembly is contained within a liquid nitrogen cryoshroud to reduce the gauge background pressure which is $< 10^{-9}$ Torr when operating in an As_4 growth ambient pressure of 10^{-6} Torr. The sensor and substrate are oriented 30° from the vertical normal of the source. The metal sources, which are high temperature liquids when evaporated, are contained within pyrolytic boron nitride crucibles in which Ta grounding posts are inserted through the bottom. This prevents contamination of the sources from the copper hearth.

With the described electron beam evaporation system AlAs, GaAs and $Al_xGa_{1-x}As$ epilayers have been grown. The response of the AlAs growth rate is shown in Fig. 2.24a. It is seen that with a constant power input, the growth rate drifts with time due to an imbalance of heating and cooling of the source. Thus, it is apparent that feedback control of the growth rate is an essential requirement. With feedback control of the electron beam power, it is seen that the growth rate remains constant at a fixed set point. There is a small and rapid deviation in the apparent evaporation rate of the sources, which is estimated to be about $\pm 2\%$. This is in part due to noise generated by the slow sampling rate (1 Hz) of the source along with 60 Hz line frequency noise. However, there is expected to be some small real deviation in the evaporation rate due to convection of the liquid metal source induced by the large temperature gradient in the crucible. A deviation of ± 1 K would account for a $\pm 2\%$ change in the vapor pressure of the source. Similar deviations in the evaporation rate of the Ga effusion cell have been observed, which appear to depend upon the height of the liquid metal

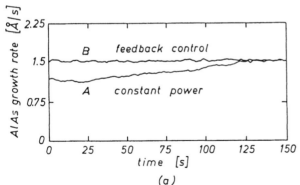

(a)

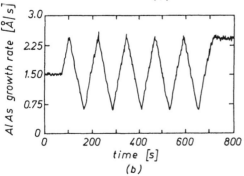

(b)

Fig. 2.24. (a) Output control signal response of the Al source for (A) constant power input from electron gun and (B) with feedback control of electron gun to maintain a fixed evaporation rate [2.47]. (b) Output control signal response of Al source to programmed sawtooth evaporation rate profile. The set point was programmed to rise and fall at the same rate over a 1 min time interval [2.47]

in the crucible. Nevertheless, the small apparent deviations in evaporation rates are negligible in comparison to the angular dependence of the beam flux, which can be time-averaged out by substrate rotation during growth.

The dynamic response of the Al source evaporation rate is demonstrated in Fig. 2.24b. The source was programmed to follow a sawtooth profile with linear increasing and decreasing evaporation rates. It is seen that the Al source easily tracks the programmed profile without phase lag and there is no overshoot of the peak set point or undershoot of the valley set point. This profile was obtained through continuous growth and would be impossible to obtain by conventional effusion cells. The electron beam sources have also demonstrated controlled evaporation rates down to 0.01 Å/s. Changes in the evaporation rate between 0.01 and 1.0 Å/s have been achieved in < 10 s.

The geometry of the electron beam source has some other advantages. There are no flux transients observed upon opening the shutter, unlike in effusion cells where radiation cooling can cause the beam flux to decrease by up to 30 % over several minutes. In addition, the electron beam source has a broad evaporation pattern, unlike effusion cells which tend to collimate the molecular beam, leading to large nonuniformity over 3 in. wafers. Indeed, within the resolution of a Dektak II profiler, there was no measurable thickness variation in a 1 μm GaAs epilayer grown on a 3 in. diameter Si (100) wafer using substrate rotation in the electron beam source MBE system. Thus, the electron beam source approach may be ideally suited for production requirements of III–V integrated circuit (IC) wafers.

Laser radiation heated sources have been applied in MBE for growing II–VI compounds [2.46, 80]. With this technique, also known as laser assisted deposition and annealing (LADA), epitaxial layers of $Hg_{1-x}Cd_xTe$ and CdTe as well as HgTe-CdTe superlattices have been grown. Knudsen cells are replaced there by targets evaporated by pulsed laser radiation. Figure 2.25 shows a simple schematic drawing of an MBE system with laser radiation heated beam source

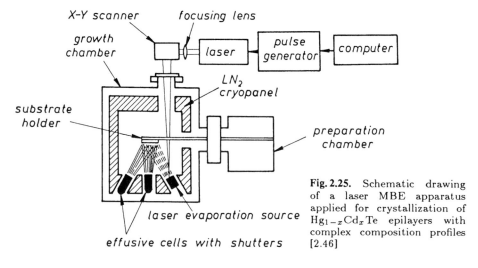

Fig. 2.25. Schematic drawing of a laser MBE apparatus applied for crystallization of $Hg_{1-x}Cd_xTe$ epilayers with complex composition profiles [2.46]

[2.46]. This system consists of a preparation chamber, a growth chamber and an Hg removal chamber specially designed for growing $Hg_{1-x}Cd_xTe$ compound layers. Three sources are used in this system: effusion cells for Te and Hg beams and a laser radiation heated source with polycrystalline CdTe target. In growing a graded $Hg_{1-x}Cd_xTe$ structure, the Te and Hg beams were kept constant. Only the CdTe beam was modulated.

Nd : YAG lasers are used because of their ruggedness and long-term stability. The 1.06 μm wavelength radiation can be transmitted through UHV viewports. Two types of Nd : YAG lasers were used for comparison: an acousto-optically Q-switched laser and a Pockels cell Q-switched laser, which have very different output characteristics. The acousto-optical Q-switched laser has a repetition rate in the kilohertz range. Energy content per pulse is about 1 mJ with pulse duration of approximately 200 ns. The Pockels cell Q-switched laser is best operated in the free oscillation mode where energy content per pulse can be as high as 800 mJ for a duration of 100 μs. The repetition rate can be varied from 0.5 Hz up to 20 Hz.

$Hg_{1-x}Cd_xTe$ epilayers were grown on (100) CdTe substrates at 180° C by simultaneously opening all three beams. The Hg beam is about three orders of magnitude more intense than the other components in order to prevent thermal decomposition of the as-grown layer. The growth rate is determined by the sum of the Te and CdTe beams, whereas the composition x is determined by the CdTe to Te beam intensity ratio.

Multiphoton absorption of the 1.06 μm radiation by CdTe causes a rapid rise in surface temperature and thus evaporation to begin. The velocity distributions of the evaporants were measured by time-of-flight analysis and were found to be Maxwell-Boltzmann. This suggests that the evaporation process was in equilibrium. Surface temperatures were determined to be in the range from 1400° C (0.6 W laser average power) to 3200° C (4.8 W laser average power). The cooling time is estimated to be around 1 ms due to the low thermal conductivity in CdTe. This time scale is comparable to or longer than the time between consecutive pulses from the acousto-optical Q-switched laser. Therefore, adjacent pulses can overlap and cause additional surface heating. In other words, the surface temperature and evaporation rate are sensitive to the scanning rate. This condition does not apply to the Pockels cell Q-switched laser, where the time between consecutive pulses is much longer than the thermal relaxation time.

There are three ways to modulate the CdTe evaporation induced by the laser irradiation. The most obvious way is to change the power density of the laser because the peak surface temperature is proportional to $P\sqrt{\tau}$, where τ is the pulse duration and P is the power density. Since the evaporation rate depends exponentially on the surface temperature, the dependence of the evaporation rate on power density is also exponential. This relationship is seen for both lasers shown in Fig. 2.26a. This dependence is so sensitive that it is very difficult to achieve precise control.

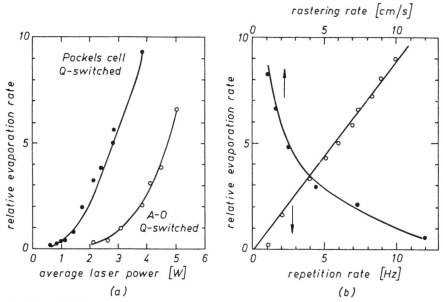

Fig. 2.26. (a) CdTe relative evaporation rate versus average radiation power for two different Nd : YAG lasers described in the text [2.46]. **(b)** Dependence of CdTe evaporation rate on the scanning rate (•) and repetition rate (o) for the Pockels cell Q-switched Nd : YAG laser [2.46]

The second approach is to control the evaporation rate by changing the laser scan rate. It relies on the heat accumulation due to pulse overlap. An example is shown in Fig. 2.26b. The dependence is nonlinear. The evaporation rate decreases slowly as the scan rate increases. This process is very slow, with a response time as long as a few seconds. The degree of overlap between consecutive pulses depends strongly on the surface morphology and is not very reproducible. Therefore, this approach is not practical.

The third approach is to modulate the CdTe evaporation by changing the laser repetition rate. This approach can only be applied to the Pockels cell Q-switched laser, where the energy content per pulse is constant except for at very slow repetition rates. The rolloff results from the setting of the internal optics in the laser cavity. For the laser unit used (JK Laser model HY750), it occurs at below 2 Hz. Figure 2.26b also shows the dependence of evaporation rate on repetition rate for a fixed laser energy. The relationship is linear down to 2 Hz and the evaporation rate is stable over an extended period of time. A random sampling of 20 evaporation rates integrated over a single rastering showed < 1 % deviation. This modulation technique has a very fast response such that at total typical conditions, each pulse evaporated only enough material to form a fraction of a monolayer surface coverage. Therefore, composition can be varied from monolayer to monolayer. The composition profile will be smooth and the resolution is limited by the interdiffusion, but not by the growth process.

The experimental results reported in [2.46] clearly demonstrate that pulsed laser-induced evaporation can be used to modulate beam intensity for thin film growth with MBE.

2.4 Beam Sources Used in GS MBE Systems

Gas sources were introduced to the UHV growth technique as early as in 1973 by *Morris* and *Fukui* [2.81]. They used arsenic AsH$_3$ and phosphine PH$_3$ cracked in a boron nitride cracker tube at 800° C as arsenic and phosphorus sources for growing polycrystalline GaAs and GaP films on Si substrates. Gas sources have been applied because in growing compounds with As and/or P as constituents, there is need of a source that provides precise control over the group V element beam intensities, at high intensity levels [2.82]. Later, gas sources with metalorganic compounds were introduced to MBE of III–V compounds [2.83] and also of some II–VI compounds [2.84].

2.4.1 Arsine and Phosphine Gas Source Crackers

Two types of crackers may be distinguished [2.13, 64, 85]: one that cracks the hydrides at pressures in the range of hundreds of Torr, called the high pressure gas source (HPGS), and one that cracks them at pressures that are probably in the range of tenths of Torr or less, the low pressure gas source (LPGS). These are illustrated in Fig. 2.27. The HPGS has two stages of cracking. At the usual

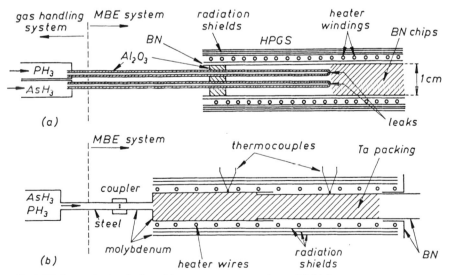

Fig. 2.27a,b. Arsine and phosphine gas source crackers: (a) high pressure gas source, and (b) low pressure gas source [2.13, 64, 85]

69

operating temperature of 900° C, AsH_3 and PH_3 are expected, on the basis of thermodynamic equilibrium, to decompose to form the tetramers As_4 and P_4 plus H_2 under the pressure conditions in the alumina cracking tube illustrated in Fig. 2.27a. These species then diffuse through a small leak in the alumina tube into the low pressure portion of the cracker where further thermal cracking to dimers is expected. Because it is presumed that the HPGS will yield species as expected from thermodynamic equilibrium, that cracker has been referred to as a thermodynamic cracker [2.13]. The LPGS is a catalytic cracker (Fig. 2.27b) and there is very little basis on which the relative amounts of the group V monomers, dimers and tetramers can be predicted.

Although the HPGS has been used for most studies of basic GS MBE, and the LPGS has been used for most studies of the GS MBE that involve modification by the use of group III metal organics, only recently a study has been done in which the beam species effusing from these sources are directly observed [2.64]. Measurements were made of the ion current resulting from a collimated and periodically interrupted beam from the crackers effusing directly into the ionization chamber of a quadrupole mass spectrometer equipped with synchronous detection by means of a phase sensitive amplifier. The mass spectrometer was otherwise completely protected by a liquid nitrogen cooled surface. The results of this modulated beam mass spectrometric study are given in Table 2.4. What is significant is that both types of crackers can yield primarily dimer species under proper experimental conditions. The beam effusing from the LPGS appears also to contain significant amounts of monomer as was claimed by *Calawa* [2.86]. These results suggest that either the HPGS or the LPGS should be satisfactory with any of the versions of GS MBE.

An extended study on a phosphine cracking source has been performed by *Chow* and *Chai* [2.12]. The thermodynamic calculations performed by these authors showed that phosphine should dissociate into the phosphorus dimer to a high degree. Furnace materials were compared by construction of all-tantalum and all-quartz cartridges which slid into the furnace hot zone. Phosphine was found to react with tantalum, but not to any great extent with quartz. The effects of baffles have been studied with the quartz cartridges. In general, baffling increases

Table 2.4. Peak ion current ratios[a] As_2^+/As_4^+ and P_2^+/P_4^+ from the cracking of AsH_3 and PH_3 with high and low pressure gas sources [2.64]

Source	T [° C]	M_2^+/M_4^+
HPGS AsH_3	900	11–130
HPGS PH_3	900	∼ 100
LPGS AsH_3	1000	7–30
LPGS PH_3	1000	22–46

[a] A range of ion current ratios is given because the ratio varies with flow through the cracker. The range studied was that useful for MBE. In separate studies pure M_2 gave no M_4^+ and pure M_4 gave a ratio $M_2^+/M_4^+ \sim 0.5$.

the cracking efficiency to over 90 % and causes the peak efficiency to occur at a lower temperature. It has been found that in accordance with theoretical predictions the phosphorus dimer is the major cracked species.

2.4.2 Gas Sources Used in MO MBE

In the MO MBE growth technique (Sects. 1.3.1 and 7.2.1), when applied to growing III–V compounds, all the sources are gaseous group III and group V alkyls [2.83]. Unlike atmospheric or LP MO CVD, in which thermal pyrolysis of group III alkyls also occurs in the gas phase, the In and Ga in MO MBE are derived by the pyrolysis of either trimethylindium (TMIn) or triethylindium (TEIn) and trimethylgallium (TMGa) or triethylgallium (TEGa) entirely at the heated substrate surface. The As_2 and P_2 are obtained by thermal decomposition of triethylphosphine (TEP) and trimethylarsine (TEAs) instead of arsine and phosphine in contact with heated Ta or Mo at 950°–1200° C, respectively, after mixing with H_2. The Ta or Mo serve both as a baffle to the gas flow and as a catalyst in dissociating the alkyls [2.87]. The choice of the group V alkyls instead of the hydrides despite the poorer purity is mainly because they are less hazardous. (Group V alkyls are extremely malodorous, with smell threshold believed to be in the ppb range, and they rapidly oxidize with a room-temperature partial pressure of ~ 20 Torr for TEP and ~ 300 Torr for TMAs.) A gas handling system similar to that employed in MO CVD with precision electronic mass flow controllers is used for controlling the flow rates of the various gases admitted into

Table 2.5. Metalorganics for semiconductor thin film deposition [2.88]

Group of metal in periodic table of elements	Compound	Symbol
IIa	Biscyclopentadienylmagnesium	Cp_2Mg
IIb	Dimethylzinc	DMZn
	Diethylzinc	DEZn
	Dimethylcadmium	DMCd
	Dimethylmercury	DMHg
	Diethylmercury	DEHg
IIIa	Trimethylaluminum	TMAl
	Trimethylgallium	TMGa
	Triethylgallium	TEGa
	Diethylgallium chloride	DEGaCl
	Trimethylindium	TMIn
	Triethylindium	TEIn
IVa	Tetramethyltin	TMSn
	Tetraethyltin	TESn
	Tetramethyllead	TMPb
	Tetraethyllead	TEPb
Va	Triethylphosphine	TEP
	Trimethylantimony	TMSb
	Trimethylarsine	TMAs
VIa	Dimethyltelluride	TMTe
	Diethyltelluride	TETe

the growth chamber. Hydrogen or argon is used as the carrier gas for transporting the low-vapor-pressure group III alkyls in order for the electronic mass flow controllers to work properly and reproducibly.

A large variety of metalorganic sources may be used for growing other compound materials with MO MBE. The metalorganics used in MO CVD are listed in Table 2.5 [2.88]. In principle, they could be applied also for growing films in an UHV environment. At present, however, the MO MBE technique has been applied only for ZnSe, among the family of II–VI compounds. This has been achieved by using an optimized cracking cell for DES (diethylselenide) [2.84], which was similar to the arsine cracking cell of *Kapitan* et al. [2.70], with minor modifications.

3. High Vacuum Growth and Processing Systems

The maturity which the MBE technique has now achieved is reflected in the demand for high throughput, high yield MBE machines. A whole set of companies currently manufacture MBE-growth and MBE-related equipment that is sophisticated in design and reliable in application. Among the largest manufacturers which share the major part of the world market, the following six may be listed: ISA RIBER and VG SEMICON in Europe, VARIAN/TFTD and PERKIN-ELMER in the USA, and ANELVA and ULVAC in Japan.

Historically, MBE technology developed in response to the increasing attention paid by the semiconductor community to GaAs devices of increasing complexity. In particular, microwave and high speed component manufacture has seen remarkable growth in order to meet dramatic increases in the market for GaAs field effect transistors and integrated circuits. The development and production of the next generation of high speed discrete and IC devices is inextricably linked to the ability to grow highly complex device structures epitaxially. Present-day MBE has become the leading edge of this technology.

Device performance is critically dependent on the electrical and morphological quality of epitaxial layers and interfaces. For example, high yield production of modulation-doped high electron mobility transistors, quantum well heterojunction lasers or superlattice avalanche photodiodes requires pure materials, abrupt and smooth interfaces and defect-free surface morphology. The performance of the MBE growth system, or more specifically, of the individual components in the MBE growth environment, has an impact on device quality and yield. The UHV environment, beam source performance, shutters, substrate manipulator, growth geometry and sample transfer all affect the integrity and reproducibility of device structures.

The modular approach to system design has been established as the most effective and flexible one. Modularity means providing the system in "building-block" units, combining the advantages of standardization and flexibility. MBE systems built in the modular technology can be customized from standard units to address the particular concern of the individual user.

The enormous potential of MBE to produce structures of unequalled complexity can only be fully realized by utilizing effective automatic control. Stringent specifications for the accuracy and reproducibility of semiconductor structures such as superlattices, with layer thicknesses down to a few atomic layers, necessarily entail the use of a real-time computer control system. Real-time control ensures precise synchronization of events during the growth process. For example, shutter status and molecular beam fluxes must be assessed and adjusted simultaneously every one or two seconds in many growth procedures, especially when complicated device structures are to be grown with MBE. This means, however, that a computer control system is an indispensable part of the MBE equipment.

The construction blocks and functional elements of modern MBE production systems available commercially will be described in this chapter, taking as examples the ISA Riber and VG Semicon MBE systems. These manufacturers supplied the most comprehensive and illustrative material with permission for publication in this book. Consequently, the technical data presented here are based on authorized information from these manufacturers. The chapter will be completed with a presentation of a new version of hot-wall epitaxy equipment, the so-called hot-wall beam epitaxy (HWBE) growth unit, followed by a detailed presentation of the FIB-MBE growth and processing machinery.

3.1 Building Blocks of Modular MBE Systems

In essence a modern, dedicated, high throughput MBE system needs to be able to reproducibly fabricate semiconductor material with purity levels of better than 10 ppb, with device-quality minority and majority carrier characteristics, and with excellent uniformity. Growth rates should be up to a few micrometers per hour with thickness control of tenths of a monolayer. The fact that this combination of material and production criteria can now be satisfied by commercially available machines is a clear indication of the present status of MBE [3.1]. In order to satisfy these demanding specifications UHV processing environments with unparalleled cleanliness are employed, with means of maintaining these ultraclean conditions during a process which involves both a relatively large heat input ($\approx 1 \, \mathrm{kW}$) and large reactive gas loads. Special combinations of inert and ultrapure materials compatible with high-temperature reactive environments have evolved, and ultraclean and controlled evaporation sources with significantly improved peformance, in terms of purity and control, over anything previously available have been developed. These, along with a number of mechanical developments in sample handling and shutter manipulation, have all contributed significantly to the development of MBE. In addition, detailed diagnostic work and analysis has given clues to the subtle and complex relationship between system design and the resulting material quality.

The main building blocks of a modern MBE production system are schematically presented in Fig. 3.1, with the principal functions indicated. These systems consist of several blocks: cassette entry stage, interstage substrate transfer system, preparation and analysis chambers, MBE deposition chamber, device structure processing chamber, and others. The modular form of the systems enables the

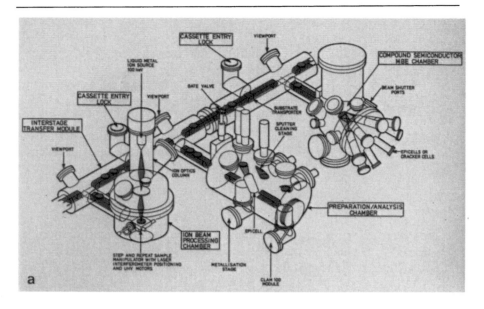

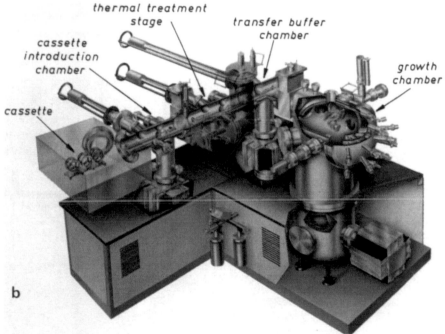

Fig. 3.1a,b. Schematic illustrations of the main building blocks of modern MBE production systems, with their principal functions indicated [Courtesy of VG Semicon (a) and ISA Riber (b)]

user to combine more than one of each of the listed blocks into final production systems [3.2, 3].

Two medium-scale modular MBE systems, designed and manufactured according to present-day criteria on controlled and reproducible UHV deposition of semiconductor films, are presented in side view and in a schematic illustration in Figs. 3.2–5. A key feature of these machines is their ability to be used for research and development purposes, as well as for small-scale production of thin film structures on substrates as large as 3 in. (7.6 cm) in diameter. The individual blocks of a typical modular MBE system will be described in detail in the following sections, with emphasis on the particular features of their construction.

3.1.1 The Cassette Entry Stage

The entry stage allows a single cassette holder to be rapidly loaded or unloaded from the MBE machine. A cassette, which may hold up to ten 3 in. diameter substrates (Fig. 3.6a) is loaded into the "cassette rapid entry lock" system where it is initially evacuated. A schematic diagram of this entry lock system is given in Fig. 3.6b. A simple linear motion drive/valve platform is operated to transfer the loaded cassette directly to the preparation/analysis chamber. The linear drive is used to index the individual substrate for identification and selection. Another design of entry lock system is given in Fig. 3.7 [3.2, 3].

Fig. 3.2. The V80H MBE system installed in a clean room (Courtesy of VG Semicon)

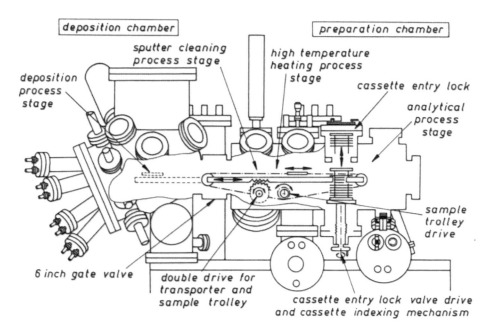

Fig. 3.3. Schematic illustration of the V80H MBE system showing in the cutaway part the sample transfer system (Courtesy of VG Semicon)

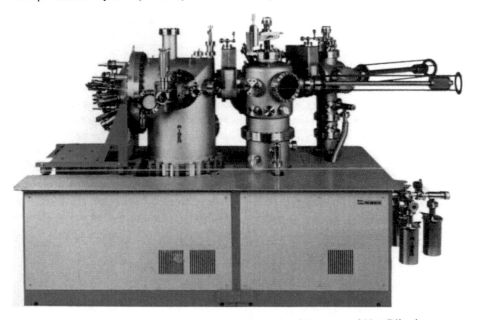

Fig. 3.4. The MBE 32 system in an R&D configuration (Courtesy of ISA Riber)

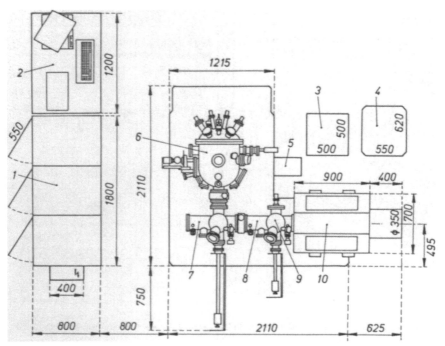

Fig. 3.5. Schematic illustration of the two-module MBE 32 P system with a glove box attached – the scale is given in millimeters (*1* electronic cabinets, *2* computer table, *3* cryopump compressor, *4* mobile rough pumping, *5* cryopump, *6* growth chamber, *7* transfer module, *8* loading and preparation module, *9* thermal treatment system, *10* glove box (Courtesy of ISA Riber)

The entry system comprises: a low volume cassette entry chamber, a rapid access entry door to load cassettes into the entry chamber, an isolation valve to isolate the entry system from the preparation and analysis chambers during load/unload procedures, a cassette transfer system to move the cassette from the entry chamber to the cassette entry position in the preparation and analysis chambers, and a ten-position aluminum substrate holder cassette with ten substrate holder face plates fabricated in molybdenum for substrate marking. An example of a substrate holder for indium-free bonding of 3 in. substrates is shown in Fig. 3.8 (see also [3.4]).

3.1.2 The Interstage Substrate Transfer System

The substrate transfer system is a reliable and practical device which conveys substrates from the cassette entry position to and from the stages in the preparation chamber (Fig. 3.3). In addition it conveys individual substrates through the isolation gate valve into the deposition chamber and thence to the deposition chamber substrate holder.

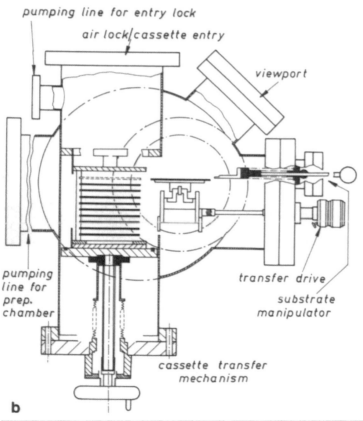

Fig. 3.6. (a) The cassette for ten 3 in. (in diameter) substrates used in the V80H MBE system. **(b)** Schematic illustration of the cassette entry lock system of the V80H MBE system (Courtesy of VG Semicon)

a

0   5   10   15   20 cms

pumping line for entry lock

air lock/cassette entry

viewport

pumping
line for
prep.
chamber

transfer drive

substrate
manipulator

cassette transfer
mechanism

b

Fig. 3.7. Side view of an entry lock system (Courtesy of ISA Riber)

The transfer system generally employs a directly coupled rack-and-pinion-translation device as, for example, in the design by VG Semicon with a substrate holder trolley shown in Fig. 3.9. The trolley can be moved in a way that allows complete flexibility in substrate transfer to any chosen process position in either the preparation chamber or the deposition chamber and thence back to the cassette. Both ISA Riber and VG Semicon trolleys are moved by rotary drive manipulators housed in the preparation/analysis chamber. Transfer forks are provided to move the substrate onto and off the substrate trolleys to the independent process stages. Separate viewports at each process stage allow complete visual access for manipulation.

Fig. 3.8. Substrate mounting block for indium-free bonding of 3 in. (in diameter) substrates which is used in the MBE 32 system (Courtesy of ISA Riber)

3.1.3 The Preparation and Analysis Stages

The preparation of the substrates, which mainly means heat treatment in suitable conditions, is an important step of the MBE process. It influences the epilayers' quality to a large extent. Often it is useful to have surface analysis capabilities, either to check the quality of the substrates prior to the deposition process, or to check the quality of grown layers.

Preparation and analysis are two different steps which can be performed in separate chambers or in a common chamber. All manufacturers offer an out-gassing stage (typically at 800° C) and surface analysis capabilities, including x-ray photoemission spectroscopy (XPS), ultraviolet photoemission spectroscopy (UPS), scanning Auger microscopy (SAM), secondary ion mass spectroscopy (SIMS) and low energy electron diffraction (LEED). See Chaps. 4 and 5.

High quality epitaxial layers grown by MBE on large-size substrates (3 in.) are now routinely processed for industrial production. Defect density is critical for the yield of the total process and should be reduced by all means. Minimizing the defect density requires starting the MBE growth on a clean and stoichiometrically perfect surface. As a necessary preliminary step to MBE deposition, the oxide layer formed in the course of the previous chemical etching step has to be removed by heating. This is a critical operation, as the oxide layer is very thin (≈ 10 Å). In addition, the surface stoichiometry has to be preserved while the removal of the oxide is under way [3.5, 6].

Numerous in situ pretreatment techniques were studied. Among them, heating under arsenic flux, in arsine or hydrochloric acid atmosphere, for example,

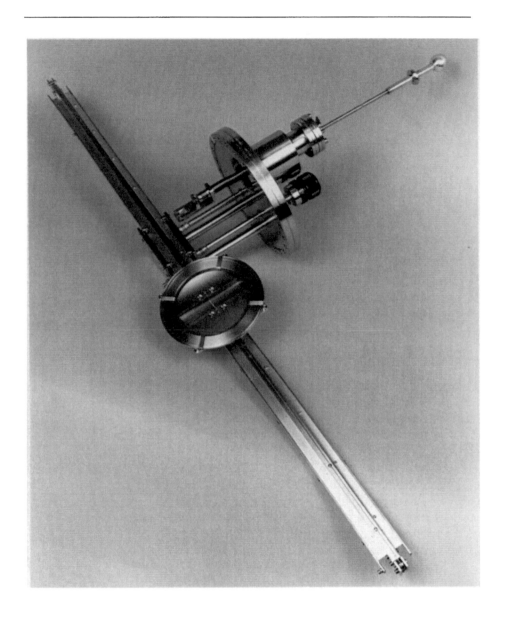

Fig. 3.9. Substrate holder trolley wire-driven on a rail from the V80H MBE machine (Courtesy of VG Semicon)

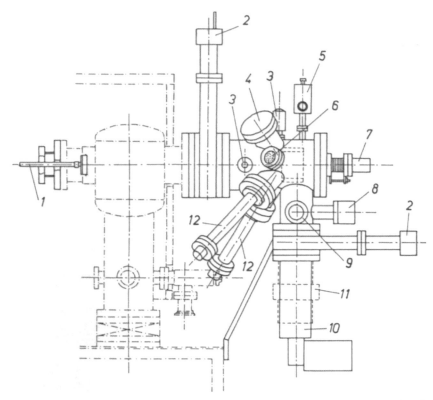

Fig. 3.10. Schematic illustration of the "Controlled Atmosphere Treatment Chamber" of the MBE 32 P system (Courtesy of ISA Riber). *1* transfer system, *2* isolation gate valve, *3* cell shutter, *4* spare port $2\frac{1}{2}$ in ID, *5* gas inlet, *6* viewport, *7* thermal treatment system, *8* Bayard-Alpert gauge, *9* roughing line, *10* cryogenic pump, *11* turbomolecular pump, *12* effusion cell port

were found to be effective for GaAs. For InP substrates, arsine, arsenic and phosphine have been investigated. When using arsenic, dimer species were found to be more efficient than tetramer species. Preventing any pollution of the MBE deposition system requires that the surface preparation stage be performed in a specific UHV system, like the one shown schematically in Fig. 3.10 and in side view in Fig. 3.11. Another design solution is possible in which the preparation stage and the analysis stage are incorporated into one preparation/analysis chamber (Figs. 3.3 and 3.12).

It is possible to site a sputter cleaning source at the outgassing stage, although sputtering is not commonly used in MBE. A metallization facility can be part of the preparation stage, although this technique is rather used after the growth process and deserves a separate chamber. It can be fitted with either a single or a multiple-hearth electron-beam evaporator with a dedicated film thickness monitor, and with several effusion sources. Additional parts may allow some simple processing equipment, e.g. plasma oxidation apparatus.

Fig. 3.11. Side view of the preparation stage shown schematically in Fig. 3.10 (Courtesy of ISA Riber)

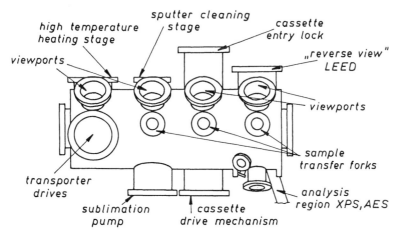

high temperature heating stage

sputter cleaning stage

cassette entry lock

viewports

"reverse view" LEED

viewports

sample transfer forks

transporter drives

sublimation pump

cassette drive mechanism

analysis region XPS, AES

Fig. 3.12. Front view of the preparation/analysis chamber of the V80H MBE system (Courtesy of VG Semicon)

3.1.4 The MBE Deposition Chamber

The deposition chamber is the most important stage in the whole MBE system. Here are located all the elements which enable controlled epitaxial growth in an extremely clean environment. These are beam sources with their individual shutters, a common cryogenic shroud filled with liquid nitrogen, a substrate manipulator with heating and continuous rotation facilities, the electron gun and phosphor screen of the RHEED system, and the pressure control unit (ion gauge or mass spectrometer).

As an example of the present-day design standards concerning MBE deposition chambers, cutaway illustrations of such chambers are shown in Fig. 3.1b (ISA Riber MBE 32) and Fig. 3.13 (VG Semicon V80H [3.7]). Both chambers comprise large-area reservoir cryopanelling to give optimum growth environment protection and to provide a high pumping speed facility at the growth position. The main chamber cryopanel employs a double skin reservoir design and is a cylindrical vessel with a hemispherical enclosure at one extremity.

In the ISA Riber MBE 32 chamber, a separate cryopanel is located around the sources, to allow sequential cooling of the main cryopanel and cell cryopanel, thus encouraging residual contaminants to freeze onto the main cryopanel, away from the sources. In the VG Semicon V80H chamber, the main panel ends with a flat separating reservoir around all the sources. Tubular entrances provide access for all the beam sources and monitoring facilities provided in the growth region. Cell/substrate geometry differs from one manufacturer to another. Cell-to-substrate distance is around 120 mm in the ISA Riber MBE 32 and 130 mm in the VG Semicon V80H.

Eight ports are provided for the individual beam shutters positioned to interrupt the beams between the sources and the substrate. The ISA Riber MBE

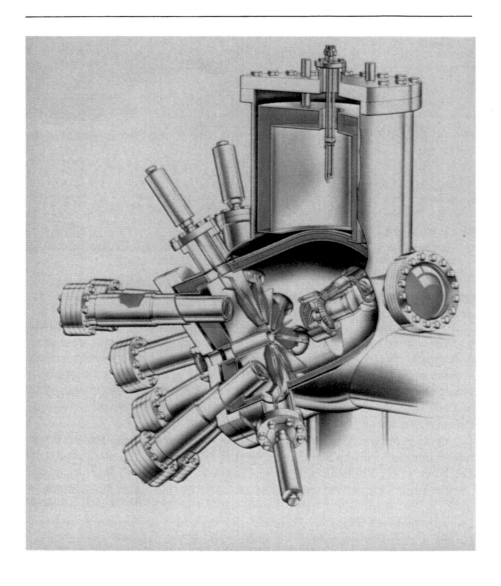

Fig. 3.13. Cutaway illustration of the deposition chamber of the V80H MBE system showing the configuration and geometry of the evaporation sources, shutters and substrate manipulator (Courtesy of VG Semicon)

32 also features a general shutter, allowing the eight beams to be interrupted at once, while the VG Semicon V80H has this facility mounted on the sample manipulator. For both systems, the sample manipulator port is on a horizontal axis and provides continuous sample azimuthal rotation and primary rotation along the axis of the manipulator.

A full complement of ports is provided for the growth monitoring facilities. There are ports for a cross-beam quadrupole mass spectrometer and a RHEED gun and phosphor screen. Also sited on the main sample manipulator port is the movable ionization gauge and main substrate shutter.

Rough pumping of the chamber is achieved using an oil-free rotary pump and sorption pumps, separated from the chamber by an all-metal gate valve. A pumping well accommodates a liquid nitrogen cooled, baffled sublimation pump, located underneath the chamber in the ISA Riber MBE 32 configuration and at the top of the chamber in the VG Semicon V80H.

However, a range of secondary UHV pumping systems can be fitted to this primary facility. A variety of vacuum pumping systems (ion pumps, turbo pumps, diffusion pumps and cryopumps) are available to meet individual processing needs, e.g., for phosphorus, mercury, sulfur, arsine or phosphine. All options give the same guaranteed critical ultraclean residual vacuum level of 5×10^{-11} Torr.

The substrate sample is fixed, during the growth process, on a horizontal axis manipulator, which provides facilities to automatically continuously rotate

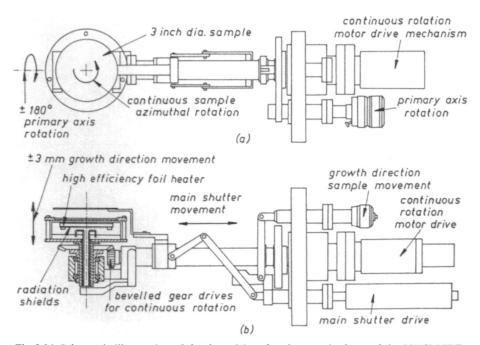

Fig. 3.14. Schematic illustration of the deposition chamber manipulator of the V80H MBE system. **(a)** Top view. **(b)** Side view (Courtesy of VG Semicon)

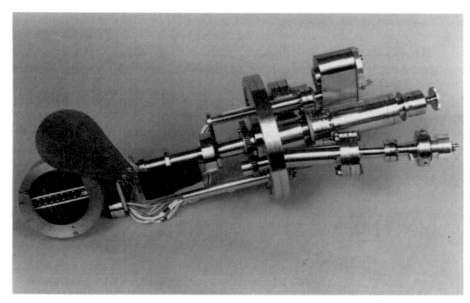

Fig. 3.15. Side view of the deposition chamber manipulator of the V80H MBE system, showing construction details of the substrate heating and rotation facilities (Courtesy of VG Semicon)

Fig. 3.16. Side view of the deposition chamber manipulator of the MBE 32 P system (Courtesy of ISA Riber)

and heat the substrate. The substrate can be positioned with respect to the beam sources at any angle between horizontal and vertical. The degrees of freedom provided by the manipulator are illustrated schematically in Fig. 3.14. A side view of the manipulator fabricated by VG Semicon is shown in Fig. 3.15. The ISA Riber MRC manipulator (Fig. 3.16) includes additional X, Y and Z motions, thus allowing the RHEED beam incidence angle to be varied.

The manipulator incorporates a non-inductively wound high efficiency Ta radiation foil (Fig. 3.17) which allows substrate heating to temperatures up to $800°$ C [3.7] with temperature feedback from a W/Re-W/Re (5 %–26 %) thermocouple.

3.1.5 Beam Sources

The basic design of the beam sources used in MBE is described in Sect. 2.3.1. It should be emphasized that the design of sources has a major impact on the output for the user. Standard effusion cells are made of refractory materials (PBN, Mo, or Ta). The heating element is a large-area foil heater, which provides a very efficient form of radiant heating. It is nearly self-supporting in the VG Semicon design, which allows minimized ceramic insulation. The ISA Riber heater is supported by PBN insulators.

Fig. 3.17. Large surface area heater of the deposition chamber manipulator of the V80H MBE system (Courtesy of VG Semicon)

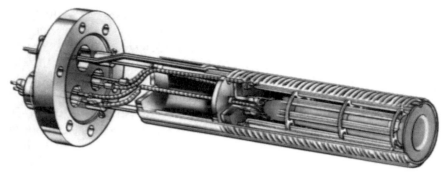

Fig. 3.18. Cutaway illustration of the standard effusion source used in the V80H MBE system (Courtesy of VG Semicon)

The crucibles are tapered with an increase in diameter towards the cell end orifice and incorporate a large flange plate at the top of the cell to effectively prevent the substrate having a direct line of sight to any of the high temperature refractory metal components.

A cutaway illustration of a standard effusion cell used in the V80H deposition chamber is shown in Fig. 3.18. The boron nitride crucible of this cell,

Fig. 3.19. A boron nitride crucible surrounded by its foil heater, which are parts of the standard effusion cell of the V80H MBE system (Courtesy of VG Semicon)

Fig. 3.20. Complete assembly of three effusion sources together with their shutters (Courtesy of ISA Riber)

surrounded by its foil heater, is shown in Fig. 3.19. A complete ISA Riber assembly of three effusion sources with their BN crucibles is shown in Fig. 3.20.

Associated with each cell is a separate refractory metal shutter. The shutters are sited to interrupt the beams between the sources and substrate and to effectively completely shield the substrate from each individual beam when closed. In the VG Semicon system (Fig. 3.21), the shutters are activated linearly, and are at an angle of 24° to each source in order to significantly reduce back-reflected radiation into the cells. This minimizes disruption of the thermal radiation in the cell during shutter operation. The shutters are driven by a magnetically coupled mechanism. An external double solenoid of an all-bakeable construction, surrounds a magnetic core inside the UHV system, and is coupled to this drive through the nonmagnetic stainless steel wall. By applying an alternating current to either solenoid the magnetic core can be moved backwards or forwards. A separate direct current is used for holding positions. This mechanism allows very rapid action (< 100 ms). The magnetic core element is effectively thermally isolated from the refractory metal shutter.

In the ISA Riber system (Fig. 3.20), the shutters are rotary activated at a distance and angle to each source in order to reduce the flux transients. They are activated by a lubricant-free rotary feed-through located on the evaporation flange. A stepping motor allows soft and fast action of the shutters (< 100 ms) with positive knowledge of the position and feedback control by the computer process control system. It also eliminates the need for elastomer shock absorbers.

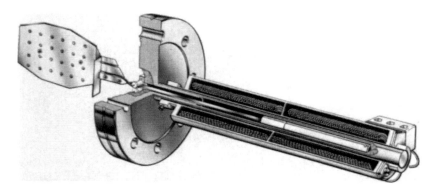

Fig. 3.21. Cutaway illustration of the shutter used with standard V80H effusion cells. The external twin solenoids allow bellows-free actuation without disruption of the UHV environment (Courtesy of VG Semicon)

In place of any of the standard effusion cells it is possible to fit cracker cells (Sect. 2.3.2). The cells are designed for dimer generation of the group V elements and also for antimony or tellurium. The cracker source comprises two distinct temperature-controlled heater zones. The first zone is a reservoir and is used for the controlled evaporation of the group V material (generally in the tetramer form). This reservoir may have a high capacity. The evaporated species are then channeled into a separate heater zone in which they undergo multiple collisions at elevated temperatures in a baffled region. In this high temperature region, dissociation of the arsenic and phosphorus to the dimer species occurs.

The ISA Riber cracker cells feature a PBN cracking stage, with tantalum baffles ensuring multiple wall collisions and enhanced catalytic cracking of molecules. Using these materials instead of graphite, which was formerly used in commercial cracker cells, a nearly 100% dimer flux (arsenic) is obtained at a cracking temperature of 800° C (1000° C for graphite).

It may be useful to have effusion sources vacuum interlocked to the deposition chamber. This allows loading or unloading of the cells without the necessity of air-exposure of the deposition chamber. A schematic illustration of the operation principle of such a load-lock effusion cell assembly is shown in Fig. 3.22,

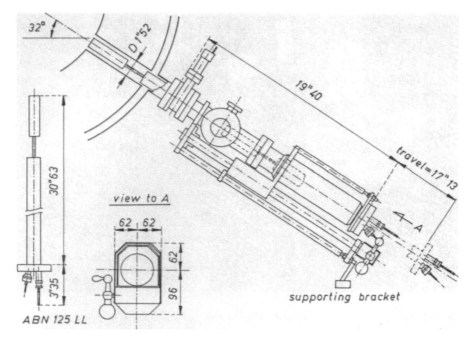

Fig. 3.22. Schematic illustration of the operating principle of the load-lock effusion cell assembly used in the MBE 32 P system (Courtesy of ISA Riber) (Where not indicated, the dimensions are given in mm)

and a front view of this assembly as attached to the deposition chamber of the MBE 32P system is shown in Fig. 3.23.

Special precautions in the MBE deposition chamber are required in the case where some of the effusion cells have to be filled with a high vapor pressure liquid metal, for example mercury. In this case (see Sect. 2.3.1 and Fig. 2.19) the cell should be filled only during the time of deposition and special pumping facilities should be added to the basic pump system. A good mercury cell has the following properties:

–   The flux should be:
    constant during a ten-hour period of growth ($> 10\,\mu$m),
    very reproducible from run to run,
    easy to change in a short period of time.
–   Since several hundred grams of Hg have to be used every day, the cell should have a large Hg capacity or should be connected to a large Hg reservoir.
–   It is also convenient if the mercury cell can be filled in situ without breaking the vacuum.

An example of a design solution suitable for the described case is shown schematically in Figs. 3.24 and 3.25, the first of which shows a constant level mercury

Fig. 3.23. Front view of the load-lock effusion cell assembly attached to the deposition chamber of the MBE 32 P system (Courtesy of ISA Riber)

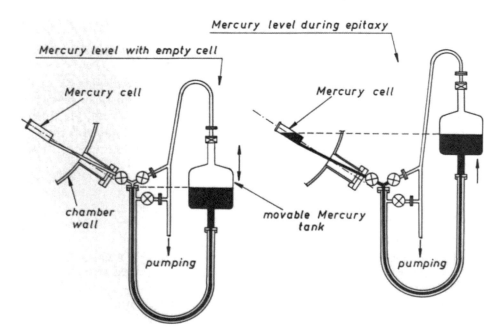

Fig. 3.24. Schematic illustration of the operating principle of the constant level mercury source, model MCL 160, used in the MBE 32 P-CMT machine. The design avoids contact of Hg with copper or other reacting materials (Courtesy of ISA Riber)

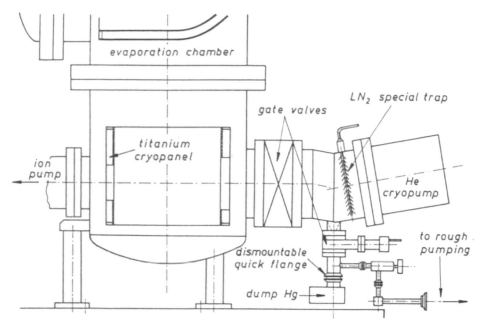

Fig. 3.25. Schematic illustration of the attachment principle of the special CMT pumping system added to the basic pumps of the MBE 32 P-CMT machine (Courtesy of ISA Riber)

source used in the MBE 32P-CMT system, and the second, the additional pumping components attached to the deposition chamber of this MBE system.

The Riber MCL 160 Mercury Cell provides a constant Hg level in a regular cell connected to a large reservoir. The Hg level is kept constant between the cell and the reservoir using only gravity. The motion of the reservoir is monitored by a sensor located in the effusion cell, in order to keep the level constant. Due to the large difference in the areas of the mercury surface in the cell and in the reservoir, $10\,\mu m$ of CMT grown at $200°$ C on CdTe of (111) B orientation changes the Hg level by only 1 mm. Thus the motion of the reservoir is not used during growth. With the help of the sensor, the level is always the same when a new experiment starts. The reproducibility of such a cell is excellent. In this effusion cell, precise control of the Hg temperature is achieved, which allows a good knowledge of the real value of the Hg flux. As regards flexibility, since the effusion cell is small, the temperature can be changed within a few minutes. This property is useful for applications which require a specific grading in doping and composition. The layers obtained are suitable for device applications.

It has been mentioned before (Sect. 1.3.1) that GS MBE has the advantage of replacing condensed phase beam sources of CPS MBE by gaseous sources such as group V hydrides cracker sources, or metalorganic gas sources. Gaseous sources designed as alternatives to the condensed phase beam sources are shown in cutaway illustrations in Fig. 3.26 [3.3,8].

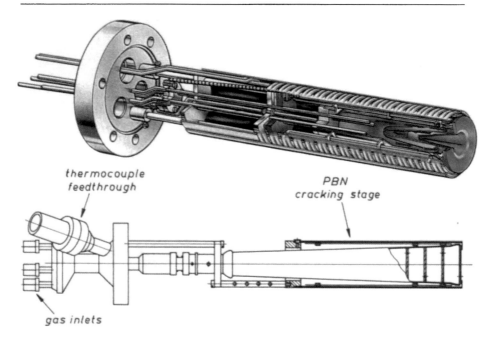

Fig. 3.26. Cutaway illustration of the hydride gas source used in the V80H MBE deposition chamber. Four inlet lines and the pyrolytic boron nitride cracker assembly are shown [Courtesy of VG Semicon (**a**) and ISA Riber (**b**)]

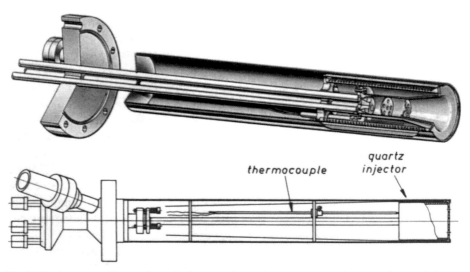

Fig. 3.27. Cutaway illustration of the metalorganic gas source showing three of four independent gas lines, the boron nitride mixing zone and the integral heater [Courtesy of VG Semicon (**a**) and ISA Riber (**b**)]

The group V hydrides source has a cracking zone held at 900°–1000° C. It has several introduction lines, which permits several gases to be introduced into the same cell, thus giving high compositional uniformity, both in space at the output of the cell and in time, due to precise flow regulation control. The VG Semicon source has four independent inlet lines, while the ISA Riber source has five.

In the VG Semicon design, each Al_2O_3 inlet line has a micromachined orifice which sustains a pressure difference between the inlet line and the UHV environment. The cell employs two cracking zones, which allow high cracking efficiency of the hydrides, thereby eliminating condensation of AsH_3 and PH_3 on the internal cryopanel surfaces. Each of the four independent lines can supply controlled group V fluxes. The gas inlet system permits rapid switching between each line, allowing rapid group V switching at the substrate. Group V alloy control is obtained by simultaneously opening two inlet lines into the cell.

In the ISA Riber design, the gases are introduced under low pressure and mixed in the cell flange, thus giving compositional uniformity. The mixed gases are introduced into a PBN cracking stage which has a near 100 % cracking efficiency. The gas mixture is then directed toward the substrate, thanks to a PBN diffuser designed to promote high uniformity at the substrate level. The five inlet lines allow the use of independently controlled group V flows or gaseous dopants.

Alternatively, two lines can be used for the same gas with different flows. The flow can be rapidly switched out of the cell by operating a three-way valve external to the cell. In this operation, the flow coming from the gas handling system is directed toward a pumping system and kept constant, so it can be redirected toward the cell without a stability problem. Due to the low pressure inside the cell, the activation of the external three-way valve stops the growth instantaneously.

Metalorganic gas sources are gas injectors and do not require any cracking zones (Fig. 3.27). The VG Semicon source has four independent gas inlet lines with micromachined orifices. The gas flows into a PBN conical crucible with perforated baffle plates to ensure thickness and compositional uniformity. An integral heater ($< 100°$ C) ensures that gas does not condense in the cell. The gas is then decomposed at the wafer surface.

The ISA Riber source has five independent low pressure gas inlet lines. The gases are mixed in the cell flange for compositional uniformity. They then go through a quartz crucible and are directed towards the substrate through a quartz diffuser, ensuring high flux uniformity at the substrate level. The source has an outgassing facility (750° C) and preheating capability (50° C) with sufficient precision to avoid organometallic condensation and decomposition.

To ensure the precision and stability of gas flow demanded by the GS MBE process, the gas handling system of a MBE machine using gas sources, instead of effusion sources, is specially designed. Each gas inlet has an associated independent control line comprising a number of isolation valves which control gas inlet, gas outlet, gas exhaust, bypass and isolation. All control lines are mounted in a gas cabinet which is easily connected to the laboratory vent system and scrubber. A dedicated exhaust pumping system is configured for fast, safe gas exhaust and ease of servicing. A vacuum trapped rotary pump forms the basis of the system with the addition of a dedicated turbomolecular pump. Specific features of VG Semicon and ISA Riber systems are:

VG Semicon

- Each gas inlet has an associated independent control line comprising a number of pneumatic valves which control gas inlet, gas outlet, gas exhaust, bypass and isolation.
- At the heart of each control line is a capacitance manometer pressure transducer and associated pressure control valves. This offers (i) sensitivity for high precision gas control, (ii) stability for day-to-day reproducibility, (iii) simple, inherently clean construction, and (iv) control of MO gas pressure without resort to a H_2 dilution-pumping system protected against excessive H_2.
- All lines are mounted in a gas cabinet which is easily connected to the laboratory vent system and scrubber.
- The valve configuration is uniquely designed for economical usage of valuable toxic gas.
- Stabilization of gas line pressure is effected using a two-valve control scheme which regulates the rate of gas flow in and out. The two-valve system facilitates rapid stabilization of gas line pressure, thus permitting ease of control for growth of compositionally varying structures.
- The dedicated exhaust pumping system is configured for fast, safe gas exhaust and ease of servicing. A vacuum ballasted, trapped rotary pump forms the basis of the system with the addition of a dedicated turbomolecular pump for metalorganic gas lines. The exhaust system is separately housed for connection to the common vent system.
- Fast gas switching is achieved by rapidly venting the gas outlet line to exhaust. Growth of complicated heterostructures is therefore achieved without the need for mechanical shutters.
- In special applications where extremely rapid gas switching is required, for example in the growth of InGaAs/InP multiple-quantum-well structures, two or more cells with liquid-nitrogen-cooled shutters can be utilized. The shutter assembly consists of a standard effusion cell shutter with a liquid nitrogen cooler can, both on a $4\frac{1}{2}$ in. flange. Primary switching is performed by the shutter, the valve system being used to commence gas flow before the shutter is opened and to shut off gas flow after the shutter is closed.

— The regulation of fluxes is based upon flow control. For group V gases, the main items are a pressure reducer and a mass flow controller. For group III metalorganic compounds, a Baratron pressure gauge and a control valve allow the bubbler pressure to be accurately adjusted. An associated mass flow controller permits precise monitoring of the flow. This system gives precise and reproducible control of all lines. Each line can be fitted with purge input (N_2/H_2), and a hydrogen purifier can be mounted in the gas cabinet.

— The gas switching is achieved using a RUN/VENT technique. A three-way valve directs the flow either toward the cell or toward a secondary pumping system, without flow disturbance. The three-way valve is located near the cell inlets, thus eliminating any transient effect. In fact there is no need to actuate the shutters during growth, even for quantum wells.

— The general design is such that the growth can be controlled down to a fraction of a monolayer, leading to very sharp interfaces.

— Thermostated baths for the metalorganics are housed in the gas cabinet. They can be either heated, or cooled/heated, depending on which compound is used, with high temperature stability.

3.1.6 Monitoring and Analytical Facilities

Full beam monitoring and structural monitoring facilities are usually fitted to the deposition chamber of a MBE machine. The monitoring system comprises a mass spectrometer, monitoring ionization gauge, and the RHEED system.

The quadrupole mass spectrometer is sited with its ion source adjacent to the substrate position and with the axis of its rods normal to the growth direction. The quadrupole assembly is enclosed in an extension of the main cryopanel reservoir. The ionization gauge is sited on the main substrate manipulator assembly. By primary rotation of the manipulator drive the ionization gauge can be sited in an equivalent substrate position. In this position the beams from the individual sources can be monitored and calibrated. If a substrate is also fitted onto the manipulator holder this will be protected from evaporation by the main shutter.

Ports are available in the deposition chamber to fit a complete glancing incidence reflection electron diffraction system. The RHEED system is sited to give structural information on the substrate and growing film in the normal growth position. Sample azimuthal rotation can optimize the pattern obtained. The RHEED diffraction pattern generated by the sample surface is displayed on a phosphor screen suitably located in the deposition chamber. This pattern may be evaluated and further processed in a specially designed optoelectronic system. As an example of such a system the video system used in Philips Research Laboratories will be described following the paper of *Bölger* and *Larsen* [3.9].

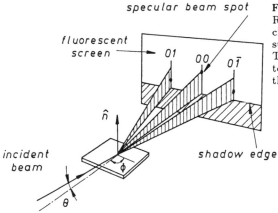

Fig. 3.28. Schematic diagram of RHEED geometry showing the incident beam at an angle θ to the surface plane; azimuthal angle φ. The elongated spots indicate the intersection of the Ewald sphere with the 01, 00, and $0\bar{1}$ rods [3.9]

The experimental geometry of RHEED is illustrated in Fig. 3.28. Electrons having energy of typically 5–50 keV are incident on the substrate at a variable glancing angle ($0 \leq \theta \leq 5°$). The diffraction of the incoming primary beam leads to the appearance of intensity-modulated streaks (or rods) normal to the shadow edge superposed on a fairly uniform background which is due to inelastically scattered electrons. It is obvious from the figure why this geometry is ideal for combination with MBE, where it is desirable to have the molecular beams impinging on the substrate surface at near-normal incidence.

Figure 3.29 shows a diffraction pattern in the $[\bar{1}10]$ azimuth of the reconstructed GaAs(001)-(2 × 4) surface (zero-order Laue zone). The fourfold periodicity is evident from the appearance of fractional-order (i.e., 1/4, 1/2, and 3/4 order) streaks between the integral-order streaks. The intensity of the streaks is strongly modulated, with the section of the highest intensity lying approximately on a semicircle. This can be understood in a simplified picture if it is assumed that the penetration of the electron beam is restricted to the very uppermost lay-

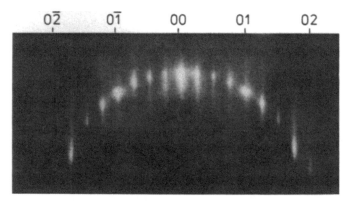

Fig. 3.29. Diffraction pattern from the GaAs(001)-(2 × 4) surface, in the $[\bar{1}10]$ azimuth, taken at an electron energy of 12.5 keV and an incident angle of 3.1° [3.9]

ers (for which there is experimental evidence [3.10]). This results in a nearly two-dimensional Laue condition with the reciprocal lattice points drawn out into lattice rods perpendicular to the surface. The intersection of these with the Ewald sphere will lie on semicircles (Sect. 4.1). In practice, the rods are broadened, resulting in elongated and/or broadened streaks and their shape can be related to disorder, domains, etc. [3.11].

The detection system should be able to resolve fine features in diffraction patterns like that shown in Fig. 3.29. The distance $d_{1/4}$ between two adjacent streaks is given by

$$d_{1/4} = Lg_{\|}/k \quad , \tag{3.1}$$

where L is the distance between the substrate and the phosphor screen, $k = (2\pi/h)\sqrt{2mE}$ is the wave vector of the primary beam of energy E, and $g_{\|} = G_{\|}/4$ is the smallest lattice vector of the 2 $\times$ 4 surface Brillouin zone (Sect. 4.1.2). Here $G_{\|} = (2\pi/a_0)(0, 1, 1)$, where a_0 is the lattice constant. For the conditions of Fig. 3.29 ($E = 12.5\,\text{keV}$, $L = 280\,\text{mm}$) one finds $d_{1/4} = 1.9\,\text{mm}$. Perpendicular to the streaks, therefore, a resolution of at least 0.2 mm (corresponding to an angular resolution of 0.04° or 0.7 mrad) is needed. Along the streaks the angular resolution $\delta\theta$ (i.e., system linewidth) determines the maximum measurable correlation length W, where $W = \pi/k\delta\theta \sin \theta$. It is therefore important to minimize $\delta\theta$. There are several other requirements. As mentioned above, the streak intensity may be highly modulated and strongly dependent on the angle of incidence. In fact, intensity variations exceeding 10^3 are found [3.12]. This necessitates a large dynamic range and high stability and reproducibility of the angular setting. Further requirements related to MBE growth are the need for fast simultaneous detection of two or more beams for establishing phase relationships and rapid successive measurements of line profiles.

In order to ensure fast data acquisition *Bölger* and *Larsen* [3.9] relied on processing the video signal from a television camera imaged on the RHEED screen. The camera was equipped with a sensitive silicon vidicon tube and a wide-aperture lens with calibrated diaphragm. The horizontal TV line scan was adjusted so as to be parallel to the diffraction streaks and perpendicular to the substrate's rotation axis defining the angle of incidence. The fluorescent screen was made from a commercially available phosphor with a grain size of about 0.1 mm (this determines the ultimate resolution of the system). It is, however, possible to produce phosphor screens having a resolution of $\sim 30\,\mu\text{m}$. Such screens are used for medical applications.

The video signal was processed by two different systems, each having its own merits. The first system, shown in Fig. 3.30, used a two-channel boxcar integrator. Each channel has a time aperture over which the signal was averaged. Proper triggering of the boxcar can place this aperture anywhere on the video signal (TV picture). TV frame-synchronization (F.S.) and line-synchronization (L.S.) pulses, taken from the video monitor, are used to trigger two pulse-delay generators for scanning purposes. Their output pulses pass through an AND

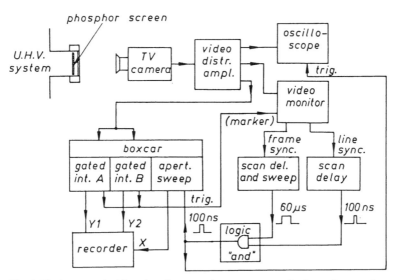

Fig. 3.30. Layout of video signal processing system employing a boxcar integrator [3.9]

gate which triggers the boxcar. The horizontal delay is thus referenced to the preceding horizontal synchronization pulse. The delayed frame synchronization pulse has to have the duration of a line sweep (i.e., $60\,\mu s$) for triggering on a single line. The aperture of the boxcar can be swept linearly over the screen in any desired direction. The positions of the two apertures are made visible by adding the aperture gates to the video signal on the monitor. The instantaneous signal along a TV line is monitored on an oscilloscope. Each unit receives the video signal from a separate output of a distribution amplifier.

An aperture of about 50 ns is used to measure the intensity along a streak. The initial L.S. delay is set so that the scan starts at the position of the straight-through beam and the stationary F.S. delay is set to the required TV line as seen on the monitor and oscilloscope. A scan is made by using either the boxcar scan (leaving one channel stationary) or the horizontal scan delay. With reasonably strong signals a scan takes about 10 s. To measure the integral streak intensity distribution, the aperture is taken to encompass the whole streak length and scanned in the vertical direction (L.S. stationary, F.S. delay scanned). For growth studies the A and B channel apertures of the boxcar can be adjusted to two different diffraction features and monitored on a two-pen recorder.

The second system is shown in Fig. 3.31. After passing through the distribution amplifier the video signal along a horizontal line scan is digitized by a Biomation 8100 transient recorder (2048 channels, 8 bits). The Biomation output is stored at the frame repetition rate in a digital multichannel averager (Tracor TN 1710). A number of frames (typically 100–500) were averaged to obtain a good signal-to-noise ratio. If desired, the background signal could be subtracted and the resulting spectrum stored on a floppy disk. The trigger timing was ob-

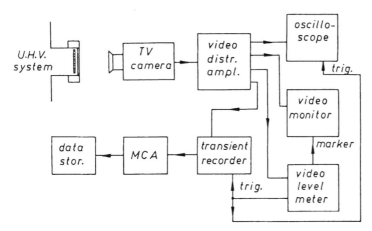

Fig. 3.31. Layout of video signal processing system employing a Biomation transient recorder and a multichannel averager [3.9]

tained here from a video level meter (Philips PM5548) giving slightly more stable triggering than in the first system. This video meter enables the position of the trigger, visualized on the monitor, to be at any point on the screen. A slight modification of the video meter allowed the trigger to be scanned across the screen with an external signal. Apart from being the source of a stable trigger, this meter directly measures the light intensity at the marker position.

Whereas the Biomation system has a higher accuracy, is faster (a few seconds per streak), and permits easy data handling, the boxcar system can scan its aperture in any direction, measures dual-point time dependence, and requires a smaller investment.

For both systems the deflection angle was calibrated by placing a piece of graph paper on the screen and making measurements (both horizontal and vertical). After marking the position of the straight-through beam an angular accuracy of $0.03°$ was obtained for $\delta\theta$. The spot resolution was limited by the camera tube ($20\ \mu m$ resolution) and the image demagnification (of 7.5) to $150\ \mu m$. The fluorescent screen was found to be linear over at least a factor of 10^3 in intensity. The camera was adjusted for linear response ($\gamma = 1$) and its black level adjusted to give a good signal at weak illumination. The lens diaphragm, used to prevent system overloading, was calibrated. Some problems were met due to pickup at the mains frequency on the camera front end. This causes two horizontal bands across the screen. External synchronization of the camera keeps these bands stationary, which is important for time-dependent studies.

A series of streak scans of the (00) beam at different angles of incidence θ taken with the Biomation system is given in Fig. 3.32. The crystal is GaAs(001) with the 2×4 reconstruction (substrate $T = 560°$ C) in the $[\bar{1}10]$ azimuth. The substrate was very carefully oriented and had a measured misorientation (angle between the surface normal and the [001] direction) of less than $0.02°$. The scans

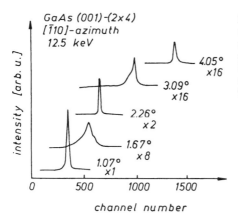

GaAs (001)-(2×4)
[Ī10]-azimuth
12.5 keV

intensity [arb. u.]

4.05°
×16

3.09°
×16

2.26°
×2

1.67°
×8

1.07°
×1

0   500   1000   1500

channel number

Fig. 3.32. Scan along the (00) rod of GaAs(001)-(2 × 4) in the [Ī10] azimuth at an electron energy of 12.5 keV for different angles of incidence. The scans were made using the processing system shown in Fig. 3.31, and averaged over 100 frames. Note the different amplification factors [3.9]

are averaged over 100 frames and presented without background subtraction. Part of the increase of the background at the end of a scan is due to the increase of the dark signal inherent to vidicon camera tubes. This effect can be eliminated by using a high-resolution CCD solid-state video camera, which is also less prone to pickup problems. The streak-profile measurement shown in Fig. 3.32 took only 4 s, increasing to about 7 s if the storage on floppy disk is included. It has been found that features in the pattern on the screen, which could be just observed by the naked eye in a darkened room, could be quantitatively measured with good signal-to-noise ratio.

The spectra in Fig. 3.32 display some of the typical features encountered. The intensity is greatly dependent on the angle of incidence and the lens diaphragm has to be adjusted in order to avoid overloading. The narrow peak, which is the elastically diffracted(00) beam, has a variable linewidth and additional features in the line profile are also seen. These features are not due to regular steps on the surface because of the very accurate orientation of the surface. Their positions also remain fixed on changing the azimuth by π. The intensity-versus-angle dependence (the rocking curve) is found to display a number of maxima which are related to primary or secondary Bragg diffraction or to surface resonances (Sect. 4.1).

The final point which should be discussed is the overall experimental resolution. The width of a sharp feature such as the peak of the line profile at 2.26° in Fig. 3.32 is about 14 channels (FWHM) as compared with 10 channels (10^{-3} rad) for the incident electron beam (VG LEG 110) and five channels for the vidicon resolution. This means that linewidth deconvolution (or a better electron gun) might be needed in linewidth studies. If the intrinsic linewidth $\delta\theta$ is attributed only to a finite correlation length W, a lower limit in the direction of the beam can be estimated from $W = \pi/k\delta\theta \sin\theta$ to be 140 nm.

Modern MBE machines are also usually equipped with apparatus for a variety of surface analytical techniques (Chap. 4), including Auger, SAM, XPS, UPS and depth profiling for compositional analysis and a reverse-view LEED

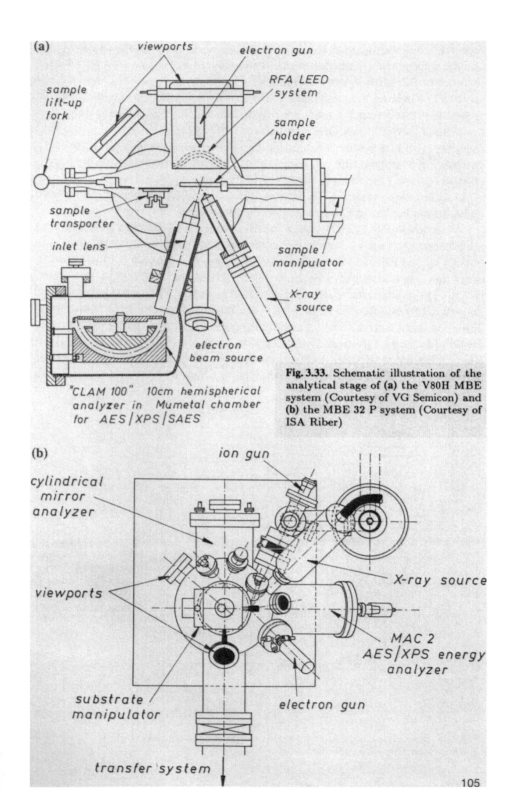

(a)

viewports

electron gun

RFA LEED system

sample lift-up fork

sample holder

sample transporter

inlet lens

sample manipulator

X-ray source

electron beam source

"CLAM 100" 10cm hemispherical analyzer in Mumetal chamber for AES/XPS/SAES

Fig. 3.33. Schematic illustration of the analytical stage of **(a)** the V80H MBE system (Courtesy of VG Semicon) and **(b)** the MBE 32 P system (Courtesy of ISA Riber)

(b)

ion gun

cylindrical mirror analyzer

X-ray source

viewports

MAC 2 AES/XPS energy analyzer

substrate manipulator

electron gun

transfer system

system [3.13] for structural analysis. These analytical units are usually sited in the analytical stage of the preparation/analysis chamber [3.3] or in a separate analysis chamber of the MBE system. In the V80H system, for example, compositional analysis is based on the so-called CLAM 100 module, which, together with the chosen radiation source, is mounted below the analysis stage (Fig. 3.33a).

The CLAM 100 module comprises a spherical sector analyzer (model 849) complete with channeltron electron multiplier and lens assembly housed in an independent Mumetal (this is a special alloy exhibiting magnetic shielding properties) chamber. Electrons are conveyed to the analyzer by a three-stage electrostatic Einzel lens assembly. The analyzer/lens assembly is mounted on a flange which includes flanged ports for the radiation sources.

A reverse-view LEED system fitted in the analysis chamber, with a viewing screen at the top of the chamber horizontally configured, is also shown in Fig. 3.33a. The LEED system employs a transparent screen and a miniature electron gun. This allows the diffraction pattern at normal incidence to be viewed for the 3 in. diameter sample from behind the screen. The optics are mounted on an FC 150 double-sided flange with all the necessary electrical connections. The minimum distance from screen to viewport is less than 70 mm. Another example of a design for an analytical chamber is the one fabricated by ISA Riber (Fig. 3.33b). More detailed information about the surface analytical tools used in present-day MBE systems will be presented in Chap. 4.

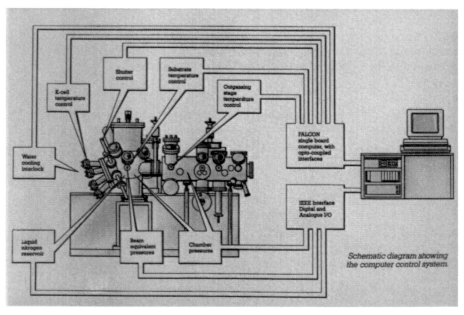

Fig. 3.34. Schematic diagram showing the computer control system of the V80H MBE machine (Courtesy of VG Semicon)

As already mentioned at the beginning of this chapter, the potential of MBE to produce device structures of great complexity may only be realized by utilizing effective automatic control. This means that a computer control system is an indispensable part of each MBE system. An example of the computer control system for a MBE machine (the V80H MBE system) is shown in Fig. 3.34.

3.2 Multiple-Growth and Multiple-Process Facilities in MBE Systems

Because of its linear design, coupled with a suitable substrate holder transfer system, a modular MBE machine can, if required, be easily extended to incorporate additional chambers, e.g., a glove box (Fig. 3.5), a second deposition chamber, or a more sophisticated surface analysis chamber including, for example, a scanning electron microscope, a secondary ion mass spectrometer, or other analytical facilities.

Two examples of extended modular MBE systems, with twin growth chambers, are shown in Figs. 3.35 and 3.36. Each of these systems may be further enlarged by adding, for example, a metallization chamber (Figs. 3.37 and 3.38) with electron-beam gun evaporators (Fig. 3.39) or an ion beam processing stage (Fig. 3.1).

Fig. 3.35. A twin growth chamber V80H MBE system (Courtesy of VG Semicon)

Fig. 3.36. The MBE 32 P Modutrac system with two identical growth chambers (Courtesy of ISA Riber)

3.2.1 The Hot-Wall Beam Epitaxy Growth System

As has been pointed out in Sect. 1.2.1, the "three-temperature method" of Günther developed historically in two directions. First we have MBE, the more sophisticated technique, a cold-wall method with a low growth rate crystallization process far from thermodynamic equilibrium. The second technique, hot-wall epitaxy (HWE), is relatively simple with a high growth rate crystallization process near thermodynamic equilibrium.

In many cases of semiconductor film crystallization procedures, a quite thick buffer layer, matching the substrate structural properties to the epilayer structure, is required. It would be reasonable to grow first the thick buffer layer with a high growth rate technique, e.g., GS MBE or HWE, and then after this, to grow the more sophisticated epilayer structure in a second growth stage by MBE. To do this, one may attach to an extended MBE system a growth stage where buffer layers may be grown on suitably prepared substrate wafers. For this final substrate preparation procedure HWE is preferred to GS MBE due to its simplicity accompanied simultaneously by high structural quality of the epilayers grown. This is especially true in the case of II–VI or IV–VI compound structures, and even more when hot-wall beam epitaxy (HWBE) is applied. This [3.14] is a new modification of the well-established HWE technique [3.15]. It is performed in an UHV growth environment that enables the advantages of the HWE technique to be combined with those of the in situ surface diagnostic facilities typical of the MBE technique.

Fig. 3.37. Side view of the MBE 32 P Modutrac metallization chamber (Courtesy of ISA Riber)

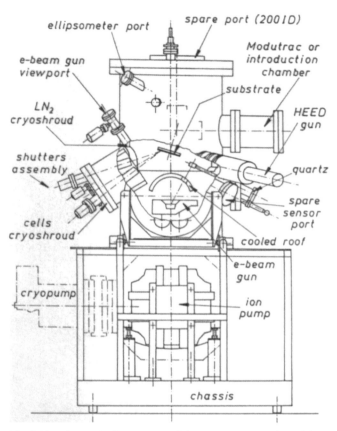

ellipsometer port   spare port (200 ID)

Modutrac or
introduction
chamber

e-beam gun
viewport

substrate

LN₂
cryoshroud

HEED
gun

shutters
assembly

quartz

spare
sensor
port

cells
cryoshroud

cooled roof

e-beam
gun

cryopump

ion
pump

chassis

Fig. 3.38. Schematic illustration of the EVA 32 UHV deposition system, which is equipped with an electron beam gun evaporator (up to 5 crucibles), effusion source (4 standard cells) and flux monitors. This deposition chamber may be applied for silicon epitaxy, metallization or superconductor deposition (Courtesy of ISA Riber)

A schematic illustration of a conventional HWE evaporation reactor is shown in Fig. 3.40. A heated wall, in the form of a cylindrical tube, is inserted between the source and the substrate. In this manner: (a) loss of evaporation material is strongly reduced, (b) the environment within the growth reactor is kept clean as compared to the rest of the vacuum system, and (c) a relative high pressure of the evaporating materials can be maintained inside the reactor. There are three independently heated ovens in the HWE reactor. These maintain the temperatures of the evaporating sources, the hot-wall tube, and the substrate, respectively. Suitable adjustment of these three temperatures enables one to achieve low supersaturation during growth with HWE, and thus effective control over the nucleation stage, and consequently high crystalline quality of the grown epilayers.

The HWBE reactor differs from the conventional HWE reactor by having the substrate holder removed from the hot-wall tube to a definite distance along the symmetry axis of the reactor. Accordingly, the assembly of the evaporating

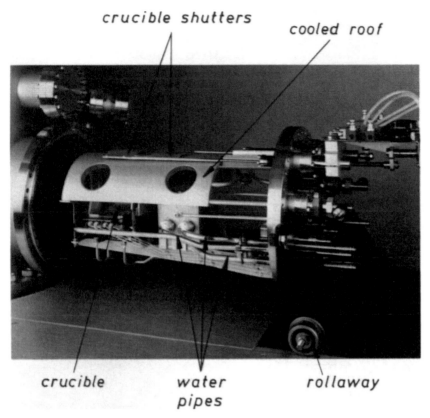

crucible shutters

cooled roof

crucible

water pipes

rollaway

Fig. 3.39. Side view of the electron-beam gun EB 22/300 of the EVA 32 UHV deposition chamber. This gun evaporates material from the 40 cm³ crucibles (Courtesy of ISA Riber)

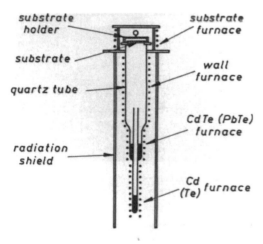

substrate holder

substrate

quartz tube

radiation shield

substrate furnace

wall furnace

CdTe (PbTe) furnace

Cd (Te) furnace

Fig. 3.40. Schematic illustration of the conventional HWE evaporation reactor used for crystallization of CdTe and PbTe films [3.15]

111

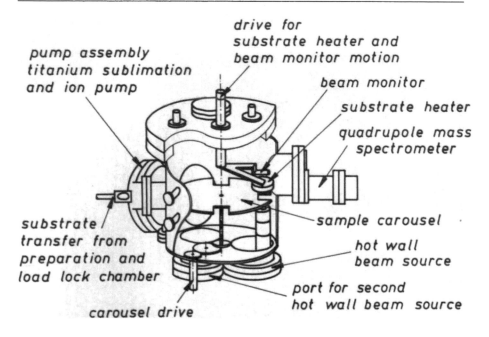

pump assembly
titanium sublimation
and ion pump

drive for
substrate heater and
beam monitor motion

beam monitor

substrate heater

quadrupole mass
spectrometer

substrate
transfer from
preparation and
load lock chamber

sample carousel

hot wall
beam source

carousel drive

port for second
hot wall beam source

Fig. 3.41. (a) Cutaway illustration of the HWBE growth chamber. **(b)** Front view of the HWBE 2500 growth chamber with transfer system (Courtesy of TOPLAB a division of Hainzl Industriesysteme)

sources together with the hot-wall tube (Fig. 2.9) creates a single channel extended beam source. The hot-wall tube acts as an equilibrating element for the reactor (Sect. 2.2.3).

Figure 3.41a shows a schematic drawing of a HWBE growth chamber and Fig. 3.41b a picture of the original apparatus. The system is designed to accept two HW-beam sources; one is shown in the figure. The substrates are transferred from a load-lock or a preparation chamber into the growth chamber through a gate valve and inserted into the carousel. Up to five films may be grown one by one by simply rotating the carousel and positioning the substrate over the beam source. Before growth is initiated the beam flux is measured with a beam monitor. During this measurement the substrate and the substrate heater are not in the growth position. Either the substrate heater or the beam monitor can be brought into the growth position, as desired. Additionally the beam may be analyzed by means of a quadrupole mass spectrometer with its cross beam ion source positioned above the beam monitor.

The HWBE-growth chamber has been successfully used for the growth of high quality CdTe layers on GaAs [3.16]. Based on this experience with HWBE-grown CdTe layers, the main features of the HWBE technique can be characterized as follows:

- High uniformity in thickness and luminescence properties, proven in the case of 1 in. CdTe layers on GaAs, can be obtained.
- Although the substrate is no longer put on top of the hot wall tube to close the reactor, high quality layers can be grown at high yield.
- A high growth rate is achieved as in the case of conventional HWE without significant loss of layer purity.
- Loss of evaporating material is minimized.
- A low source temperature is sufficient to establish an appropriate particle flux.

3.2.2 Focused Ion Beam Technology

As mentioned before (Sect. 1.3.4), fabrication of semiconductor device structures entirely in a high vacuum environment is a trend in the technology of present-day microelectronics and optoelectronics. For this purpose, welding of the MBE growth technique and the focused ion beam processing technique in one fabrication scheme is especially useful (Fig. 3.1). Lithography is the key patterning step in all current integrated circuit fabrication. Resist is spun on a wafer, baked, and exposed in an intricate pattern, usually by ultraviolet light, although x-ray and electron beams are beginning to play a role. After development and baking the surface is left partly covered by an inert organic film that "resists" various treatments to which the bare surface is subjected. The treatment may, for example, be material removal by a wet chemical etch or by a gaseous plasma; doping by ion implantation (broad beam); or addition of material by evaporation (lift-off).

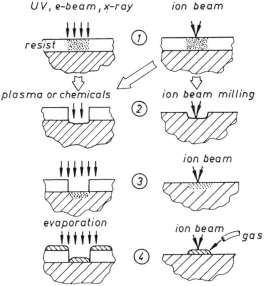

UV, e-beam, x-ray   ion beam

resist

①

plasma or chemicals   ion beam milling

②

ion beam

③

evaporation   ion beam

④   gas

Fig. 3.42. Schematic illustration of the advantages of the FIB technology compared to conventional lithographic technology [3.17]. Conventional resist-based fabrication is shown in the left, i.e., resist is exposed and developed, then material is removed from the unprotected area or the unprotected area is implanted, or material is deposited on the exposed area. The focused ion beam can expose resist, but more importantly, it can remove material, implant, or deposit in a fine pattern *without* the use of resist or a mask

1 — lithography
2 — material removal
3 — doping (implantation)
4 — deposition of material

This patterned alteration of a surface using lithography is a multistep process, and treats the whole wafer in the same way.

Ion beams focused to submicrometer diameters offer a radical departure from this conventional fabrication routine [3.17]. Resist may be eliminated and the dose of ions can be varied as a function of position on the wafer. This is illustrated in Fig. 3.42. Focused ion beam (FIB) fabrication is a serial process, and the wafer is exposed point by point. Needless to say, this is slow and some of the applications can be considered only over very limited sample areas.

Ions of keV energies incident on a solid surface produce a number of effects: several atoms are sputtered off, several electrons are emitted, chemical reactions may be induced, atoms are displaced from their equilibrium positions, and ions implant themselves in the solid, altering its properties. Some of these effects, such as sputtering and implantation, are widely used in semiconductor device fabrication and in other fields. Thus the capability to focus a beam of ions to submicrometer dimensions, i.e., dimensions compatible with the most demanding fabrication procedures, is an important development. Activities in the focused ion beam field have been spurred on by the invention of the liquid metal ion source and by the utilization of focusing columns with mass separation capability. This has led to the use of alloy ion sources, making available a wide range of ion species, in particular the dopants of Si and GaAs. The ability to sputter and also to induce deposition by causing breakdown of an adsorbed film has produced an immediate application of focused ion beams to photomask repair.

Interest in FIB is growing rapidly. The following range of specifications of the FIB system have been reported: accelerating potential 3–200 kV, ion current density in the focal spot up to 10 A/cm^2, beam diameters 0.05–1 μm, deflection accuracy of the beam over the surface $\pm 0.1\ \mu$m, and ion species available Ga, Au, Si, Be, B, As, P, etc. Some of the applications which have been demonstrated or suggested include mask repair; lithography (to replace electron beam lithography); direct, patterned, implantation doping of semiconductors; ion-induced deposition for circuit repair or reviewing; scanning ion microscopy; and scanning ion mass spectroscopy [3.17].

A focused ion beam system can be thought of as composed of three main parts: the source of ions, the ion optical column, and the sample displacement table (Fig. 3.43). A detailed understanding of these parts is too intricate for the

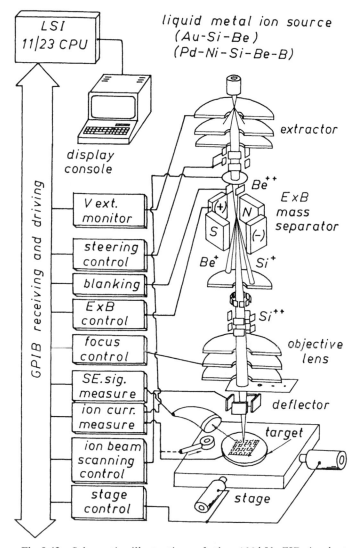

Fig. 3.43. Schematic illustration of the 100 kV FIB implanter with an ion beam controlling/writing system [3.18]

Table 3.1. System parameters of a 100 kV FIB implanter [3.18]

Accelerating voltage	40–100 kV
Ion source	Field-emission-type liquid metal
Source available	Si, Be, B
Lifetime	> 700 h (Au-Si-Be source)
	> 100 h (Pd-Ni-Si-Be-B source)
Ion species separation	$E \times B$ filter
Resolution	$\Delta M/M = 3\%$
Minimum beam size	0.1 μm
Current density	1 A/cm^2 (Si 160 keV)
Current stability	1% per hour
Writing field size	400 × 400 μm
Address size	14 × 14 bit
Dwell time	0.5–1024 μs
Sample stage movement	± 10 mm (x, y)
	0–90° (tilt)
	0–360° (rotation)
Accuracy	± 10 μm (x, y)
Sample size	up to 2 in.
Vacuum, sample chamber	3 × 10^{-10} Torr (bakeable to 150° C)
Focusing system	5 × 10^{-7} Torr

scope of this section. Instead, the principles of operation of a 100 kV FIB implanter with an ion beam controlling/writing system will be described and the current best performance and the potential application possibilities will be discussed [3.18].

Figure 3.43 shows a block diagram of the 100 kV FIB implanter with an ion beam controlling/writing system [3.19]. Table 3.1 summarizes system parameters of this implanter.

The accelerating voltage applied between the emitter tip and the target is 40–100 kV. The field emission ion source is charged with Au-Si-Be or Pd-Ni-Si-Be-B alloyed liquid metal, so Be (p-type dopant), Si (n-typed dopant), and B (dopant for isolation) ions can be emitted together. If the electric and magnetic fields are appropriately set, a desired ion beam can pass through the $E \times B$ mass separator undeflected [3.20]. Adjusting the objective lens focuses the selected ion beam onto the semiconductor target. The electrostatic deflector permits the FIB to be scanned over several-hundred-micrometer squares of the target. The dipole electrostatic blanking plates are located just above and below the $E \times B$ mass separator. While the vacuum level of the focusing column is 5 × 10^{-7} Torr, the vacuum level of the sample chamber, which is isolated from the column by a pinhole, is better than 5 × 10^{-10} Torr.

The diameter of the focused ion beam is less than 0.1 μm, and the current density of the target ion current is up to 1 A/cm^2. Stability of the target current is better than 1% for 1 h, and the lifetime of the Au-Si-Be liquid metal ion source exceeds 700 h. Maskless ion implantation doping can thus be performed under stable and reproducible conditions.

To arbitrarily fabricate the geometry of p- and n-type pattern doping regions in both the depth (z) and lateral (x, y) directions with the maskless FIBI process, the system has various functions for controlling the maskless ion doping process as well as for arbitrary-shape ion beam writing. As shown in Fig. 3.43, the extractor, ion beam steering quadrupole, $E \times B$ mass separator, objective lens (OL) electrode, ion current, secondary electron signal, electrostatic deflector, and sample stage are controlled or measured semiautomatically by an LSI 11/23 controlling interface. Thus ion species, ion energy, and beam diameter are freely and quickly selected, and beam position is recorded during ion beam writing so that impurity doping patterns formed in the semiconductor substrate can be precisely aligned with each other. The ion beam can be scanned on 16 000 points in the x and y directions over a 400 μm square field with the incident axis 7° off at a working distance of 40 mm. The sample stage is used to move a sample substrate within a 20 mm square. Ion dose is determined by controlling dwell time (0.5–1000 μs) and by choosing frame counts of the scanned FIB, considering ion current and beam diameter. One of the most significant benefits of maskless ion implantation is the rapid and arbitrary selection of ion dopant species during the process [3.18].

The described FIB implanter may be coupled with a MBE growth chamber via a sample transfer vacuum stage into one growth-processing system, as is shown in Fig. 1.13. As in the process drawn in the upper part of this figure, the purpose of this system is to arbitrarily fabricate p- and n-type pattern doping geometries controlled in depth and lateral directions with fine resolution over GaAs and AlGaAs multilayers. To realize this structure, actual ion beam writing in a MBE-FIBI process is performed as follows. Etched fiducial marks are located at the corners of a 400 μm square field on the GaAs substrate as shown in Fig. 3.44. When an MBE layer is grown on the substrate, the mark grooves are replicated on the layer. Ion beam writing is performed on the grown layer with reference to the mark positions of the newly replicated grooves.

In this case, however, the appearance of the mark groove changes as MBE GaAs grows because of the anisotropic growth of crystallographic orientation [3.21]. This is expected to influence the alignment accuracy of mark position detection as the MBE layer grows thicker. Figure 3.44b shows scanning ion microscopy (SIM) images of the registration marks produced by dry etching the GaAs substrate to 2 μm depth followed by 1 μm thick MBE layer growth. Electron emission from the edge of the side wall is especially high because the concentration of penetrated ions is higher at the side wall than at the plane surface. Thus, two intensive peaks are observed, corresponding to both side walls [3.22]. Figure 3.44c shows the detected signal in $x[1\bar{1}0]$ and $y[\bar{1}\bar{1}0]$ directions as a function of relative ion beam position. It can be seen that the signal shapes differ because of slight distortion of the marks due to anisotropic MBE growth on the edges and corners of the mark grooves. From these data, the coordinates of the center of the mark are found automatically. Positional reproducibility of this mark detection technique is better than 0.5 μm even after the slight distortion of marks by the 1 μm thick MBE growth.

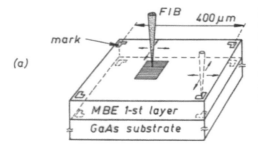

Fig. 3.44. (a) Method of mark position detection over the multicrystal layer. **(b)** SIM image of registration mark covered with $1\,\mu$m thick MBE layer and **(c)** corresponding SIM signals for mark position detection in different directions ([1$\bar{1}$0] and [$\bar{1}\bar{1}$0] orientations, respectively) [3.18]

(a)

(b)

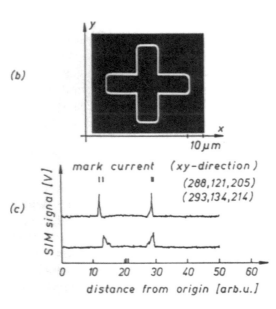

(c)

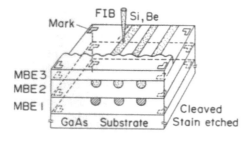

Fig. 3.45.
Cross-sectional view of double-layered Be (p-type) and Si (n-type) impurity doping regions produced with a $7 \times 10^{13}\,\mathrm{cm}^{-2}$ dose with the FIBI system together with a MBE deposition chamber [3.18]

Using this mark detection technology, a patterned Be- and Si-doped MBE-grown GaAs multilayer structure is fabricated [3.18]. As shown in the upper part of Fig. 3.45, a finely focused (0.2 μm diameter) Si ion beam (160 keV) was referenced to etched registration marks and implanted into n-type (100) MBE-grown GaAs (5×10^{15} cm^{-3}) to form impurity doping line patterns. Then the sample was covered by a 1 μm thick epitaxial layer followed by Be implantation (80 keV) with identical patterns. The sample was implanted with doses of 5×10^{13} to 7×10^{15} cm^{-2}. After the ion implantation, a 0.5 μm thick GaAs layer was grown on the implanted layer by MBE, and a 1000 Å thick CVD SiO$_2$ film was deposited for the encapsulation. Then the sample was annealed at 850° C for 20 min. To characterize the lateral and depth profiles of the implanted region, the sample was cleaved at the center of the line patterns and was chemically stainetched [3.19].

The microphotograph in Fig. 3.45 shows the cross-sectional view of the double-layered, finely p- (80 keV Be) and n- (160 keV Si) doped regions with a dose of 7×10^{13} cm^{-2}. It should be noted that the upper and lower doping regions are well aligned, i.e., the relative positional shift between the identical patterns in adjacent layers was found to be within 0.3 μm regardless of ion species or layer. This alignment technology is a prerequisite for precisely fabricating pattern-doped multilayer crystal growth structures.

Another approach to the ion beam doping technology for MBE-grown semiconductor films is the low-energy ion doping procedure [3.23]. The ion source used in this technology consists of a single chamber which combines the functions of an effusion cell, a vapor transport tube, and a glow discharge or electron beam ionizer [3.24]. Typical ion energies are in the range 0.5–3 keV. The low-energy ion doping processes will be described in more detail in Sect. 6.5.4, because they belong in fact to in-growth doping techniques for MBE-grown films or layered structures.

4. In-Growth Characterization Techniques

The experimental arrangement of MBE is unique among epitaxial thin film preparation methods in that it enables one to study and control the growth process in situ in several ways. In particular, reflection high energy electron diffraction (RHEED) allows direct measurements of the surface structure of the substrate wafer and the already grown epilayer. It also allows observation of the dynamics of MBE growth [4.1–20]. The forward scattering geometry of RHEED is most appropriate for MBE, since the electron beam is at grazing incidence (Fig. 3.28), whereas the molecular beams impinge almost normally on the substrate. Therefore, RHEED may be called an in-growth surface analytical technique.

Two more surface sensitive diagnostic facilities may be incorporated into the MBE arrangement as in-growth analytical techniques. These are ellipsometry [4.21–25] and ultraviolet photoelectron spectroscopy (UPS) [4.26, 27]. In both of these techniques neither the probing beam nor the elements of the apparatus used block or disturb the flow of molecular beams in the deposition chamber. In both cases light is directed at grazing angles toward the substrate, and thus, it intersects the molecular beams nearly perpendicularly. Consequently, light beams used in these techniques do not disturb the MBE growth process, while delivering useful information about it [4.23, 26].

In modern multichamber MBE systems other surface diagnostic facilities are also often incorporated. Most frequently the following analytical techniques are used:

- Auger electron spectroscopy (AES), for determining surface chemical composition of the substrate wafer or the MBE-grown epilayer [4.28–30], and for determining the bulk chemical composition of the grown structures in the case of a depth profiling procedure [4.31, 32]. The AES technique may be used in MBE systems in two possible modes. The simpler one is AES point analysis (Fig. 3.33a), while the more sophisticated mode is scanning Auger microscopy (SAM).
- Secondary ion mass spectroscopy (SIMS), for determination of the chemical composition of the outermost atomic layers of the grown structure [4.33, 34]. Compared to AES, the main advantages of SIMS are high absolute sensitivity for many elements and high mass resolution. The quadrupole mass spectrometer included in the SIMS system can also be utilized for effective diagnostics of the ambient gas in the deposition chamber.

- X-ray photoelectron spectroscopy (XPS) [4.35] and angle-resolved ultraviolet photoelectron spectroscopy (ARUPS) [4.27], mainly for studying the electronic structure of the epilayer surface, or the energy band distribution at heterointerfaces [4.36–39].
- Scanning electron microscopy (SEM) [4.40, 41], for displaying the structure of the film (or substrate) surface.

All of the above-mentioned analytical techniques are postgrowth in situ techniques. They give important information about the films or surfaces grown with MBE, however, they cannot be used directly to control the growth process itself.

A comprehensive presentation of all of the listed surface analytical techniques far exceeds the scope of this book. Therefore, only the two most important in-growth characterization techniques, namely, RHEED and ellipsometry, will be discussed in this chapter in detail. The postgrowth characterization methods, in situ and ex situ in character, will be discussed in the next chapter.

4.1 RHEED

In the conditions which are usually employed in MBE a high energy beam of electrons in the range 5–40 keV is directed at a low angle (1°–3°) to the surface (Fig. 3.28). The de Broglie wavelength of these electrons is therefore in the range 0.17–0.06 Å and the penetration of the beam into the surface is low, being restricted to the outermost few atomic layers [4.42]. Generally, the wavelength λ corresponding to an accelerating voltage V (expressed in volts) is given to a good approximation by [4.43]

$$\lambda \approx \frac{12.247}{\sqrt{V(1 + 10^{-6}\,V)}}\ [\text{Å}]\ . \tag{4.1}$$

Most geometrical aspects of the diffraction pattern can be interpreted on the basis of a limited penetration scattering model, i.e., a model which is kinematic in the diffraction sense [4.43]. A detailed analysis of the diffracted beam intensity, particularly the way in which it changes with incident angle (Θ in Fig. 3.28), should in principle permit an exact determination of the complete structure of the surface unit cell [4.42]. A set of data concerning the GaAs (001) surface has already been published [4.14], and a theoretical treatment also exists [4.2]. However, it is very clear from many experiments [4.14, 15, 44] that the incident beam penetrates deeper into the solid and, consequently, very strong multiple scattering effects (dynamical effects in the diffraction sense) dominate the diffraction process. These dynamical processes transfer the scattered intensity between beams, i.e., between the fractional order beam and the specular beam, or between the surface resonances and specular beam. This means that one has to be cautious when interpreting the intensity and apparent width of diffracted features [4.42].

4.1.1 Fundamentals of Electron Diffraction

The corpuscular nature of electrons is quite apparent in certain processes, however, the wave-like nature of these microparticles is clearly shown in scattering by crystals. In view of the undulatory character of electrons, the details of their interaction with matter can be determined by the characteristic wavelength of the electron de Broglie waves and the atomic amplitudes [4.45]. Consequently, the fundamental concepts of the theory of scattering and structure analysis of crystals are based on Fourier series and integral formalism.

In the first approximation of electron diffraction theory, the methods of calculation in the structure analysis of crystals are based on the kinematic theory of diffraction. The basic assumption of this theory may be formulated as follows: since the absolute magnitude of atomic amplitudes of scattering of electrons is very small, the intensities of the scattered beam will, in the presence of a limited number of scattering centers (atoms), be small compared with the intensity of the primary beam. Thus, it is possible to ignore the loss of energy of the primary beam in the course of its "expenditure" in the formation of coherently scattered radiation. One may also ignore the coherent scattering of secondary beams, which, acting in their turn as primary beams, give rise to new diffracted beams, and so on. As the volume of coherent scattering increases, i.e., the number of scattering centers increases, the intensities of the secondary beams will increase, and the description of the processes taking place must be based on the dynamical theory of scattering. In this theory, energy interrelationships are taken into account and, generally speaking, all beams are taken as qualitatively equivalent to the primary beam and to one another.

In the kinematic theory the problem of scattering is analyzed as follows. A plane monochromatic wave incident upon a specimen gives rise to an elementary secondary wave in each element of its volume. The amplitude of this scattered wave will, naturally, be proportional to the scattering power of the given element of volume. Scattering of electrons is brought about by the electric potential of the scattering volume $\varphi(r)$, where r is the position vector in the volume. Thus the scattering power of an element of volume dv_r is proportional to $\varphi(r)dv_r$. The incident wave $\exp(ik_i \cdot r)$ with a wave vector of magnitude $|k_i| = 2\pi/\lambda$ reaches different points r of the volume in different phases and consequently the secondary waves $\exp(ik \cdot r)$ arising from these points also have different phases. The overall phase shift in the secondary wave depends also on the direction of its wave vector k and equals $(k - k_i) \cdot r$. In the primary wave the phase shift is equal to the projection of the vector r on the direction k_i, i.e., to $k_i \cdot r$, and in the scattered wave it is equal to the projection of r on the direction k, i.e., to $k \cdot r$. To obtain the amplitude of scattering from the whole volume through which $\varphi(r)$ is distributed, the elementary waves $\exp[i(k - k_i) \cdot r]dv_r$ arising from all its elements with an amplitude proportional to $\varphi(r)$ must be summed (integrated). Taking $k - k_i = s$ and integrating over the whole volume of the specimen, one finds that the amplitude of scattering f_0 is equal to [4.45]

$$f_0(s) = \frac{2\pi m_e e}{h^2} \int \varphi(r) \exp(is \cdot r) dV_r \quad , \qquad (4.2)$$

where m_e and e are the mass and elementary charge of the electron ($m_e = 9.1096 \times 10^{-31}$ kg, $e = 1.6022 \times 10^{-19}$ C), respectively, and h is Planck's constant ($h = 6.6262 \times 10^{-34}$ J s).

For the sake of simplicity, only the integral part of (4.2) will be used hereafter as the amplitude of scattering, and this will be written as $f(s)$, thus, $f_0(s) = (2\pi m_e e/h^2)f(s)$. An important feature of (4.2) is that in its mathematical form it represents a Fourier integral. Thus, all the fundamental premises of the theory of scattering and of structure analysis may be obtained from the theory of Fourier integrals and from Fourier series.

The vector s in (4.2) has a dimension which is inverse to that of the vector r used for measuring distances within the specimen. The values of the amplitude $f(s)$ may be regarded as the distribution of magnitude f in the space of vector s — known as reciprocal space — in exactly the same way as the magnitude $\varphi(r)$ is regarded in the real space of the specimen. The concept of reciprocal space and the vector s plays an important role in the theory of diffraction. For a given position of the specimen, i.e., when k_i is fixed relative to it, the end of vector s can lie only on the sphere with a radius equal to $|k_i|$. This is the sphere 1 shown in Fig. 4.1a. This statement is based on the kinematic approximation of diffraction theory; the loss of energy (the change in the value of the vector k_i) of the primary beam, in the course of diffraction, is ignored ($|k| = |k_i|$).

The sphere 1 is called the Ewald sphere or the sphere of reflection since, in diffraction from crystals, its position determines the formation of discrete reflections. At the point of an electron diffraction pattern to which the vector k is

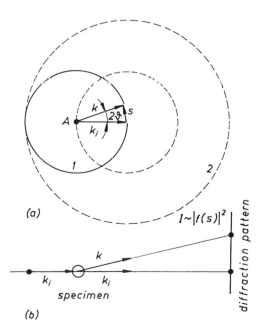

(a)

(b)

$I \sim |f(s)|^2$

diffraction pattern

specimen

Fig. 4.1. (a) Spheres of reflection (the Ewald sphere) (1), and limitation (2) in reciprocal space. (b) Formation of the diffraction pattern corresponding to (a) [4.45]

123

directed, one generally finds an intensity $I(k)$ corresponding to the square of the amplitudes $f(s)$ cut out from reciprocal space by the Ewald sphere. Figures 4.1a and b show how the mathematical concepts of reciprocal space and the Ewald sphere are realized in the corresponding diffraction pattern. By changing the direction of vector k_i (by a change in the direction of the incident beam or rotation of the specimen), i.e., by moving its origin along the sphere A (Fig. 4.1a), sphere 1 can be made to pass through any point within sphere 2, the so-called limiting sphere. The angle of scattering 2ϑ between vectors k and k_i is determined by the expression

$$|s| = |k - k_i| = \frac{4\pi \sin \vartheta}{\lambda} \quad , \quad \text{since} \tag{4.3}$$

$$|k| = |k_i| = \frac{2\pi}{\lambda} \quad .$$

A series of $f(s)$ values may be obtained experimentally by rotating the specimen relative to the primary beam. The operation corresponds to the intersection of various regions of reciprocal space by the Ewald sphere.

Knowing $\varphi(r)$, one can calculate $f(s)$ for any s from the formula

$$f(s) = \int \varphi(r) \exp(is \cdot r) dv_r \quad , \tag{4.4}$$

i.e., the diffraction pattern may be calculated if the structure of the specimen is known. The amplitude of scattering $f(s)$ is the "image" in reciprocal space corresponding to the specimen $\varphi(r)$. The above Fourier integral (4.4) has the property of inversion:

$$\varphi(r) = \frac{1}{8\pi^3} \int f(s) \exp(-is \cdot r) dv_r \quad , \tag{4.5}$$

which allows one to calculate $\varphi(r)$ if $f(s)$ is known. This means that a picture of the scattering object can be worked out from the diffraction pattern; this is the main problem in structure analysis.

There are, however, the following important limitations. Firstly, since the observed intensities are given by $I \sim |f(s)|^2$, analysis of the diffraction pattern in general yields only $|f(s)|$ values, i.e., values of the moduli of the structure amplitudes, without any indication of their phases. Secondly, the set of these values is limited since the maximum observable values of $|s|$ amount to $|2k|$ (Fig. 4.1a, limiting sphere 2). If the function $f(s)$ decreases sufficiently rapidly, i.e., if beyond the limits of the limiting sphere $f(s) \to 0$, then the second limitation disappears. This is precisely the position in electron diffraction where the radius of the limiting sphere is very large (since the wavelength is short). The presence of a second limitation − the cutting off of $f(s)$ for large s values − precludes the calculation of $\varphi(r)$ from (4.5) with absolute precision, since not all the necessary values of $f(s)$ will enter the equation.

Using the Fourier transform operator F one may write (4.4, 5) as

$$F(\varphi(r)) = f(s) \quad , \tag{4.6}$$

$$F^{-1}(f(s)) = \varphi(r) \quad . \tag{4.7}$$

A crystal represents a three-dimensional periodic distribution of scattering material. When a periodic function is substituted in the integral (4.4) for $\varphi(r)$, the set of values for $f(s)$ will no longer be continuous.

For the one-dimensional case and a period of 2π, (4.4) will take the form

$$\frac{1}{2\pi} \int_0^{2\pi} \varphi(x) \exp(ihx) dx = \Phi_h \quad , \tag{4.8}$$

where h represents an integer. Thus, the function $\varphi(x)$ is characterized by a series of discrete Fourier coefficients Φ_h. If the period of this function is a, then

$$\Phi_h = \frac{1}{a} \int_0^a \varphi(x) \exp[i(2\pi h/a)] dx \quad . \tag{4.9}$$

Let us now find the distances between the points having weights Φ_h in reciprocal space. For this purpose, let us compare (4.4) and (4.9). The term $2\pi h/a$ in (4.9) corresponds to s in (4.4), i.e., for a periodic function of period a, the values of $f(s)$ differ from zero only when $s = 2\pi h/a$:

$$f(s) \sim \Phi_h = \Phi(2\pi h/a) \quad , \tag{4.10}$$

and occur at separations of $2\pi/a$.

Similarly, in the three-dimensional case the Fourier coefficients are

$$\begin{aligned}
\Phi_{hkl} &= \frac{1}{a \cdot b \times c} \int_0^a \int_0^b \int_0^c \varphi(xzy) \exp\left[2\pi i\left(\frac{hx}{a} + \frac{ky}{b} + \frac{lz}{c}\right)\right] dx\, dy\, dz \\
&= \frac{1}{\Omega} \int \varphi(r) \exp(2\pi i r \cdot G) dv_r \quad . \tag{4.11}
\end{aligned}$$

Here h, k and l are integers. In reciprocal space $2\pi h/a$, $2\pi k/b$ and $2\pi l/c$ are components of vector s. Consequently, in the case of scattering from a nonperiodic object (atom, molecule, etc.), the distribution of the amplitude $f(s)$ in reciprocal space is continuous, i.e., scattering with any given intensity is possible in any direction, whereas in the case of scattering by crystals, only certain definite directions of diffracted beams are possible. A scattering amplitude exists only for the above-mentioned values of the components of vector s, where h, k and l (the indices for amplitude Φ_{hkl}) are integers. The distribution of points at which the scattering amplitude differs from zero and takes on the value Φ_{hkl} is periodic in reciprocal space and forms what is called the reciprocal lattice (Fig. 4.2). Each point of this kind, i.e., each hkl point, is characterized by a vector of the reciprocal lattice:

$$\frac{s}{2\pi} = G = ha^* + kb^* + lc^* \quad , \tag{4.12}$$

having its origin at the point (0,0,0). For orthogonal unit cells, a comparison of

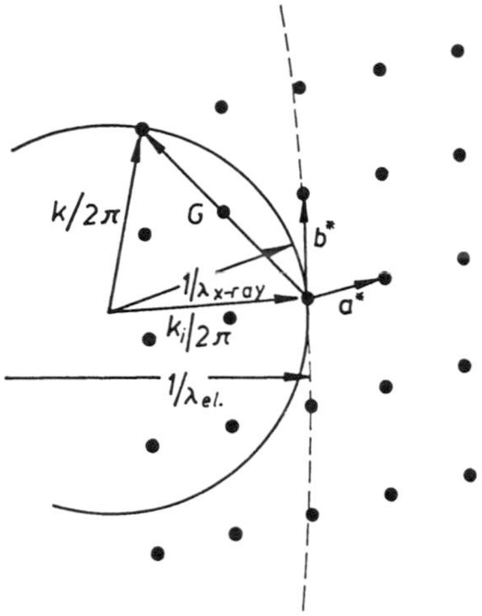

Fig. 4.2. Reciprocal lattice and Ewald spheres for x-ray and electron diffraction [4.45]

this equation with the exponent in (4.11) shows that

$$a^* = \frac{1}{a} \quad , \quad b^* = \frac{1}{b} \quad , \quad c^* = \frac{1}{c} \quad .$$

Thus the conditions of diffraction by a crystal are given by

$$\mathbf{s} = 2\pi \mathbf{G} \quad ; \quad \mathbf{k} = \mathbf{k}_0 + \mathbf{s} = \mathbf{k}_0 + 2\pi \mathbf{G} \quad . \tag{4.13}$$

This relation determines the possible directions of the beams diffracted by the crystal. However, for any given position of the crystal, only those beams arise which correspond to the intersection of points on the reciprocal lattice with the Ewald sphere. This is shown in Fig. 4.2. Since $\mathbf{G} = \mathbf{s}/2\pi$, the radius of the Ewald sphere with respect to the reciprocal lattice is $1/\lambda$.

Consequently, the conditions for the production of diffraction beams depend on the orientation of the crystal relative to the primary beam, and the radius $1/\lambda$ of the sphere. In x-ray diffraction the wavelength λ, equal to approximately 1–2 Å, is comparable in magnitude to the dimensions of unit cells (of the order of 5–10 Å), and the sphere has a significant curvature relative to the planes of the reciprocal lattice (sphere $1/\lambda_{x-ray}$ in Fig. 4.2). Hence an x-ray diffraction pattern obtained from a fixed single crystal will, in the most favorable case, show only a few reflections. In electron diffraction the wavelength λ is very small (of the order of 0.05 Å) and the radius of the sphere is large ($1/\lambda_{el}$ in Fig. 4.2). Over a considerable area of reciprocal space the section of the Ewald sphere is almost flat. If this flat region is made to correspond to a given plane of the reciprocal lattice then all the points belonging to this plane will appear in the electron diffraction pattern (those, for example, intersected by the broken line in Fig. 4.2). The pattern is thus a plane section of the reciprocal lattice of the crystal on a particular scale.

Let us now determine the relationship between the structure of the direct lattice of a crystal having the unit cell vectors $\mathbf{a}$, $\mathbf{b}$, $\mathbf{c}$ and its reciprocal lattice having unit cell vectors a^*, b^*, c^*. Let us consider first of all the simple relationship for the one-dimensional case where $a^* = 1/a$ and $G_h = ha^*$:

$$ha^*(a/h) = 1 \quad . \tag{4.14}$$

The expression $(hx/a + ky/b + lz/c)$ in the exponent of (4.11) is the expanded form of the scalar product of vector r, determining the position of point $r(x, y, z)$ in the unit cell of the crystal, and the vector G, (4.12). If r equals one of the axial vectors a, b or c, for instance $r = a(a, 0, 0)$, then, by analogy with (4.14), the following three scalar products will be obtained:

$$G_{hkl} \cdot \frac{a}{h} = 1 \quad ; \quad G_{hkl} \cdot \frac{b}{k} = 1 \quad ; \quad G_{hkl} \cdot \frac{c}{l} = 1 \quad . \tag{4.15}$$

These are the three Laue conditions.

Since $G_{100} = a^*$, $G_{010} = b^*$, and $G_{001} = c^*$, it follows from this equation that $a \cdot a^* = b \cdot b^* = c \cdot c^* = 1$, and that products of the type $a^* \cdot b = a^* \cdot c$, etc., are all equal to zero. This means that various vectors of the direct and the reciprocal lattices are mutually perpendicular: $a^* \perp b$, $a^* \perp c$, etc. The unit cell volume is determined by the mixed product of axial vectors: $a \cdot b \times c = \Omega$. Comparison of the expression for Ω with the relationship $a \cdot a^* = 1$ clearly shows that $a^* = b \times c/\Omega$.

Three vectors, a/h, b/k, c/l, determine the position of the crystal lattice plane (hkl) traced through their end points, as shown in Fig. 4.3. Subtracting in pairs the relationship in (4.15), three scalar products of the type

$$\left(\frac{a}{h} - \frac{b}{k} \right) \cdot G_{hkl} = 0 \tag{4.16}$$

are obtained, from which it follows that G is perpendicular to the vectors which form the sides of the triangle in Fig. 4.3, lying in the plane (hkl), i.e., G is perpendicular to this plane. The distance d_{hkl} from this plane to the origin of the coordinates or to the neighboring similar plane is called the interplanar distance. It is easily seen (Fig. 4.3) that d_{hkl} is a projection of vector a/h, or b/k, or c/l on the direction G_{hkl}, i.e., that any of the relationships in (4.15) lead to

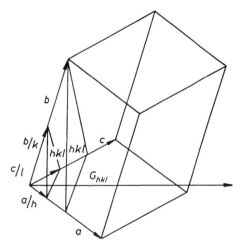

Fig. 4.3. Relation between the reciprocal lattice vector G_{hkl}, the axial vectors a, b, c of the unit cell and the (hkl) plane of the crystal. In this diagram the Miller indices of the first plane are equal to $h = 4$, $k = 2$, $l = 3$ [4.45]

127

$$|G_{hkl}|d_{hkl} = 1 \quad ; \quad G_{hkl} = 1/d_{hkl} \quad . \tag{4.17}$$

Hence a general conclusion may be formulated as follows: vector G_{hkl} of the reciprocal lattice is perpendicular to the (hkl) plane of the crystal lattice, and its absolute magnitude is the reciprocal of the interplanar distance d_{hkl}.

It was stated earlier that for orthogonal cells $a^* = 1/a$, etc., and $d_{100} = a = (G_{100})^{-1} = (a^*)^{-1}$. For nonorthogonal cells this is not so. For instance, in a monoclinic lattice, with the angle between the a and the c axes $\beta \neq 90°$, $d_{100} = a \sin \beta$ and $a^* = (|G_{100}|)^{-1} = (a \sin \beta)^{-1}$. For triclinic cells these relationships become rather complex [4.45]. Figure 4.4 illustrates the relationship between the reciprocal lattice and the planes of the direct lattice for a two-dimensional nonorthogonal cell.

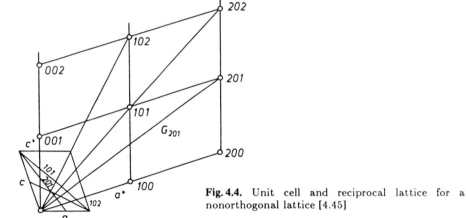

Fig. 4.4. Unit cell and reciprocal lattice for a nonorthogonal lattice [4.45]

From Fig. 4.1a it can be seen that $(\sin \vartheta)/\lambda = s/4\pi$. From (4.13) or directly from Fig. 4.2 it follows that, for diffraction by a crystal, $(\sin \vartheta)/\lambda = |G|/2 = 1/2d$. Thus the angles of scattering by a crystal are determined by

$$\lambda = 2d \sin \vartheta \quad . \tag{4.18}$$

This is the well-known Bragg equation.

Using the Bragg equation and the Ewald sphere construction, one may deduce the atom periodicities in the solid surface region by measuring the diffraction spot spacings. The relation between the atom arrangements in the solid and the observed diffraction pattern on the fluorescent screen can be described as in Fig. 4.5 [4.46] and expressed as

$$d = \frac{2\lambda L}{D} \quad , \tag{4.19}$$

where L is the distance between the sample and the fluorescent screen and D is twice the distance between the diffraction spots and the center of the screen. The interplanar distance d shown in Fig. 4.5, equal to the reciprocal of the absolute

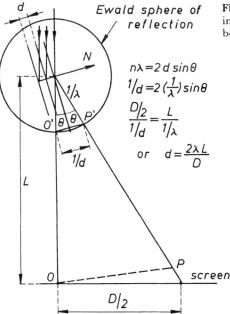

Fig. 4.5. Schematic illustration of how the interplanar distance in a crystal lattice may be measured with RHEED [4.46]

$$n\lambda = 2d\sin\theta$$

$$\frac{1}{d} = 2\left(\frac{1}{\lambda}\right)\sin\theta$$

$$\frac{D/2}{1/d} = \frac{L}{1/\lambda}$$

$$\text{or} \quad d = \frac{2\lambda L}{D}$$

magnitude of the reciprocal lattice vector G^*, see (4.17), may be expressed in the simplest case of a cubic lattice structure in terms of the Miller indices hkl as

$$d_{hkl} = \sqrt{(h/a)^2 + (k/a)^2 + (l/a)^2} \quad , \tag{4.20}$$

where a is the lattice constant. The different crystallographic interplanar spacings for different orientations of the sample are therefore equal to $a/\sqrt{2}$ and $a/\sqrt{3}$ for d_{110} and d_{111}, respectively. From the described example it is evident that RHEED may easily be used for this kind of crystallographic measurement [4.46].

4.1.2 Origin of RHEED Features

The diffraction process in RHEED is in general not a true reflection, as was assumed in the previous section, i.e., the scattered beam does not leave the scatterer from the same surface where the incident beam entered. Most surfaces are rough and the diffraction pattern is produced in transmission through the surface asperities (Fig. 4.6a). The diffraction patterns obtained in transmission-reflection diffraction and in true reflection diffraction (Fig. 4.6b) frequently differ and require separate treatment [4.43]. In the first case the diffraction pattern exhibits many spotty features, while in the second case the features are in the form of elongated streaks.

For diffraction on smooth surfaces the RHEED pattern is usually constituted by elongated streaks which are normal to the shadow edge [4.47], as shown in

(a)   (b)

Fig. 4.6a,b. Modes of electron diffraction: **(a)** Transmission-reflection diffraction, resulting in many spotty features. **(b)** True reflection diffraction, resulting in streaky features

Figs. 3.28 and 3.29. The simplest explanation is to consider that the penetration of the electron beam into the solid surface is in this case restricted to the uppermost layer of the crystal [4.48].

The conditions for constructive interference of the elastically scattered electrons may be inferred using the Ewald construction in the reciprocal lattice. In the case where the interaction of the electron beam is essentially with a two-dimensional atomic net, the third dimension in real space is missing, and therefore the third dimension in reciprocal space is not defined, see (4.12). Consequently, the surface layer is represented in the reciprocal space by rods in a direction normal to the real surface (Fig. 4.7). It should be noted that the reciprocal lattice rods have finite thickness due to lattice imperfections and thermal vibrations, and that the Ewald sphere also has finite thickness, due to electron energy spread and to beam convergence. The radius of the Ewald sphere is very much larger than the separation of the rods. This can be verified from a simple calculation of the wavelength of an electron with an energy of 5 keV. Using (4.1) it follows that the radius of the Ewald sphere is $1/\lambda = 36.6\,\text{Å}^{-1}$. If the surface lattice net has a lattice constant a of 5.65 Å (unreconstructed GaAs), then the distance between adjacent rods in reciprocal space $(2\pi/a)$ will be $1.1\,\text{Å}^{-1}$. As a result, the intersection of the sphere and rods occurs some way along their length, resulting in a streaked, rather than a spotty, diffraction pattern.

Keeping in mind that the RHEED pattern from a flat, exactly oriented low index surface results from the intersection of the Ewald sphere and the reciprocal lattice rods, one may apply the appropriate Ewald sphere construction to explain the diffraction pattern observed from definite solid surfaces.

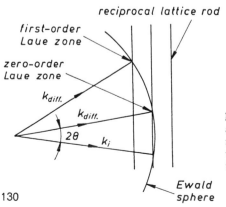

reciprocal lattice rod

first-order
Laue zone

zero-order
Laue zone

$k_{diff.}$

$k_{diff.}$

2θ k_i

Ewald
sphere

Fig. 4.7. Ewald sphere construction appropriate for a qualitative explanation of the formation of streaks in RHEED patterns. The wave vector k_i is very large compared with the interrod spacing [4.47]

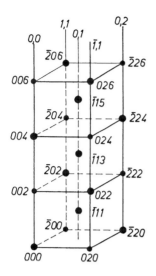

Fig. 4.8. A section of reciprocal space of the sphalerite lattice indexed with two-dimensional notation for the reciprocal lattice rods and with three-dimensional notation for the bulk lattice points [4.14]

Let us consider first some general examples, which concern diffraction effects which can be attributed to both two- and three-dimensional diffraction. In doing this we will use, following *Larsen* et al. [4.14], two-dimensional indices to describe reciprocal lattice rods, and three-dimensional indices to define directions and diffraction conditions in the crystal. A section of reciprocal space of the sphalerite lattice (this is the lattice structure of most of the semiconductors grown with MBE at present) indexed with both notations is shown in Fig. 4.8.

Because the structure of a real surface is always different from the structure of a bulk truncated material due to rearrangements of the surface atoms, the first layer has a different periodicity, which is still strongly correlated to the bulk symmetry. If the surface reconstruction is 2 × 2 (for details see Sect. 4.1.3) then the lattice vector of the surface in real space is twice as long as the bulk lattice vector. Therefore, the corresponding reciprocal lattice vector is only one-half the undisturbed reciprocal lattice vector, see (4.12). Consequently, between the zero-order and the first-order Laue zones (Fig. 4.7) additional rods will appear in the so-called half-order Laue zone. In the case of the 2 × 4 surface reconstruction the Laue zone will be divided into four subsections in one direction and into two subsections in the perpendicular direction.

Several important diffraction possibilities are illustrated in Fig. 4.9, using the conventional construction based on the intersection of the Ewald sphere with the reciprocal lattice section containing the reciprocal rods of the zero-order Laue zone (Fig. 4.7). The direction of the rod is along the surface normal and a wave vector in this direction is denoted $k_\perp$ and similarly a reciprocal lattice vector is denoted $G_\perp$. The spacing between rods is given by surface parallel reciprocal lattice vectors $G_\parallel$ while surface parallel wave vectors are designated by $k_\parallel$. It is usually assumed that the most important diffraction conditions are primary Bragg effects, as shown in Figs. 4.9a and b, which are assumed to be single scattering

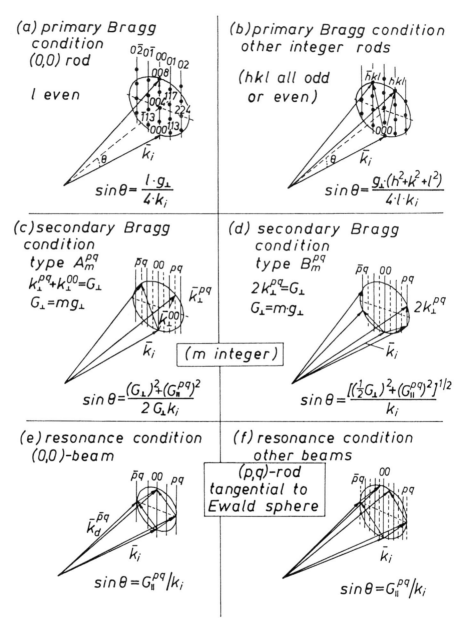

(a) primary Bragg condition (0,0) rod

l even

$$\sin\theta = \frac{l \cdot g_\perp}{4 \cdot k_i}$$

(b) primary Bragg condition other integer rods

(hkl all odd or even)

$$\sin\theta = \frac{g_\perp \cdot (h^2 + k^2 + l^2)}{4 \cdot l \cdot k_i}$$

(c) secondary Bragg condition type A_m^{pq}

$k_\perp^{pq} + k_\perp^{00} = G_\perp$

$G_\perp = m g_\perp$

$$\sin\theta = \frac{(G_\perp)^2 + (G_\parallel^{pq})^2}{2 \, G_\perp k_i}$$

(d) secondary Bragg condition type B_m^{pq}

$2 k_\perp^{pq} = G_\perp$

$G_\perp = m \cdot g_\perp$

$$\sin\theta = \frac{[(\frac{1}{2}G_\perp)^2 + (G_\parallel^{pq})^2]^{1/2}}{k_i}$$

(m integer)

(e) resonance condition (0,0)-beam

$$\sin\theta = G_\parallel^{pq}/k_i$$

(f) resonance condition other beams

(p,q)-rod tangential to Ewald sphere

$$\sin\theta = G_\parallel^{pq}/k_i$$

Fig. 4.9a–f. Ewald sphere construction illustrating various diffraction conditions, specified by the equation given in each panel. The smallest reciprocal lattice vector along the surface normal (the [001] direction) is $g_\perp = 4\pi/a$, where a is the lattice constant. The surface parallel reciprocal lattice vector $G_\parallel^{pq}$ defines the lattice rod indexed by p and q, $G_\parallel^{pq} = 2\sqrt{2}\pi\sqrt{p^2 + q^2}/a$. Here p, q are integers for an unreconstructed surface, but they can also be fractions for a reconstructed surface. Fractional order rods are indicated as dashed lines. (**a, b**) Primary Bragg reflections specified by hkl. (**c, d**) Secondary Bragg diffraction conditions. The subscript m refers to the momentum $G_\perp$ given in units of $g_\perp$. The dashed rods indicate fractional-order rods. (**e, f**) Surface resonance conditions [4.14]

events. The equations show the relationship between the angle of incidence and the wave vector of the incoming beam for these conditions. In Fig. 4.49b beams $\bar{h}kl$ and hkl are excited when $\bar{h}kl$, hkl and 000 all lie on the Ewald sphere and corresponding intensity maxima are expected in the elastically diffracted beams.

Intensity enhancement can arise from wave-vector matching to bulk reciprocal lattice vectors perpendicular to the surface (Figs. 4.9c and d). These conditions have been described by *Maksym* and *Beeby* [4.2]. It is also possible to scatter into beams propagating parallel to the surface, i.e., having zero wave vector normal to the surface ($k_\perp = 0$). For this condition the reciprocal lattice rods are tangential to the Ewald sphere, as illustrated in Fig. 4.9e. The diffracted beam remains parallel to the surface and can couple out again by diffraction back to the 00-beam or back to other beams as shown in Fig. 4.9f for a reconstructed surface. At the surface the potential is strongly dependent on the depth and only in special cases may enhancement effects be observed [4.49]. We refer to these conditions (Figs. 4.9e and f) as surface resonances, following to *Larsen* et al. [4.14]. They are expected to be extremely sensitive to surface morphology and reconstruction (Sect. 4.1.3).

All the diffraction conditions illustrated in Fig. 4.9 are quite general and independent of any surface reconstruction. The list, however, is not exhaustive but merely serves to illustrate the occurrence of multiple scattering events.

The incoming beam can also be diffracted by scattering with a surface reciprocal lattice vector $G_\|$. This is the case illustrated in Fig. 4.10 for a 2 × 4 surface reconstruction, which explains the diffraction pattern shown in Fig. 3.29.

An additional effect inherent in RHEED which may be explained with the Ewald sphere construction is refraction of electrons when they cross the potential barrier of the surface and enter the solid. This effect is frequently neglected but is particularly important in understanding intensity variations of the RHEED pattern features. As a result of refraction, electrons change their energy by an amount determined by the inner potential V_0. For a perfect surface the surface parallel component of the wave vector is conserved when the electrons cross the surface, i.e., $k_\|$ is a good quantum number. From conservation of energy considerations it follows that the angle between the surface and the incident beam in vacuum,

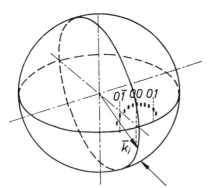

Fig. 4.10. Ewald sphere construction for the GaAs(001)-(2 × 4) reconstructed surface in [$\bar{1}$10] azimuth, illustrating the arc of the short streaks corresponding to the intersection of the sphere with the $0\bar{1}$, 00, 01 and associated fractional-order rods. Higher-order Laue zones are omitted for clarity [4.14]

Ewald sphere

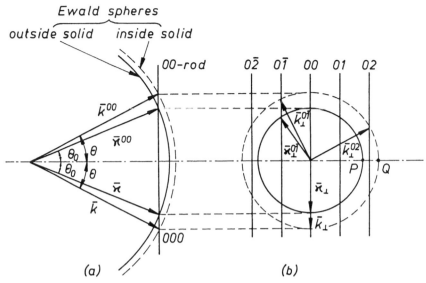

Fig. 4.11. Ewald sphere construction illustrating refraction effects. The inside and outside angles θ and θ_0, respectively, are related by $\cos\theta_0 = \sqrt{1 - V_0/E}\,\cos\theta$ [4.14]. Here the electron wave vector is denoted $\bar{\kappa}$ in the vacuum, and $\bar{k}$ inside the solid

θ_0, is related to the equivalent angle inside the solid, θ, by

$$\cos\theta_0 = \sqrt{\frac{1 - V_0}{E}}\,\cos\theta \quad, \tag{4.21}$$

where E is the primary beam energy and the inner potential V_0 is negative, measured with respect to the vacuum level. The angle inside is therefore always greater than the angle outside. The effect that this may have on the angular position of diffracted beams and their emergence from the solid is illustrated in Figs. 4.11a and b, respectively. In Fig. 4.11b both emergent (0,1) and nonemergent (0,2) beams are shown. When the 02 rod is tangential to the Ewald sphere corresponding to outside the solid (at P) it is at the horizon, i.e., the beam is just emergent. This is the condition shown in Fig. 4.9e with $V_0 = 0$. When the 02 rod is tangential to the inside solid Ewald sphere at Q, the beam is nonemergent and runs parallel to the crystal surface.

4.1.3 RHEED Data from Reconstructed Semiconductor Surfaces

From many experimental data [4.50] it is evident that in all real systems atoms at and near a surface do not exhibit the same arrangement as in the bulk. The simplest rearrangement is surface relaxation, whereby the topmost layers retain the bulk symmetry, but the atomic distances perpendicular to the surface are different from the bulk value [4.51]. Surface reconstruction is a stronger disturbance giving rise to rearrangements of the topmost layers into symmetries different from the respective bulk truncation. The structural rearrangements may be exemplified by:

1. The relaxed 1 × 1 GaAs(110) cleavage surface. Here the surface topology is unchanged from that of the ideal bulk-terminated surface.
2. The dimerized Si(100)-(2 × 1) surface. Here the topology is changed, but the one-to-one correspondence between the atoms of the reconstructed and the ideal surface is maintained.
3. Si(111)-(7 × 7) surface. Here the removal of atoms from the surface occurs, which results in a loss of the one-to-one correspondence with the ideal surface [4.52].

Much interest has been shown in the (001) and (111) surfaces of GaAs and related compounds prepared by MBE. These polar surfaces may have an excess of Ga or As atoms, depending on growth conditions or postgrowth treatment. They have associated with them a large number of reconstructions in which the surface atoms reorder themselves to produce a periodicity different from that of the underlying crystal. Such reconstructions are usually denoted by a convention described by *Wood* [4.53], and we shall follow it in this chapter. A surface structure denoted by GaAs(001)-($m \times n$) means that a GaAs crystal is oriented with the [001] direction normal to the surface and has a surface structure, due to reconstruction, whose unit mesh is $m \times n$ times larger than the underlying bulk structure. Such surface meshes may be centered, in which case the notation would be GaAs (001)-$c(m \times n)$. If the mesh is rotated so that its principal axes are not aligned with those of the underlying bulk then the rotation angle is also included, e.g., GaAs(111)-($\sqrt{19} \times \sqrt{19})R23.4°$.

The reconstructed surface layer produces a reciprocal lattice consisting of rods normal to the surface, which can be examined using low-energy electron diffraction (LEED) with the incident beam normal to the surface, or RHEED, where grazing incidence is used. Both these techniques have been used, but LEED, due to its very geometry [4.54], is unsuitable for use during MBE growth and must be considered as a postgrowth analysis technique. It does, however, allow the complete symmetry of the surface to be observed in a single pattern, and as such is a useful complement to the RHEED technique, which may be employed during, as well as after growth in an MBE chamber.

Cho [4.55, 56] was the first to use MBE to prepare GaAs surfaces prepared under well-defined conditions and used RHEED to investigate reconstructions occurring on ($\bar{1}\bar{1}\bar{1}$) and (001) surfaces and how they related to growth conditions and subsequent processing. Two reconstructions were found on the ($\bar{1}\bar{1}\bar{1}$) surface, a ($\bar{1}\bar{1}\bar{1}$)-(2 × 2) and a ($\bar{1}\bar{1}\bar{1}$)-($\sqrt{19} \times \sqrt{19})R \pm 23.4°$, the occurrences of which could be controlled by the substrate temperature and the molecular fluxes incident on the surface. Observation of the diffraction pattern during layer growth showed that after depositing only 20 monolayers the pattern became much more streaky than that obtained from the prepared substrate, implying that in the initial growth phase the impinging atoms fill in the depressions on the substrate to form a very smooth surface. The diffraction pattern was characteristic of the $(\sqrt{19} \times \sqrt{19})R \pm 23.4°$ reconstruction, but as the substrate temperature was lowered, keeping the incident molecular fluxes constant, the pattern slowly deteriorated and gave way to one characteristic of a (2×2) surface structure. The transition temperature of the reconstruction transformation was found to be a reproducible function of deposition rate, higher transition temperatures corresponding to higher rates. It was found that if a layer was grown with a (2×2) reconstruction with a substrate temperature above 400° C then the diffraction patterns changed to indicate a $(\sqrt{19} \times \sqrt{19})R \pm 23.4°$ surface as soon as the beams were interrupted with a shutter, thus indicating that postgrowth analysis of MBE surfaces does not necessarily indicate the surface present during deposition. These results, together with the fact that As is preferentially desorbed from a GaAs surface [4.57], led Cho to suggest that the (2×2) surface was As-rich and the $(\sqrt{19} \times \sqrt{19})R \pm 23.4°$ surface was Ga-rich. These reconstructed surfaces, being stable and having an excess As or Ga, are often referred to as being "As-stabilized" or "Ga-stabilized", respectively.

Similar work carried out on GaAs(001) surfaces [4.56] showed similar As- or Ga-rich associated reconstructions. The diffraction patterns were interpreted as GaAs(001)-$c(2 \times 8)$ for the As-stabilized surface and GaAs (001)-$c(8 \times 2)$ for the Ga-stabilized surfaces and are often denoted as $c(2 \times 8)$As and $c(8 \times 2)$Ga, respectively. These patterns were obtained during growth and could be interchanged reversibly by varying the As_2/Ga flux ratio with constant substrate temperature or by varying the substrate temperature with constant flux ratio. The importance of being able to set up and maintain a known reconstruction pattern during growth was demonstrated by *Cho* and *Hayashi* [4.57], who showed that Ge, when evaporated along with the As_2 and Ga to act as a controlled dopant, was incorporated principally as a donor when incident on a growing As-stabilized surface and as an acceptor when incident on a Ga-stabilized surface.

The transition from a $c(2 \times 8)$ to $c(8 \times 2)$ reconstruction on the (001) surface is complex and involves many intermediate structures, some of which may be mixtures. *Larsen* et al. [4.58] have made a detailed study of the As-rich $c(4 \times 4)$ surface of GaAs. Using RHEED and angle-resolved photoelectron spectrosocpy (ARPES) it was concluded that the $c(4 \times 4)$As reconstruction results from an As overlayer on top of the usual GaAs surface, the "excess" As atoms being chemisorbed and trigonally bonded. Figure 4.12 shows possible models for the $c(4 \times 4)$As surface, indicating how different coverages can result in the same surface structure. It is assumed that the As-As bond lengths are the same as in amorphous As but that the bond angles have distorted. The two coverages shown are not proposed as separate models: *Larsen* et al. [4.58] suggest that combinations of both structures may coexist, thereby maintaining the surface symmetry over a large range of surface As coverage. *Massies* et al. [4.59] found the work function of the $c(4 \times 4)$As surface to be 0.3 eV lower than that of the (2×4)As surface and also suggest that the $c(4 \times 4)$As reconstruction may encompass a range of As coverage [4.60].

136

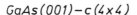

GaAs(001)–c(4x4)

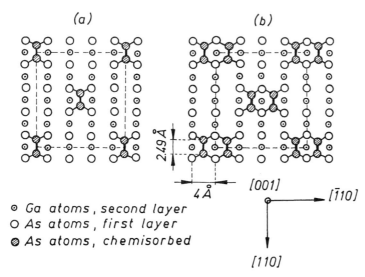

(a)   (b)

⊙ Ga atoms, second layer
○ As atoms, first layer
◎ As atoms, chemisorbed

2.49 Å

4 Å

[001]

[ī10]

[110]

Fig. 4.12. Possible models for the c(4 × 4) GaAs surface, based on a trigonally bonded excess As layer: **(a)** An additional 25 % As coverage; **(b)** an additional 50 % As coverage [4.58]

It is evident that LEED is able to distinguish between (4 × 2)Ga and c(8 × 2)Ga (001) surfaces and between (2 × 4)As and c(2 × 8)As (001) surfaces. It has been claimed by *Neave* and *Joyce* [4.61] that the two As structures cannot be distinguished by normal RHEED techniques, and the same is also true of the two Ga-stabilized structures. They claimed that in order to distinguish between the centered and noncentered structures it is necessary to obtain a diffraction in a ⟨110⟩ azimuth which includes a half-order Laue zone. Their reasons for this are illustrated in Fig. 4.13, which shows reciprocal lattice sections for both c(4 × 2)Ge and c(8 × 2)Ge structures together with the expected theoretical RHEED patterns in different azimuths. As one cannot be certain that structures present after growth (i.e., those analyzed by LEED) are the same as those present during growth, they argue that the existence of centered structures during growth cannot be demonstrated with any certainty, and reference is therefore made only to (2 × 4)As and (4 × 2)Ga as the normal reconstructions during growth. This idea may be challenged, since if the surface reconstruction were perfect then one would not expect the half-order lines in the zero-order Laue zone in the ⟨1̄10⟩ azimuth for the centered structure, whereas they should appear for the other (indeed this can be inferred from Fig. 4.13). *Cho* [4.62] clearly implies that he can distinguish the two structures, as does *Ludeke* [4.63], who has produced well-defined RHEED patterns from the two surfaces showing clear differences [4.64].

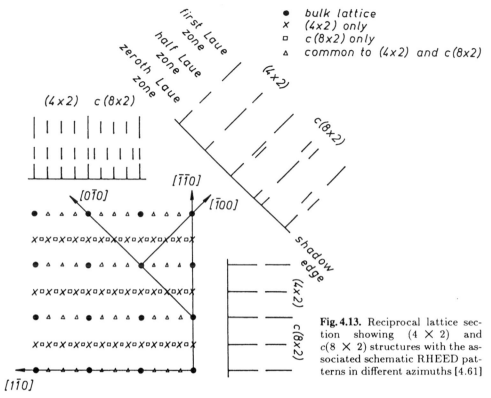

Fig. 4.13. Reciprocal lattice section showing (4 × 2) and c(8 × 2) structures with the associated schematic RHEED patterns in different azimuths [4.61]

So far most studies on surface reconstruction performed with RHEED have been carried out on GaAs surfaces [4.8, 65, 66]. However, results from other III-V semiconductor compounds, namely, InAs [4.67–69], InP [4.70, 71], and AlAs [4.72], show similar reconstructions, with similar dependences on incident fluxes and substrate temperatures [4.64].

4.1.4 RHEED Rocking Curves

There is a considerable body of evidence [4.2, 14, 15, 44, 73] which would suggest that the assumption of kinematic behavior of the RHEED process is unrealistic for typical conditions used in MBE experiments (10–20 keV primary beam energy, 1–4° incident angle). It is a comparatively simple matter to establish the nature of the diffraction process by the measurement of so-called rocking curves, i.e., the variation of diffracted intensity in the specular spot in the 00 and other rods with the incidence angle of the primary beam, at some fixed azimuth. This is best illustrated by reference to Fig. 4.14, which shows how the specular beam intensity varies with incidence angle for three different azimuths of the GaAs(001)-(2 × 4) arsenic-stabilized surface [4.14]. If the scattering were a simple kinematic process the results for the three azimuths would all be similar and show maxima only at the Bragg conditions for the (001) set of crystal planes [4.42].

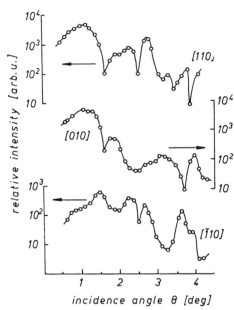

Fig. 4.14. Specular beam intensity as a function of angle of incidence (rocking curves) for the GaAs(001)-2 × 4 surface for the three principal azimuths. Primary beam energy: 12.5 keV [4.42]

Fig. 4.15. Rocking curves for 00 rod in the [110] azimuth for three differently reconstructed GaAs(001) surfaces. The calculated positions of the primary Bragg peaks with a 14.5 eV inner potential are indicated. Primary beam energy: 12.5 keV [4.15]

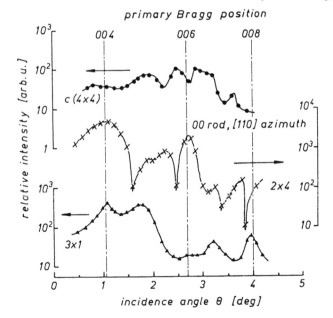

Further evidence for the multiple scattering nature of the problem comes from the observation that the oscillations of intensity of different beams in a diffraction pattern (Sects. 1.3.2 and 4.1.5) often exhibit a complex phase relationship, i.e., the oscillation of the specular beam is often nearly 180° out of phase with the oscillation of the other beams [4.74]. Figure 4.15 shows rocking curves for the specular spot intensity along the 00 rod in the [110] azimuth for three differently reconstructed GaAs(001) surfaces, $c(4 \times 4)$, (2×4) and (3×1) [4.15]. All measurements of diffracted intensities were made at substrate temperatures (T_S) and As_2 fluxes (J_{As_2}) where these structures were stable, i.e., typically for $c(4 \times 4)$, $T_S = 475°\,C$, $J_{As_2} = 5 \times 10^{14}$ molecules cm^{-2}s^{-1}, for (2×4), $T_S = 565°\,C$, $J_{As_2} = 2 \times 10^{14}$ molecules cm^{-2}s^{-1}, and for (3×1), $T_S = 630°\,C$, $J_{As_2} = 2 \times 10^{14}$ molecules cm^{-2}s^{-1}. There was no Ga flux, however, i.e., there was no growth. It has been found that these curves could be accurately reproduced in two separate systems using two quite different intensity measurement techniques [4.14]. There was an equivalent amount of structure in rocking curves measured during growth, but they were not so reproducible.

The highly dynamical nature of diffraction is immediately apparent, since for a kinematic process peaks would occur only at the primary Bragg angles marked in Fig. 4.15. A detailed discussion and interpretation of these and other rocking curves is presented in [4.14]. We will confine ourselves here to a brief summary of the essential features.

The problem central to RHEED interpretation is that the streaks or elongated spots that result from diffraction by the surface unit cell have their intensities modulated by scattering of electrons from the underlying three-dimensional lattice. An additional, but often neglected complication is caused by refraction. As the electrons cross the surface potential barrier they gain energy by an amount V_0, the "inner potential", and are therefore refracted as they enter the solid. The change in wave vector brought about by this gain in energy must be taken into account for the Ewald construction (Fig. 4.11). For a primary (three-dimensional) Bragg condition in the solid, for example, the angles of the beam with respect to the surface for, say, the 00 rod can be estimated from (4.21). This condition then gives a maximum in the intensity of the specular spot which is scattered at an angle θ_0 with respect to the surface. Similar maxima can be expected when points (hkl) on other reciprocal lattice rods lie on the Ewald sphere.

This condition will also lead to secondary Bragg maxima in the intensity of diffracted beams as the result of dynamical coupling between beams excited in the solid and the outgoing specular beam. It will occur, for example, whenever the Ewald construction produces two or more strong diffracted beams. These beams can then be considered as primary beams inside the solid which diffract electrons back into the outgoing specular beam, i.e., the process can be considered as a double diffraction involving the bulk and surface of the crystal and the same refraction effects occur as with primary Bragg diffraction. Maxima in the intensity of a specular beam may therefore be expected whenever a Bragg condition is met by a primary or secondary beam.

There is a third set of conditions, much more closely related to the surface region, which can give rise to maxima in the rocking curves. There are surface resonances which occur when the Ewald sphere is tangential to a reciprocal lattice rod and a wave is excited which travels parallel to the surface. A practical difficulty is the choice of the inner potential for electrons which do not completely enter the bulk solid, since this will determine the radius of the Ewald sphere. The problem has been discussed by *Ichimiya* et al. [4.75].

Finally, conditions can arise which are effectively a combination of surface resonance and secondary Bragg diffraction, in which a strong outgoing (hkl) reflection is excited, but which cannot emerge from the crystal because of refraction. This beam can then form a trapped resonance condition, and flux can be transferred between it and any outgoing beam.

It is quite clear that diffraction from a GaAs surface is a multiple-scattering process and it is therefore unlikely that treatment of the intensity oscillation phenomenon based on a kinematic approximation will be valid.

4.1.5 RHEED Intensity Oscillations

It is possible to explore the growth dynamics of MBE by monitoring temporal variations in the intensity of various features in the RHEED pattern (Sect. 1.3.2) [4.65]. Since 1983 it has become apparent that when MBE growth is initiated, the intensity of RHEED features shows an oscillatory behavior which is directly related to the growth rate [4.6, 7]. This has now become routinely used to calibrate beam fluxes and control alloy composition and the thickness of quantum wells and superlattice layers [4.15, 42, 76]. Data relating to surface diffusion [4.77–80], crystal growth mechanisms [4.81–90], and dopant incorporation [4.91] have also been obtained from studies of RHEED intensity oscillations.

The occurrence of intensity oscillations with growth has become a commonly acknowledged and experimentally well-evidenced fact. One may, however, create conditions during the growth such that no such oscillations are seen [4.77]. Generally, when the substrate temperature is increased, a condition is reached when oscillations become less apparent or absent. *Neave* et al. [4.77] used the disappearance of RHEED oscillations to obtain a measurement of the surface diffusion length of gallium on a terraced GaAs(001)-(2 × 4) reconstructed surface. The electron beam was directed parallel to the step edges. If the substrate temperature and incident fluxes were such that the surface diffusion length was less than the mean terrace length, then new growth centers appeared on the terraces and these were accompanied by the customary periodic changes of the RHEED intensity. The substrate temperature was increased until oscillations ceased, as shown in Fig. 4.16. From these results a surface diffusion energy of 1.3 ± 0.1 eV was obtained. Note, however, that at $T_S > 590°$ C in this particular experiment, whilst growth occurs it does not result in RHEED oscillations. Under these conditions, growth proceeds by the addition of atoms to the step edges. Absence of RHEED oscillations when growing material may indeed be a desirable situation from the standpoint of electrical and optical quality, since it implies that growth is occurring at step edges and will give fewer imperfections than for essentially random nucleation on terraces. These remarks apply only for the situation where layer growth occurs. The other situation in which RHEED oscillations are not seen is when gross three-dimensional growth

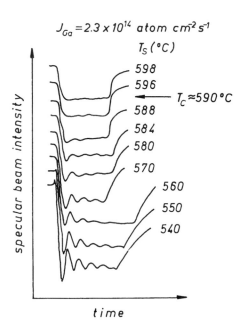

$J_{Ga} = 2.3 \times 10^{14}$ atom cm^2 s^{-1}

T_S (°C)

598
596
$T_c \approx 590\,°C$
588
584
580
570
560
550
540

specular beam intensity

time

Fig. 4.16. An example of the transition from oscillations to a constant response as a function of substrate temperature, with constant gallium flux [4.77]

occurs and this is easily determined by the spotty pseudotransmission nature of the RHEED patterns.

Let us now discuss the most important features that are seen in studies of RHEED intensity oscillations [4.42]. These are: damping of the oscillations when growth proceeds, increase of oscillation amplitude caused by dopant beam, recovery effects, initial transient behavior, and frequency doubling.

In the step edge scattering model presented in Sect. 1.3.2 and illustrated in Fig. 1.9, changes in RHEED intensity observed during growth, have been associated with changes in the step edge density. The familiar damping of oscillations therefore implies that some equilibrium step density is being achieved, i.e., some equilibrium terrace length, which will be governed by the surface diffusion length under the prevalent growth conditions, will be attained. Generally, the damping is most rapid for higher substrate temperatures at which the equilibrium terrace length will be reached most quickly.

However, there are other factors which affect the damping, the most important of which is the variation of growth rate across the part of the substrate which is sampled by the incident RHEED beam. This can give rise to complicated "beating" of oscillations [4.91]. It is important to emphasize that the conditions under which oscillations are observed should be noted and recorded.

The oscillation amplitude can increase when a dopant beam is turned on [4.6]. This implies that the fluctuations in step density have increased and it would be consistent with ideas that preferential nucleation is occurring on the terraces in the presence of dopant atoms. *Iimura* and *Kawabe* [4.92] have published RHEED oscillation data for beryllium doping in which they demonstrate such behavior, i.e., the beryllium surface concentration, even when $\sim 0.02\%$ of a monolayer, controls the growth front morphology.

It is also a generally known fact that if GaAs has been grown for several layers such that oscillations are well damped and very small, large amplitude oscillations will be seen when an aluminum source shutter is opened. This may in part be due to the change in group III-V flux ratio, but it is probably most influenced by the presence of aluminum. On the basis of the relative cohesive energies of AlAs and GaAs one may expect the activation energy for surface diffusion of Al to be higher than Ga on GaAs. This will lead to a shorter diffusion length and therefore shorter terrace lengths accompanied by an increase in the step density. A note of caution is necessary here. The actual change in the intensity when the aluminum shutter is opened must also be influenced by the diffraction conditions. The surface reconstructions are different for AlAs and GaAs and very little is known about intermediate AlGaAs alloy surface reconstructions. It may therefore be premature to offer a detailed interpretation of the behavior observed by *Yen* et al. [4.84] when growing GaAs/AlAs superlattices, although their general conclusions about interface quality, which they infer from the decay of oscillations, may be valid.

When growth is terminated by closing the group III source shutter the intensity of the specular RHEED beam recovers the original value it had prior to growth. This has been widely interpreted as a recovery of flatness following cessation of the growth [4.6, 7, 9]. This has resulted in the concept of the "growth interrupt" to improve the interfacial quality of multi-quantum-well structures [4.93–96]. There is some evidence that such growth interrupts do result in significant narrowing of the low temperature photoluminescence linewidths, and hence, it is inferred, in improvement of the interface flatness [4.93, 97–99]. The recovery period is not simple and it can usually be separated into fast and slow processes [4.9, 100]. *Lewis* et al. [4.82] have studied the recovery behavior for a wide range of conditions, including the effect of terminating growth at different points on the oscillation cycle. They have interpreted the fast process as a rapid smoothing of the growth front profile and the slow process as a recovery of long-range order, i.e., rearrangement of terraces and/or the reduction of one-dimensional disorder. *Joyce* et al. [4.44] have also shown that the recovery behavior depends on the point of the intensity oscillation at which growth was terminated. Figure 4.17 shows a set of data for which the fast initial stage is a reduction of the intensity, followed by a slow increase. In this particular instance, which has much in common with transient effects at the initiation of growth, these authors have associated the fast process with a change of the surface reconstruction and the slow change with the rearrangement of the terraces, etc. The particular form of recovery shown in Fig. 4.17 is most noticeable for the [$\bar{1}$10] azimuth, which has the highest sensitivity to changes in surface reconstruction because of the strong multiple scattering associated with the 1/4 order beams for the static arsenic-stabilized (2×4) surface.

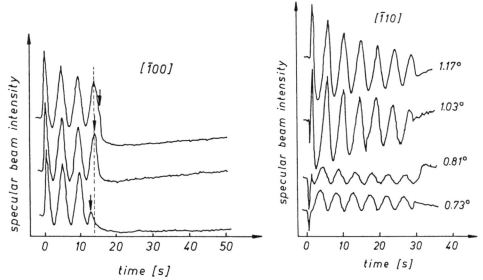

Fig. 4.17. The effect of termination of growth on the RHEED specular intensity oscillations for a GaAs(001)-(2 × 4) surface; [$\bar{1}$10] azimuth; 1.49° angle of incidence; 00 beam at 12.5 keV primary beam energy; $T_S = 580°$ C; $J_{As_2} \simeq 1.0 \times 10^{14}$ molecules cm^{-1}s^{-1}, $J_{Ga} \simeq 1.5 \times 10^{14}$ molecules cm^{-1}s^{-1}. Growth was terminated, i.e., the Ga shutter was closed, at the positions indicated. Note the initial rapid decrease in the intensity [4.44]

Fig. 4.18. Some examples of the initial transient behavior for RHEED oscillations observed for the specular 00 beam along the [$\bar{1}$10] azimuth at the angles of incidence indicated. Primary beam energy: 12.5 keV [4.42]

There are many examples of published RHEED oscillations which show an initial abrupt increase of intensity which does not always fit in with the subsequent series of regular oscillations [4.7, 101]. *Briones* et al. [4.101] suggested that this transient increase is caused by an initial smoothing of steps already present on the GaAs(001)-(2 × 4) surface. They observed the effect under well-defined conditions along the [110] and [010] azimuths, but could not record any effect along the [$\bar{1}$10] direction. They also presented some optical scattering evidence for steps running parallel to the [$\bar{1}$10] direction and showed that the transient behavior depends on the arsenic flux. Their explanation has much to commend it and it may be consistent with the model proposed by *Däweritz* [4.102, 103] for steps along the [$\bar{1}$10] being responsible for the stability of the (2 × 4) reconstruction.

Dobson et al. [4.42, 44] have examined the transient behavior for several azimuths under controlled flux of gallium and arsenic and they have observed a strong dependence on the diffraction condition, i.e., the angles of incidence and azimuth. Contrary to Briones et al. they have observed the transient most clearly along the [$\bar{1}$10] azimuth, as shown in Fig. 4.18. In view of this strong dependence on the diffraction condition, *Dobson* et al. [4.42] reject models which are based on step smoothing, and they favor a model based on a transient change of surface reconstruction. When the gallium shutter is opened, the surface stoichiometry will be instantaneously changed and one can expect the surface to adjust its reconstruction from (2×4) towards the rather ill-defined (3×1) structure. Such a change produces quite dramatic changes in the specular beam rocking curve [4.14]. The $[\bar{1}10]$ azimuth is the direction most sensitive to changes in reconstruction since the strength of the 1/4 order features is considerably weakened and the multiple scattering contribution to the specular beam intensity is changed. These authors therefore associate this transient with a change in reconstruction. They also expect the effect to be present to some extent along other azimuths.

From the practical point of view, the most important point is to appreciate that this transient behavior is significant if RHEED oscillations are to be used to control the shutters or beam fluxes during MBE growth. It is recommended [4.42] that a diffraction condition is chosen such that this effect is absent, e.g., choose the [010] azimuth with angle of incidence $\sim 1°$ at 10–12 keV primary beam energy.

The last feature of the RHEED oscillations which will be discussed here is the frequency doubling effect [4.42]. There are several examples in the literature of the appearance of harmonics in the oscillations [4.15, 19, 104, 105]. *Sakamoto* et al. [4.105] showed that for the growth of Si on Si(001) when the beam is along the [110] azimuth, oscillations corresponding to biatomic layer growth occur, whereas other azimuths showed monoatomic layer growth. The same group [4.19] has suggested that this can be explained by the growth of alternating layers of (2×1) and (1×2) reconstruction – the reconstruction features exhibiting maxima every biatomic layer. In order to establish this, it was necessary to prepare a Si(001) surface which exhibited a single (2×1) domain structure. *Aarts* et al. [4.106] did not detect any such bilayer oscillations, and it may be that the presence or absence of the effect depends on the original substrate preparation.

For III-V semiconductors the usual oscillation period corresponds to the growth of a single molecular (Ga+As) layer, i.e., $a_0/2$ for the (001) oriented surface, where a_0 is the lattice constant. However, under certain diffraction conditions oscillations corresponding to periods which are apparently $a_0/4$ can be observed [4.104], as shown in Fig. 4.19. At first sight it is tempting to attribute this behavior to the completion of successive layers of gallium and arsenic rather than complete GaAs layers. However, because of the specific diffraction condition dependence of this behavior, *Joyce* et al. [4.104] do not believe this explanation is correct, and they have suggested [4.104] that the second harmonic results from a superposition of the elastic specular scattered intensity and the diffuse scattering, as shown in Fig. 4.20. The diffuse scattering is generally increased if the step edge density increases, whilst the specular intensity is reduced. Diffuse scattering in this sense refers to directional scattering of electrons that form Kikuchi bands [4.43] or diffuse Bragg features. The latter usually results from the participation of phonons in the scattering (i.e., thermal diffuse scattering), but scattering by

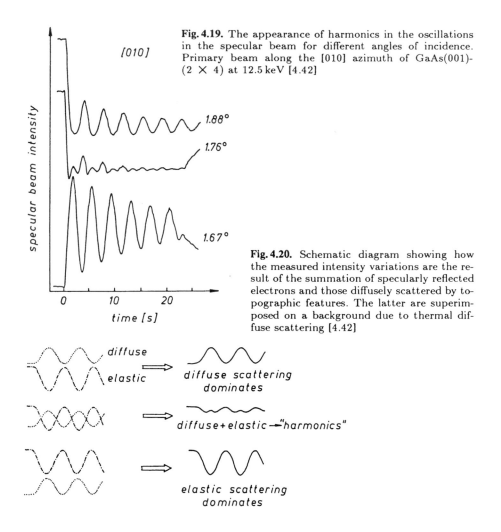

Fig. 4.19. The appearance of harmonics in the oscillations in the specular beam for different angles of incidence. Primary beam along the [010] azimuth of GaAs(001)-(2 × 4) at 12.5 keV [4.42]

Fig. 4.20. Schematic diagram showing how the measured intensity variations are the result of the summation of specularly reflected electrons and those diffusely scattered by topographic features. The latter are superimposed on a background due to thermal diffuse scattering [4.42]

surface irregularities can also provide the parallel momentum or wave vector for a directional diffuse component. Diffuse scattering is very difficult to quantify at present, but it can be stated with certainty [4.42] that it is diffraction condition dependent.

Another very important point arises from this argument. If conditions are chosen such that the diffuse scattering completely dominates over elastic specular scattering then an explanation exists of why, under some diffraction conditions, the RHEED intensity shows an increase in the intensity for the first half period and the peaks are apparently 180° out of phase with the more familiar case. This is nicely illustrated in Fig. 4.19 if the behavior for angles of incidence of 1.67° and 1.88° is compared. The angle of 1.67° corresponds to a deep minimum of intensity on the [010] rocking curve, so one may expect the elastic specular intensity to be very low.

The considerations of this section may be completed with the following three statements [4.42]:

1. RHEED intensity oscillations provide a convenient method of determining the thickness of layers or beam fluxes used in MBE provided that the conditions for their observation are carefully selected.
2. RHEED intensity oscillations provide a means of studying the mechanisms of crystal growth and of determining the kinetics of surface diffusion under the conditions of MBE.
3. In interpreting the RHEED intensity oscillations the diffraction conditions which have been used should be precisely specified and taken into consideration in a suitable way.

The reader interested in more detailed information on RHEED applications to surface analysis is referred to [4.107].

4.2 Ellipsometry

The principles of ellipsometry, as applied to materials characterization, have been widely discussed by *Aspnes* [4.108, 109] and others [4.110–113], so they will be presented below only briefly. This convenient surface-sensitive analytical technique has been applied in MBE systems only recently [4.22–25], however, one may expect that the already demonstrated feasibility of application of ellipsometry as a MBE in-growth diagnostic tool will cause its wider spread.

The computational difficulties associated with reducing raw data have been largely eliminated by the widespread availability of fast, cheap, dedicated minicomputers with adequate storage capacity to perform the necessary calculations. This has stimulated the development of a number of fast automatic ellipsometer systems, which eliminate the operator as an essential feedback element in the null-balancing process, and which can achieve over an order of magnitude improvement in precision relative to the usual divided-circle limit of $\pm 0.01°$.

Some ellipsometric systems operate at high optical efficiencies with relatively little loss of flux, therefore having the advantage of high sensitivity. This permits the use of relatively weak tunable continuum sources, such as arc lamp–monochromator combinations, instead of lasers or spectral line sources. As a result, these systems may be used for spectroscopic purposes where ellipsometric data are measured as continuous functions of the wavelength or energy of light.

These advances have transformed ellipsometry from a specialized laboratory tool to a commercially viable technique, and have stimulated a renewed interest in the field [4.21, 24, 111, 114].

4.2.1 Fundamentals of Ellipsometry

In an ellipsometric measurement, polarized light is incident on the sample. The light is reflected from the sample and the ellipsometer detects the change in the state of polarization of the light. Usually linearly polarized monochromatic light is incident at some definite angle of incidence onto the sample. After reflection from the sample, light turns out to be elliptically polarized. If the parameters characterizing the substrate ambient and light beam are known, then optical characteristics of the film and their changes during MBE growth can be studied in situ according to the changes of the reflected wave polarization.

The fact that ellipsometry measures the polarization state, rather than the intensity as in reflectometry, makes it inherently a high-precision technique which is relatively unaffected by experimental difficulties caused by source intensity fluctuations or light-scattering defects, which may occur even on the most carefully prepared surfaces. The capability of measuring phase changes directly gives ellipsometry great sensitivity to the presence of thin films on reflecting surfaces, which may be detected to average thicknesses of the order of hundredths of a monolayer [4.108].

Let us consider propagation and polarization of light in a uniform, isotropic medium with a complex dielectric function

$$\varepsilon^* = \varepsilon^*(\omega) = \varepsilon_1 + i\varepsilon_2 \quad , \tag{4.22}$$

where ω is the angular frequency of the light wave. Light propagation is described by the wave equation [4.115], which in cgs units has the form [4.108]

$$\left[\nabla^2 - \frac{\varepsilon^*(\omega)}{c^2} \frac{\partial^2}{\partial t^2} \right] E^*(r, t) = 0 \quad , \tag{4.23}$$

where $E^*(r, t)$ is the electric field vector of the light wave, and c is the velocity of light in free space ($c = 2.9979 \times 10^8 \, \mathrm{m \, s^{-1}}$). In terms of a local orthogonal coordinate system whose z axis is parallel to the propagation direction, $E^*(r, t)$ has the form of a plane wave:

$$E^*(r, t) = (E_x^* e_1 + E_y^* e_2) \exp(ik^* z - i\omega t) \quad , \tag{4.24}$$

where E_x^* and E_y^* are the complex amplitudes of $E^*(r, t)$, and e_1 and e_2 are the base vectors along the x and y axes of the coordinate system. By (4.23) the allowed values of the propagation vector k^* are given by the dispersion equation

$$\left(\frac{ck^*}{\omega} \right)^2 = \varepsilon = (n^*)^2 \quad \text{where} \tag{4.25}$$

$$n^* = n^*(\omega) = n + ik \tag{4.26}$$

is the complex refractive index. In ellipsometry, the convention is usually used according to which $\varepsilon(\omega)$ lies in the upper half plane [therefore the $+$ sign stays in the expression (4.22)] and $n^*(\omega)$ lies in the first quadrant [4.108]. The real and imaginary parts of ε^* and n^* are related explicitly by

$$\varepsilon_1 = n^2 - k^2 \quad , \tag{4.27}$$

$$\varepsilon_2 = 2nk \quad , \tag{4.28}$$

$$n = \sqrt{\frac{\varepsilon_1 + \sqrt{\varepsilon_1^2 + \varepsilon_2^2}}{2}} \quad , \tag{4.29}$$

$$k = \frac{\varepsilon_2}{2n} \quad . \tag{4.30}$$

In the absence of an unpolarized component, as will be assumed here, the polarization state of $E^*(r, t)$ is completely specified by the relative values of the complex quantities E_x^* and E_y^* and may be represented geometrically by the Poincaré sphere, Stokes parameters, or as the complex number E_y^*/E_x^* [4.108, 110]. The three limiting cases are: linearly polarized light (Im $\{E_y^*/E_x^*\} = 0$), right circularly polarized light ($E_y^* = -iE_x^*$), and left circularly polarized light ($E_y^* = iE_x^*$).

The intensity I of the light wave (the electric plane wave) is in free space given by

$$I = \tfrac{1}{2} c\varepsilon_0 (|E_x^*|^2 + |E_y^*|^2) \quad , \tag{4.31}$$

where ε_0 is the permittivity of free space ($\varepsilon_0 = 8.854 \times 10^{-12}\,\mathrm{CV^{-1}m^{-1}}$).

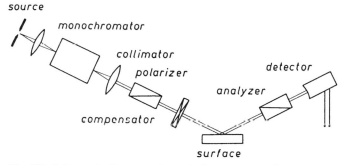

Fig. 4.21. Schematic diagram of a typical arrangement for spectroscopic ellipsometry [4.108]

The manipulation and measurement of the polarization state for most ellipsometers is performed using the linear optical elements shown in Fig. 4.21. The source, monochromator, and collimator provide quasi-parallel, quasi-monochromatic flux. The polarization state of the flux prior to reflection from the sample surface is determined by the polarizer and compensator (the compensator is optional in some systems). The polarization state after reflection is measured by a second polarizer, the analyzer, and the resultant intensity converted into electrical form by the detector.

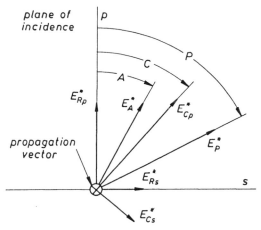

plane of
incidence

p

E_{Rp}^*

E_A^*

E_{Cp}^*

P

C

A

propagation
vector

E_P^*

E_{Rs}^*

s

E_{Cs}^*

Fig. 4.22. Diagram of field components and azimuth angle conventions for ellipsometer system of Fig. 4.21, drawn for convenience for grazing-incidence reflection. E_P^*: electric field vector transmitted by the polarizer; E_{Cp}^*, E_{Cs}^*: electric field components parallel and perpendicular, respectively, to the compensator major axis; E_{Rp}^*, E_{Rs}^*: electric field components parallel and perpendicular, respectively, to reflecting surface; E_A^*: electric field vector transmitted by the analyzer. The azimuth angles P, C, and A locating E_P^*, E_C^* and E_A^*, respectively, relative to the plane of incidence are positive as shown [4.108]

Ellipsometric data may be obtained either by adjusting the optical elements for zero transmitted flux (null ellipsometry) or by analyzing the time dependence of the transmitted flux that results from the periodic variation of one of the elements (photometric ellipsometry) [4.108]. In either case, the general equation for the complex amplitude of the electric field vector of the transmitted flux is the same. This is usually calculated using Jones matrices [4.115], but the result can be obtained directly by calculating the effect of each linear element of the measuring system upon the orthogonal components of the field in the frame of reference of the element. Using the conventions given in Fig. 4.22, it is straightforward to show that the complex field amplitude transmitted by the analyzer is given by

$$
\begin{aligned}
E_A^* = E_P^* [&r_p t_p \cos C \cos (P - C) \cos A \\
&- r_p t_s \sin C \sin (P - C) \cos A \\
&+ r_s t_p \sin C \cos (P - C) \sin A \\
&+ r_s t_s \cos C \sin (P - C) \sin A] \quad ,
\end{aligned}
\tag{4.32}
$$

where the quantities in (4.32) are defined as follows: E_A^* and E_P^* are the complex amplitudes of the linearly polarized plane wave components transmitted by the analyzer and polarizer prisms, respectively; A and P are the azimuth angles which locate the planes of polarization of E_A^* and E_P^*, respectively, measured from the plane of incidence, and are positive for clockwise rotation looking in the direction that the light is propagating (right-hand rule); C is the azimuth angle which locates the major axis of the compensator, in the same manner as A and P; t_p and t_s are the complex transmittance (transmitted field/incident field) coefficients

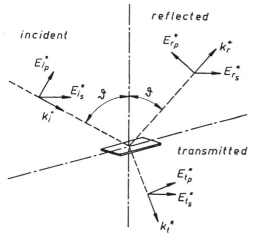

incident

$E_{i_p}^*$

$E_{i_s}^*$ ϑ

k_i^*

reflected

$E_{r_p}^*$ k_r^*

ϑ $E_{r_s}^*$

transmitted

$E_{t_p}^*$

$E_{t_s}^*$

k_t^*

Fig. 4.23. Electric field direction conventions used in calculating the complex reflectances at a plane-parallel interface in the two-phase (ambient–optically thick substrate) approximation; ϑ is the angle of incidence [4.108]

of the compensator for field components parallel and perpendicular, respectively, to its major axis. Finally, r_p and r_s are the complex reflectance coefficients of the reflecting surface for field components parallel and perpendicular, respectively, to the plane of incidence. The signs of r_p and r_s depend upon the convention chosen for the incident and reflected fields. The convention used here is the one convenient for ellipsometry, and is shown in Fig. 4.23, where for simplicity an optically thick substrate has been assumed. In terms of incident and reflected complex field amplitudes E_{ip}^*, E_{is}^* and E_{rp}^*, E_{rs}^*, respectively,

$$r_p = \frac{E_{rp}^*}{E_{ip}^*} \quad \text{and} \tag{4.33}$$

$$r_s = \frac{E_{rs}^*}{E_{is}^*} \quad . \tag{4.34}$$

The absolute reflectance coefficients, R_p and R_s, giving the change in intensity of light wave upon reflection, are related to r_p and r_s by

$$R_p = \frac{I_{rp}}{I_{ip}} = |r_p|^2 \quad \text{and} \tag{4.35}$$

$$R_s = \frac{I_{rs}}{I_{is}} = |r_s|^2 \quad . \tag{4.36}$$

For many applications not involving absolute intensity measurements, it is convenient to express (4.32) in the form

$$\begin{aligned}
E_A^* = E_P^* r_s t_p [\varrho^* \cos C \cos (P - C) \cos A \\
- \varrho^* \tau^* \sin C \sin (P - C) \cos A \\
+ \sin C \cos (P - C) \sin A \\
+ \tau^* \cos C \sin (P - C) \sin A] \quad ,
\end{aligned} \tag{4.37}$$

151

where ϱ^* and τ^* are complex reflectance and transmittance ratios of the reflecting surface and compensator, respectively, and are defined by the expressions

$$\varrho^* = \frac{r_p}{r_s} = e^{i\Delta} \tan \psi \quad \text{and} \tag{4.38}$$

$$\tau^* = \frac{t_s}{t_p} \sim e^{i\delta} \ . \tag{4.39}$$

The complex reflectance ratio is often expressed in terms of the angles ψ and Δ defined in (4.38). The complex transmittance ratio, on the other hand, is expressed in terms of the relative phase shift, δ, defined by (4.39) since usually $|t_s/t_p| \cong 1$. The major axis of the compensator is the fast (slow) axis if $\delta > 0$ ($\delta < 0$). If $|\delta| = 90°$, the compensator is a quarter-wave plate.

The parameters that are obtained from an ellipsometric measurement are the real angles Δ, ψ (most often), and δ (in special cases), depending on the experimental configuration and the mode of operation of the ellipsometer. These parameters may, however, also be calculated theoretically for a given reflection angle, wavelength and known (or given by the assumed model) materials parameters. For example, Δ and ψ are expressed in the basic equation of ellipsometry for an ambient–one-overlayer–substrate system [4.110, 112, 113] by

$$e^{i\Delta} \tan \psi = \frac{(r_p^{01} + r_p^{12} e^{i2D})(1 + r_s^{01} r_s^{12} e^{i2D})}{(1 + r_p^{01} r_p^{12} e^{i2D})(r_s^{01} + r_s^{12} e^{i2D})} \ , \tag{4.40}$$

where r_p, r_s are the amplitude reflection coefficients defined by (4.33) and (4.34), respectively. The superscripts stand for the particular interface in the system (01 for the ambient–overlayer, and 12 for the overlayer–substrate interface). Here D is a phase shift and can be expressed as [4.112, 113]

$$D = \frac{d}{\lambda} 2\pi \sqrt{(n^*)^2 - (n_a^*)^2 \sin^2 \vartheta} \ , \tag{4.41}$$

where n_a^* and n^* are the refractive indices of the ambient and the overlayer, respectively. Figure 4.24a shows the reflection geometry of the considered ambient–one-overlayer–substrate system. From one ellipsometric measurement the two

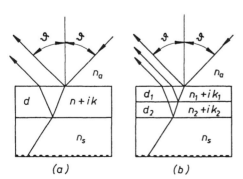

(a)   (b)

Fig. 4.24.
Light reflection (a) from an ambient–thin-film–substrate system, and (b) from an ambient–two-layer-film–substrate system; d, d_1, and d_2 are the layer thicknesses; n_a and n_s are the real refractive indices of the ambient and the substrate; $n + ik$, $n_1 + ik_1$, and $n_2 + ik_2$ are the complex refractive indices of the layers [4.112]

angles Δ and ψ are obtained. Therefore, two unknown parameters can be determined. In the simplest case the unknown parameters are the thickness d and the real refractive index n of the film.

A more difficult problem arises when there is a many-layer sample on the substrate. The calculated $\tan \psi$ and $\cos \Delta$ values are now obtained by using (4.40) m times recursively, where m is the number of layers on the substrate. Figure 4.24b shows the example of two layers.

In a spectroscopic ellipsometric measurement the reflection angle ϑ is usually fixed and the wavelength changed. This increases the data that can be obtained. However, the changing wavelength also makes it necessary to introduce the wavelength dependence of the refractive index. A more accurate analysis also requires that the imaginary part of the refraction index not be neglected.

Hence in a more exact analysis, where the reflecting system is modeled by the two-overlayer–substrate configuration shown in Fig. 4.24b, one has instead two unknown parameters (d and n) the unknown thicknesses d_1 and d_2 and two wavelength-dependent complex refractive indices $n_1(\lambda)+ik_1(\lambda)$ and $n_2(\lambda)+ik_2(\lambda)$. One should note that in the two-layer model the choice has to be made how to divide the analyzed film into the two parts [4.112].

In certain applications requiring high accuracy, the simple relation given by (4.32) or (4.37) is not adequate. Correction terms may be necessary to include the effect of residual strain birefringence in polarizers, compensators, and cell windows, optical activity in polarizers or compensators, and intrinsic anisotropies such as those occurring in uniaxial, biaxial, or optically active reflecting surfaces. These terms can become quite large when operating near normal modes of the system, i.e., for A, P, or $C \cong 0$, or $|\delta| \cong 0$ or π. For anisotropic reflecting surfaces, three complex reflectance coefficients, $r_s \rightarrow r_{ss}$, $r_p \rightarrow r_{pp}$, and $r_{sp} = r_{ps}$, must be determined, since in general a plane-polarized wave incident as either a p or s mode will generate both p and s components in the reflected wave. The ratios of any two of the three complex reflectance coefficients can be determined by comparing ellipsometric measurements performed at different adjustments, angles of incidence, or at different orientations of the reflecting surface.

4.2.2 Ellipsometric Systems Used for In-Growth Analysis in MBE

In order to exemplify how ellipsometry may be used to study the MBE process in the in-growth mode, two ellipsometric systems will be described. The first uses a laser as the light source [4.116], while the second [4.22, 23] is a spectroscopic ellipsometer using an arc lamp as the source.

The monochromatic ellipsometer [4.25] developed for incorporation into the MBE system as an automatic measuring unit consists of two parts: the polarizer part and the analyzer part. It is schematically illustrated in Fig. 4.25. The polarizer part consists of a He-Ne laser (1), a circular polarizer (2), a polarizer prism (3), modulative (4) and compensative (5) Faraday effect cells [4.115], and a

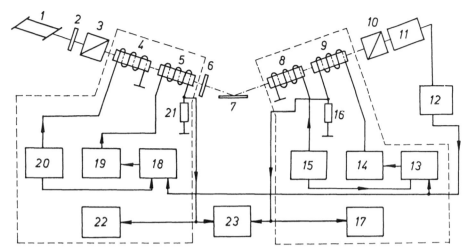

Fig. 4.25. Construction elements of the automatic monochromatic ellipsometer designed for incorporation into a MBE system [4.116]. *(1)* He-Ne laser, *(2)* circular polarizer, *(3)* polarizer prism, *(4,8)* modulative Faraday effect cells, *(5,9)* compensative Faraday effect cells, *(6)* compensator, *(7)* sample, *(10)* analyzer prism, *(11)* photomultiplier with a wide-band amplifier, *(12)* selective amplifiers unit, *(13,18)* synchronic detectors, *(14,19)* compensation current sources, *(15,20)* modulation current sources, *(16,21)* measuring resistors, *(17,22)* x-y plotters, *(23)* digital recording unit

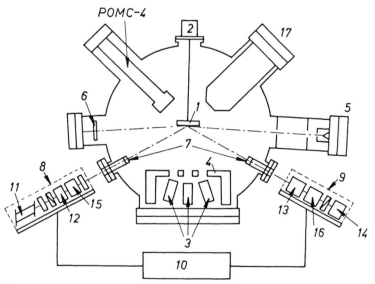

Fig. 4.26. Schematic of the MBE deposition chamber with the optical parts of the automatic monochromatic ellipsometer [4.116]. *(1)* substrate holder, *(2)* deposition chamber manipulator, *(3)* effusion cells, *(4)* cryopanel, *(5)* RHEED electron gun, *(6)* luminescent screen, *(7)* optical leads providing minimal distortion of the light beam polarization, *(8)* polarizer part of the ellipsometer (see Fig. 4.25), *(9)* analyzer part of the ellipsometer (see Fig. 4.25), *(10)* recording unit. Numbers *11,12,15,13,16* and *14* correspond to nr. *1,4,5,8,9* and *11* respectively in Fig. 4.25

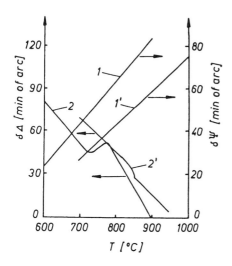

Fig. 4.27. Dependences of the ellipsometric angles $\delta\Delta$ and $\delta\psi$ on silicon crystal temperatures [4.118]; curves 1, 2 are for the vicinal face with 5° deviation from (111) to (110); curves 1', 2' are for the singular (111) face

compensator (6). The analyzer part consists of modulative (8) and compensative (9) Faraday effect cells, an analyzer prism (10) and a photomultiplier (11) with a wide-band amplifier. Figure 4.26 shows how the ellipsometer is combined with the MBE deposition chamber.

The described ellipsometer has been used for studying roughness and superstructural reconstruction of the silicon (111) surface (curves 1' and 2' in Fig. 4.27) and the silicon vicinal surface characterized by 5° deviation from (111) towards (110) orientation (curves 1 and 2 in Fig. 4.27), as a function of the sample temperature. The reversible transition from the ordered step system with a height of two interplanar distances ($2d_{111}$) to ordered steps with the height of one interplanar distance [111] taking place on a clean silicon surface near the (111) pole at $750 \pm 25°$ C [4.117] has been studied experimentally using ellipsometry [4.118] (see curves 1, 2 in Fig. 4.27). The known phase transition Si(111)-$(7 \times 7) \rightleftharpoons$ Si(111)-(1×1) has also been studied and is presented in Fig. 4.27 as curves 1', 2'. The in situ measurements have shown that the surface roughness change at the first transition by the magnitude $\Delta h = d_{111} = 0.314$ nm led to a $\delta\Delta$ magnitude change (curve 2) of 30' at $T = 750°$ C, and the calculation of this parameter gives the change $\delta\Delta = 35'$. The $\delta\Delta = 5-6'$ jump is observed at the second phase transition for a device sensitivity of 5–15' (curve 2').

The spectroscopic ellipsometer [4.22, 23] has been used for studying MBE of cadmium mercury telluride (CMT). The optical system of this spectroscopic ellipsometer is shown in Fig. 4.28. The growth chamber used in this study was a Riber 2300 P system. The standard viewport permits one to observe the sample in the growth position with an angle of incidence of about 70°. This angle is very favorable for ellipsometry, but no viewport existed that it was feasible to use for collecting the reflected beam. So, a modification of the well-known spectroscopic ellipsometer with rotating polarizer [4.119] was introduced by adding a mirror working at quasi normal incidence (Fig. 4.28). This mirror returns the light and

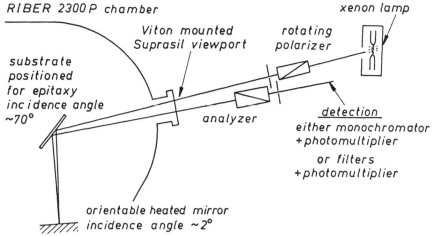

Fig. 4.28. Optical path diagram of the spectroscopic ellipsometer with rotating polarizer (schematic) [4.22]

after a second reflection on the sample the light exits via the same viewport through which it entered the chamber. The input and output beams are separated by the beam being not exactly normally incident on the mirror. The angle of incidence is kept small enough that this reflection has practically no effect on the polarization of the light, and then, to first order, no correction is necessary to take this reflection into account. First and second reflections are kept symmetrical, as shown in Fig. 4.29, so the incidence angles on the sample are exactly the same. We have to assume homogeneity of the sample because the incidence points (A and B) are slightly different. Assuming homogeneity, the detected signal can be analyzed like the signal in a standard configuration, but the measured parameters, which are usually $\tan \psi$ and $\cos \Delta$ (where ψ and Δ are ellipsometric angles), are now $\tan^2 \psi$ and $\cos 2\Delta$.

The accuracy of the described ellipsometer was checked on a well-known system like silicon, and after finding good agreement with published data, the measurement technique was applied to CMT [4.22]. For this new material, first the evolution of the dielectric function (4.22) of CMT layers with substrate temperature was established for different compositions of this material. Good agreement was found between the results obtained by ellipsometry and the previously published data. Furthermore, it could be shown [4.22] that spectroscopic ellipsometry permits the experimental determination of the composition of the CMT system, and the initial growth rate in the MBE process.

Measurement of the composition of CMT with spectroscopic ellipsometry is illustrated in Fig. 4.30. At each step of CMT growth, it is possible to spectroscopically measure the ellipsometric angles. After about 1000 Å CMT deposition, the signal results only from the top homogeneous layer and the ellipsometric data can be inversed using a single two-phase model (Fig. 4.24a), so that one obtains

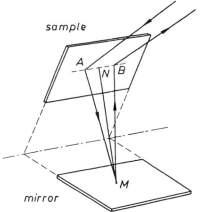

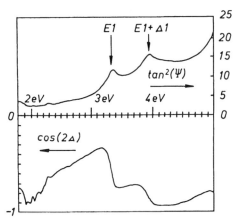

Fig. 4.29. Details of the optical path on the sample and the mirror. The mirror is adjusted so that AB is perpendicular to MN, the normal to the mirror [4.22]

Fig. 4.30. Typical ellipsometer spectra: in situ measurement of CdTe MBE layer [4.22, 23]

the dielectric function of the layer. This dielectric function can be fitted by a theoretical function calculated using summation of harmonic oscillator functions. The energies of the transitions of the oscillators depend on the percentage of cadmium. The energy of the optical transition E_1 can be determined with good precision, and a calibration curve has been established giving the correspondence between the measured energy and the cadmium percentage measured by infrared transmission. At room temperature, E_1 varies from 2.1 eV for HgTe to 3.2 eV for CdTe. With a precision of 10 meV in the determination of the transition, one obtains the composition with an absolute precision of 1 % of cadmium. Even at growth temperature (180° C), broadening of the transition does not prevent a determination with that precision.

The measurement of the initial growth rate is illustrated in Fig. 4.31. Plotting the ellipsometric angles at a fixed wavelength versus time during the deposition

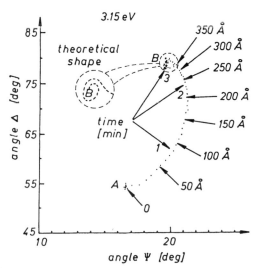

Fig. 4.31. Measurement at fixed light energy: 3.15 eV. The evolution from A(CdTe) to B(CMT) is labelled in the (ψ, Δ) plane either by the deposition time or by the layer thickness. A comparison of the two labels gives the growth rate (here about 150 Å/min). After about 350 Å, the theoretical shape (spiral) of the deposition can no longer be distinguished in the figure [4.22]

process, one obtains curves which can be parametrized either by the time or by the deposition layer thickness. Comparison of the two labels determines the growth rate. After about 1000 Å (CMT) deposition, the ellipsometric signal does not evolve any more (there is no longer an interface reference for the thickness because of the absorption of light in the layer before the interface). Thus growth rate measurements are limited to initial growth rate, which is, however, very useful for multilayers or quantum wells [4.23].

Using spectroscopic ellipsometry the interdiffusion processes between CdTe and HgTe layers may also be studied. Details of these studies are given in [4.23].

5. Postgrowth Characterization Methods

Devices grown by MBE include complex multilayer structures, besides layers of metals and dielectrics grown on single-crystal substrates. The technology of ion implantation even allows buried layer structures to be created. Two recently published more complicated structures are shown schematically in Fig. 5.1: (a) separate absorption grading and multiplication avalanche photodiode (SAGM APD) [5.1], (b) graded-index (GRIN) separate-confinement heterostructure (SCH) multi-quantum-well (MQW) laser [5.2].

The device performance is critically dependent on the electrical and morphological quality of epitaxial layers and interfaces. To point out the critical areas of a multilayer structure we take as an example the modulation doped high electron mobility transistor (HEMT), which is already being produced in large

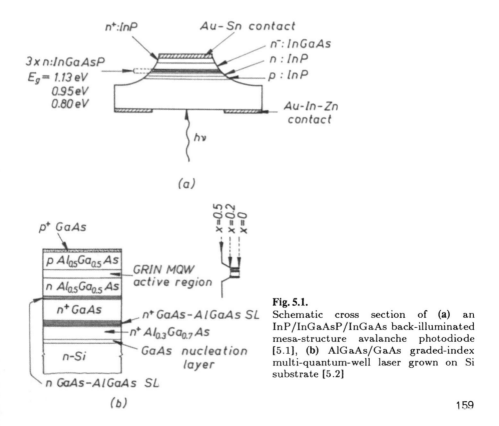

Fig. 5.1.
Schematic cross section of **(a)** an InP/InGaAsP/InGaAs back-illuminated mesa-structure avalanche photodiode [5.1], **(b)** AlGaAs/GaAs graded-index multi-quantum-well laser grown on Si substrate [5.2]

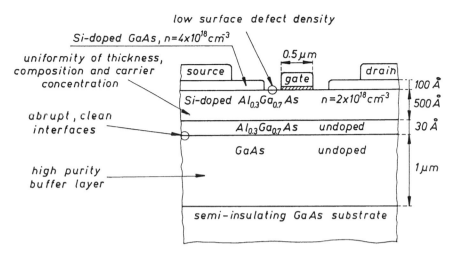

low surface defect density

Si-doped GaAs, $n=4\times10^{18}\,cm^{-3}$

uniformity of thickness,
composition and carrier
concentration

source   gate   drain $0.5\,\mu m$

Si-doped $Al_{0.3}Ga_{0.7}As$ $n=2\times10^{18}\,cm^{-3}$

$100\,\AA$
$500\,\AA$

abrupt, clean
interfaces

$Al_{0.3}Ga_{0.7}As$ undoped $30\,\AA$

GaAs   undoped

high purity
buffer layer

$1\,\mu m$

semi-insulating GaAs substrate

Fig. 5.2. Schematic diagram of a modulation doped high electron mobility transistor

numbers. Figure 5.2 shows a cross section of a HEMT structure. Indicated are the regions with the required critical properties for reliable operation of such a device.

High purity GaAs is needed as a buffer layer to ensure low deep-level densities for high electron mobility in the high frequency device. The abrupt and clean interface provides low electron scattering and long-term device stability. The homogeneity of the thickness and alloy composition as well as a controllable carrier density is necessary for an acceptable uniformity of device characteristics. Finally, a low surface defect density results in a uniform gate metalization and consequently in a higher level of integration.

However, the interaction between the layers and the substrate during growth and the course of high temperature processing steps or long-term device operation cannot always be predicted and often results in a failure of the device. Thus there is a clear need for a variety of precise structural, optical and electrical information about these multilayer systems.

Materials scientists have responded to this with an enormous number of sophisticated analytic tools. It would take us far beyond the scope of this book if we were to cover all the different techniques of characterization available so far. Therefore we have to restrict ourselves to some fundamental methods. Following the example of a HEMT given above, the most important physical parameters which should be detected in a characterization laboratory in the close vicinity of a MBE machine are:

— crystalline perfection of the individual epilayers
— doping profiles and defect distribution within layers
— structural parameters and electrical properties of the interfaces
— potential barriers between the layers.

In the first section of this chapter we will give a short survey of the most important characterization methods to guide the reader to the technique appropriate to a particular analysis problem. In the following sections we will concentrate on some methods which are indispensable in every MBE laboratory. We will discuss them in the order of structural, optical, electrical and more sophisticated techniques.

5.1 Survey of Postgrowth Characterization Methods

Crystal growers working with MBE, as well as device designers, are mainly concerned with materials properties such as carrier concentration, compensation, deep-level distribution, carrier lifetime and diffusion length. Of special interest are nondestructive methods particularly suitable for very thin ($\leq 1\,\mu$m) layers. Most of the experimental techniques are well established nowadays and are usually referred to by acronyms, which are of course familiar to all the materials scientists working in the field but may be somewhat confusing for a crystal grower. Therefore we start this section with a table of important characterization methods (in alphabetical order), including the most commonly used acronyms and the main references where a basic description of the method can be found (Table 5.1). Especially emphasized are those methods which are indispensable in a MBE laboratory and which will be described in the following sections. Table 5.2 acts as a guide to the variety of experimental techniques, listing the material

Table 5.1. Survey of characterization methods, acronyms and main references

Method	Acronym	References
Capacitance Voltage characterization	*C-V*	[5.3]
Deep Level Transient Spectroscopy	DLTS	[5.4]
Auger Electron Spectroscopy	AES	[5.5]
Electron Beam Induced Current	EBIC	[5.6]
Electron Energy Loss Spectroscopy	EELS	[5.7]
Ellipsometry	Ellipsometry	[5.8]
Electron Microprobe Analysis	EMA	[5.9, 10]
Etch-Pit Density	EPD	[5.11]
Electron Spectroscopy for Chemical Analysis	ESCA	[5.12]
Hall effect	Hall	[5.13]
Infra-Red Spectroscopy	IRS	[5.14]
Current(*I*)-Voltage characteristics	*I-V*	[5.3]
Photo-Luminescence	PL	[5.15]
Reflection High Energy Electron Diffraction	RHEED	[5.16]
Rutherford Back-Scattering	RBS	[5.17]
Scanning Electron Microscopy	SEM	[5.18]
Secondary Ion Mass Spectroscopy	SIMS	[5.19]
Transmission Electron Microscopy	TEM	[5.20]
Ultraviolet Photoelectron Spectroscopy	UPS	[5.21]
X-ray Photoelectron Spectroscopy	XPS	[5.12]
X-ray Diffraction	XRD	[5.22]

Table 5.2. Characterized features and possible techniques

Object feature being characterized		Method
Structural	Surface morphology	SEM, RHEED, TEM
	Crystalline quality	XRD, TEM, RBS
	Dislocation density	TEM, PL, EPD
	Interface roughness	TEM, PL, RBS
	Layer thickness (quantum well)	IRS, Ellipsometry, PL
	Strain or stress	XRD, PL, RBS
	Layer composition	EMA, Ellipsometry, XRD, AES, SIMS
	Abrupt interfaces	I-V, C-V, TEM, PL
Electrical	Carrier density and mobility	Hall effect, C-V
	Deep defect distribution	DLTS
	Band discontinuity at the interface	EBIC, UPS, I-V, XPS, PL
	Degree of compensation	Hall effect, PL
	Two-dimensional electron gas	Hall effect, PL
Optical	Radiative transitions	PL
	Complex refractive index	Ellipsometry
	Band gap and composition	PL, Absorption

properties under investigation. The second column gives the possible techniques that can be applied to solve a particular problem. Of course this listing is incomplete, due to the enormous amount of work which is being done with more and more complicated equipment.

5.2 Auger Electron Spectroscopy

In Auger electron spectroscopy (AES) a focused beam of electrons in the energy range 2–20 keV [5.23] irradiates the sample. Atoms up to a depth of about 1 μm are ionized in an inner core level, e.g., the K level (Fig. 5.3), and subsequently deexcite by an electron falling from a higher level L_1 with the energy balance removed by a third electron from level L_3. This last electron, the Auger electron, is emitted with an energy E_A, defined by [5.24]

$$E_A = E_K(Z) - E_{L_1}(Z) - E_{L_2}(Z + \Delta) - \Phi \quad , \tag{5.1}$$

where Z is the atomic number of the atom and Φ is the work function of the surface (Fig. 5.3). The third term on the right hand side of (5.1) has an extra component Δ included to take account of the fact that the atom is in a charged state when the final electron is ejected. Experimentally Δ is found to have a value between 1/2 and 3/2 [5.25]. An Auger electron is described by the transitions which take part in its production. An ABC Auger electron is one which results from ionization of the A electron shell, with an electron from shell B falling into the hole created and an electron from orbital C being ejected. The energies of the Auger electron transitions are plotted as a function of atomic number

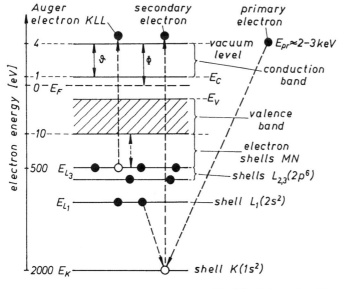

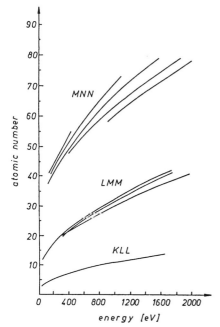

Fig. 5.3. Schematic illustration of the Auger electron generation process for the transitions *KLL*. The numerical values of the electron energy indicated bear no relation to any specific atom

Fig. 5.4. The principal Auger electron energies for three different Auger processes [5.24]

in Fig. 5.4. The threefold splitting of the *LMM* Auger electrons is due to the three possible ionization energies of the *L* shell (L_1, L_2, L_3 see Fig. 5.3). The same holds for the splitting of the *MNN* Auger electrons. The most intense Auger peaks are observed for transitions *ABC* where $B = C = A + 1$, i.e., *KLL*, *LMM*, *MNN*, etc., which reflects the fact that electron interactions are strongest between electrons whose orbits are close together.

Auger electrons can be produced in any atom that has been ionized by the removal of an inner shell electron. This ionization may be caused by bombarding the atom with electrons, photons or ions, provided that the energy is greater than the binding energy of the electron which must be removed in the initial ionization. It therefore follows that Auger peaks will be observed in XPS in which the surface is exposed to x-rays. These peaks are often regarded as a nuisance by operators of XPS instruments, but in fact much useful information can be obtained from them. However, it is usual in Auger electron spectroscopy to use an electron beam to produce the initial ionization in the surface atoms. This has the advantage that the incident beam may be focused, thus giving good spatial resolution. In addition, the primary electron beam energy need not be known accurately. It is sufficient simply for the primary electron energy to be greater than about four times the energy of an Auger electron. This follows from the fact that the ionization cross section is greatest when the ratio of primary electron energy to the Auger electron energy is ~ 3 and falls slowly as this increases [5.26]. Neither is it necessary to have an electron beam with a sharply defined energy spread; in fact it can be an advantage to have ill-defined incident beam energy, since otherwise loss peaks become very sharp and this may obscure Auger peaks of interest.

From the relation (5.1) and Fig. 5.4 it is evident that Auger electron energy is a characteristic feature of the atom for a given Auger transition. This is the reason for applying AES in analysis of the chemical composition of materials. The determination of the E_A energy is therefore fundamental in AES. Initially AES was developed using the three and four grid LEED optics as retarding field energy analyzers (RFA) [5.27]. As AES became appreciated as a useful surface analytical technique, energy analyzers were designed specifically to detect Auger electrons. The first of these analyzers was known as the cylindrical mirror analyzer (CMA) [5.28].

In the CMA shown schematically in Fig. 5.5a the Auger electrons excited from the sample pass through an entrance slit in the center cylinder of two concentric cylinders [5.24]. A negative potential V is applied to the outer cylinder and electrons which have the correct energy are deflected through the exit slit and collected using an electron multiplier. By sweeping V over the desired range an Auger spectrum can be generated. This type of analyzer has excellent collection efficiency, a high transmission value, and, because only those electrons with the desired energy pass through the analyzer, the signal-to-noise ratio is very high.

The best energy resolution, and consequently the highest efficiency in gaining chemical information from the sample, can be obtained, however, with the hemispherical analyzer (HSA) (see Figs. 3.33a and 5.5b). The HSA is operated in two modes: fixed analyzer transmission (FAT) and fixed retard ratio (FRR). In the FAT mode the electrons are retarded to a fixed energy prior to entering the analyzer and thus the resolution is constant throughout the spectrum. With the FRR mode the resolution is a similar function to that in the CMA analyzer, although the absolute value is much less in the HSA. This type of analyzer is almost always used to detect photoelectron peaks; Auger peaks which are present

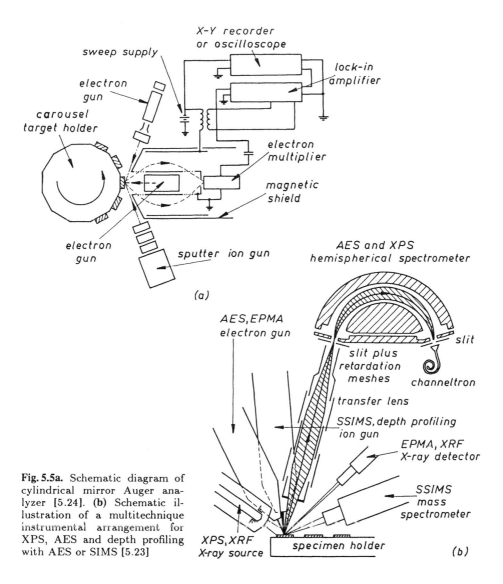

electron gun

X-Y recorder or oscilloscope

sweep supply

electron gun

lock-in amplifier

carousel target holder

electron multiplier

magnetic shield

electron gun

sputter ion gun

(a)

AES and XPS hemispherical spectrometer

AES, EPMA electron gun

slit

slit plus retardation meshes

channeltron

transfer lens

SSIMS, depth profiling ion gun

EPMA, XRF X-ray detector

SSIMS mass spectrometer

XPS, XRF X-ray source

specimen holder

(b)

Fig. 5.5a. Schematic diagram of cylindrical mirror Auger analyzer [5.24]. **(b)** Schematic illustration of a multitechnique instrumental arrangement for XPS, AES and depth profiling with AES or SIMS [5.23]

in XPS spectra can be studied in this system. The Auger spectra can be generated in the absence of photoelectron peaks by bombarding the surface with an electron beam. A typical arrangement is shown schematically in Fig. 5.5b. Here the electron beam is incident on the sample surface at a glancing angle. The Auger electrons are focused using an electrostatic lens system and pre-retarded prior to passing through the entrance slits of concentric hemispherical analyzers. Again, electrons are deflected by a negative potential on the outer hemisphere to pass through the exit slit and are collected using a channeltron electron multiplier. These analyzers produce the best resolution Auger spectra, but the collection and transmission efficiency is considerably less than the CMA. As a result the spectra take longer to obtain and it is often difficult to use these analyzers for kinetic studies. However, they should yield useful chemical information from the Auger peaks, which, when combined with the superior spatial resolution of electron-excited Auger spectroscopy, will make these very powerful analytical tools.

It is interesting to compare the different types of spectra one can obtain from the instruments described above [5.24]. In modern analyzers it is common to record spectra in the $N(E)$ versus E mode using multichannel analyzers, then to store the data on either magnetic tape or floppy disk systems and then to use the on-line data handling facilities to process the raw data. Reproduced in Fig. 5.6 is a spectrum recorded in this way from a fracture face of a sample of Nimonic 80A. The spectrum (a) is the raw data in the $N(E)$ form and this has been differentiated using the data handling facilities and reproduced as spectrum (b) $[dN(E)/dE]$. LMM transitions from nickel and titanium together with KLL transitions from oxygen and carbon can be readily identified [5.24].

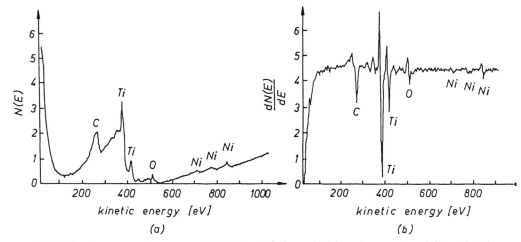

Fig. 5.6. Auger spectrum recorded in the $N(E)$ mode **(a)** and computer differentiated $[dN(E)/dE]$ **(b)** for a Nimonic 80A fracture face [5.24]

Spectra differ considerably from one analyzer type to another. This is amply illustrated in Fig. 5.7, where the LMM transitions from chromium have been recorded using a retarding field analyzer in a, a cylindrical mirror analyzer in b, and a hemispherical analyzer in c. All spectra were recorded under similar conditions, i.e. recording times of ~ 1000 s and the best possible resolution consistent with good signal-to-noise ratio. These were recorded from clean chromium metal but it is difficult to distinguish the chromium LMM transition from the oxygen KLL transition using the RFA. There is no problem of this nature with the CMA but the resolution is such that the peak width is ~ 3 eV. The HSA, however, resolves considerable fine structure, which cannot be detected using

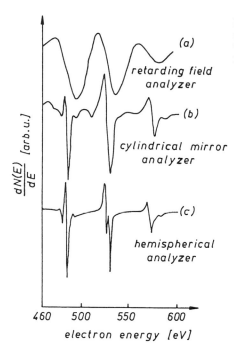

Fig. 5.7. Chromium Auger *LMM* peaks recorded using analyzers described in the text. The different energy resolution of different analyzers is evident [5.24]

the CMA analyzer. To give spectrum c, the HSA is operating at a resolution $\sim 0.1\,\mathrm{eV}$.

5.2.1 Chemical Composition of Solid Surfaces

The Auger electrons have a very short range before inelastic scattering occurs and it is this that makes the spectra characteristic of the outermost atom layers. The characteristic information length is given by the inelastic mean free path λ, which depends on the Auger electron energy E_A and the matrix material. An analysis of many experimental data [5.29] gives approximately

$$\lambda = 538 \frac{a_A}{E_A^2} + 0.41 a_A \sqrt{a_A E_A} \qquad (5.2)$$

for elements and

$$\lambda = 2170 \frac{a_A}{E_A^2} + 0.72 a_A \sqrt{a_A E_A} \qquad (5.3)$$

for inorganic compounds, where E_A is in electron volts and λ and a_A are in nanometers. Here a_A is the average monolayer thickness defined by

$$\varrho n \mathcal{N}_A a_A^3 = M \quad , \qquad (5.4)$$

where ϱ is the density, $\mathcal{N}_A$ is Avogadro's number, M is the molecular weight, and n is the number of atoms in the molecule. Equations (5.2) and (5.3) are not

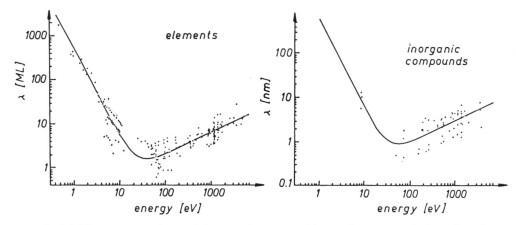

Fig. 5.8. Electron mean free path in pure elements and inorganic compounds (after [5.29])

of the precise form expected theoretically [5.30] but are an approximate fit to the experimental data, as shown in Fig. 5.8.

Recently a new general formula was proposed [5.31] for determining the inelastic mean free path (in angstroms)

$$\lambda = \frac{E}{E_p^2 \beta \ln(\gamma E)} \quad [\text{Å}],$$

(5.5)

where E is the electron energy (in electron volts) and

$$E_p = 28.8 \sqrt{\frac{\varrho N_V}{M}} \quad [\text{eV}]$$

(5.6)

is the electron plasma energy. Here N_V is the total number of valence electrons per atom or molecule, the density ϱ is expressed in $g\,cm^{-3}$ and M is the atomic or molecular weight. For a metal, N_V is computed from the number of $4s$ and $3d$ electrons. Obviously the parameters β and γ determine the magnitude and energy dependence of the inelastic mean free path. The following relations between the parameters β and γ and common material constants such as ϱ, N_V and M were found empirically [5.31]:

$$\beta = -2.52 \times 10^{-2} + \frac{1.05}{\sqrt{E_p^2 + E_g^2}} + 8.10 \times 10^{-4} \varrho$$

(5.7)

and

$$\gamma = 0.151 \varrho^{-0.49},$$

(5.8)

where E_g is the band gap energy for nonconductors.

Figure 5.9 shows a curve, labelled TPP, which was calculated with (5.5–8). For comparison a curve, labelled SD, based on (5.2) is also drawn in Fig. 5.9. In the energy range above 1000 eV the SD curve deviates clearly from the exper-

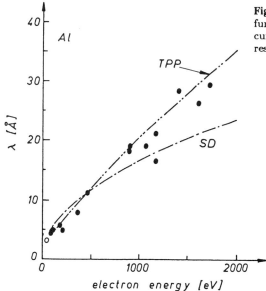

Fig. 5.9. Inelastic mean free path as a function of energy for aluminum. SD curve: results from [5.29]; TPP curve: results from [5.31]

imental data, whereas the TPP curve fits the data points reasonably well. Thus this new set of formulas (5.5–8) is believed to be useful for determining the inelastic mean free path as a function of energy for a given material and the material dependence for a given energy.

The emitted spectrum shows the Auger electron peaks on a background comprised of cascade secondary (Fig. 5.3) and inelastically scattered primary electrons [5.23], as shown in Fig. 5.6a. The intensity of the peaks shown in this figure arises first from the ionization cross section for the inner electron shell of the relevant atoms (in the case of the KLL process shown in Fig. 5.3 this is the K shell) and the subsequent cascade to eject Auger electrons. The ionization cross section increases from zero at the threshold ionization energy and reaches a maximum at some three times this energy. Additional ionization occurs from the backscattered primary electrons, and this causes the total ionization of the atomic level to remain approximately constant at higher energies. This explains why the primary electron beam energy does not have to be known accurately in AES.

Since AES was primarily developed as a tool to measure amounts of atoms at solid surfaces, considerable effort has gone into making the technique quantitative [5.12, 23, 24]. The most successful method of determining the concentration of electrons in a surface is based on obtaining sensitivity factors I^∞ using pure elements. In this method [5.24] the AES peak height or peak area from a particular element is measured with the spectrometer used for analysis, ensuring the cleanest obtainable form of that element. This is repeated for all elements which may be encountered in the sample to be analyzed. It should be emphasized that the sensitivity factors for the specific elements are characteristic features of the spectrometer used. If the spectrometer is changed, the sensitivity factors have to be determined again. Having measured the sensitivity factors of the spectrometer for the relevant elements one can determine the atomic concentration X_A of element A in the sample surface with a homogeneous distribution of atoms, measuring the Auger signals I from the elements in the sample and using the formula

$$X_A = \frac{(I_A/I_A^\infty)}{\sum_i (I_i/I_i^\infty)} \quad . \tag{5.9}$$

This formula is correct when the Auger signals measured for the elements in a definite sample are independent of the composition of the sample (are matrix independent [5.23]). However, because both the backscattering contribution to the Auger electron intensity and the inelastic mean free path are matrix sensitive, a better equation for quantitative Auger analysis is [5.23]

$$X_A = \frac{(I_A/I_A^\infty)}{\sum_i (F_{iA} I_i/I_i^\infty)} \quad , \tag{5.10}$$

where F_{iA} is a calculable Auger matrix factor, for which tables were first given by *Hall* and *Morabito* [5.32]. On the basis of their approach a more accurate estimate of F_{iA} has been given by *Ichimura* and *Shimizu* [5.33]. Fortunately, matrix effects are generally weak and may be easily predicted [5.23]. For instance, a low atomic number element in a high atomic number matrix will have an enhanced Auger signal due to the backscattering effect. The sensitivity factors for the principal lines of the elements are tabulated with respect to the silver line "Ag(351 eV)" for a standard experimental configuration [5.34].

In quantitative analysis, the attenuation of the escaping Auger electrons by the different layers has to be taken into account [5.35], together with the different escape depths of Auger electrons with different energies [5.7]. Let us consider, as an example, a binary compound AB for which we will assume a uniform distribution of each of the elements A and B in the volume of the sample [5.36]. The composition of this compound is determined by the atomic concentration of the element A, defined by

$$c_A \equiv \frac{n_A}{n_A + n_B} \quad , \tag{5.11}$$

where n_A and n_B are the surface concentrations of atoms A and B, respectively. The sensitive layers of the sample, illustrated in Fig. 5.10 using the rigid sphere model, contribute to the total Auger signal of the element (A or B) according to the following expressions:

First layer: $I_A^{(1)} = I_0 S_A n_A \quad , \tag{5.12}$

where I_0 is the incident electron beam intensity, S_A is the sensitivity factor for

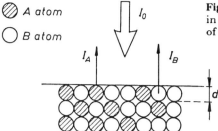

A atom

B atom

I_0

I_A I_B

d

Fig. 5.10. Schematic illustration of a binary alloy in the rigid sphere model [5.36]. The explanation of the symbols used is given in the text

the element A and n_A has the same meaning as in (5.11).

Second layer:
$$I_A^{(2)} = I_0 S_A n_A \exp\left(-\frac{d}{\lambda_A}\right) \ . \tag{5.13}$$

Here the exponential term describes the attenuation of the Auger signal caused by inelastic scattering in the layer of thickness d; λ is the mean free path of the element A Auger electrons.

Third layer:
$$I_A^{(3)} = I_0 S_A n_A \exp\left(-\frac{2d}{\lambda_A}\right) \ , \tag{5.14}$$

kth layer:
$$I_A^{(k)} = I_0 S_A n_A \exp\left(-\frac{(k-1)d}{\lambda_A}\right) \ . \tag{5.15}$$

Summing over all the layers, the total Auger signal of the element A may be determined by

$$I_A^\infty = I_0 S_A n_A \sum_{k=1}^\infty \exp\left(-\frac{(k-1)d}{\lambda_A}\right) \ . \tag{5.16}$$

This expression may be rewritten in the form

$$I_A^\infty = \frac{I_0 S_A n_A}{1 - \exp(-d/\lambda_A)} = I_0 S_A n_A G_A \ , \tag{5.17}$$

because it contains the sum of a geometrical progression.

With analogous symbols the Auger signal due to the element B is given by

$$I_B^\infty = \frac{I_0 S_B n_B}{1 - \exp(-d/\lambda_B)} = I_0 S_B n_B G_B \ . \tag{5.18}$$

Now one can replace the concentrations n_A and n_B in (5.11) by the relevant values determined from (5.17) and (5.18). The following expression then results for the composition of the compound:

$$c_A = \frac{R}{R + G_A/G_B} \ , \quad \text{where} \tag{5.19}$$

$$R = \frac{I_A^\infty/S_A}{I_B^\infty/S_B} \ , \quad \text{and} \quad \frac{G_A}{G_B} = \frac{1 - \exp(-d/\lambda_B)}{1 - \exp(-d/\lambda_A)} \ .$$

In the last equation, R is the weighted peak-to-peak height ratio between the Auger lines of the two elements, and the ratio G_A/G_B depends on the mean free paths of both the elements. If the Auger electron energy E_A of these electrons is not too different ($\Delta E_A = E_A^A - E_A^B \sim 0.30 E_A^A$) then the ratio G_A/G_B can be considered to be equal to unity. However, even in this simplified scheme an accuracy of typically about 20 % in determining the composition of the compound is achieved [5.36].

5.2.2 Sputter Depth Profiling

For the sputter depth-profiling analysis of thin films the ion gun is used to remove the surface at rates of up to 2 μm/h, but typically an order of magnitude less than this. For AES depth profiles the electron beam is placed in the middle of the ion beam crater and, if the system alignment is suitable this latter crater may be limited in size to 100 μm or less. For AES analysis at beam currents above 0.5 μA and spot sizes of a few micrometers, the AES analysis may be made whilst sputtering. However, at low beam currents, where electron counting techniques are required, the ion beam must sometimes be switched off during analysis since the ion beam itself generates secondary electrons which add large unwanted background to the spectra.

When presenting the most important problems related to AES sputter depth profiling we will follow the considerations of *Seah* [5.23]. The first problem is the quantification of the sputter depth. In determining the depth sputtered away by the ion beam impinging on the sample surface one may use laser interferometry [5.35] or other optical methods. However, if the initial surface is not smooth and flat, one must estimate the depth sputtered from the ion sputtering conditions.

If monoenergetic argon ions of current density J_i are used to sputter a target with a sputtering yield of S atoms per ion, the rate of removal is given by $\dot{Z}_i$ where

$$\dot{Z}_i = \frac{J_i S M}{e \varrho \mathcal{N}_A n} \quad . \tag{5.20}$$

Here M is the molecular weight of the material with n atoms per molecule, e the electronic charge, ϱ the density and $\mathcal{N}_A$ Avogadro's number. The dot over the sputtered depth Z means the time derivative. Thus, for a given material, the removal rate may be determined if J_i and S are known. Values of S for the pure elements, calculated according to the theory of *Sigmund* [5.37], are shown in Fig. 5.11. In (5.20) it is assumed that the ion beam is incident along the surface normal. If this is not the case the removal rate will change, but not strongly for angles below 45° since J_i reduces to $\cos \theta$ and is cancelled out by S increasing very roughly as $\sec \theta$ [5.38].

The next problem connected with AES depth profiling is the depth resolution ΔZ achieved in the profiling process. The convention used to define ΔZ is the depth over which the intensity from a step-function profile drops from 84 %

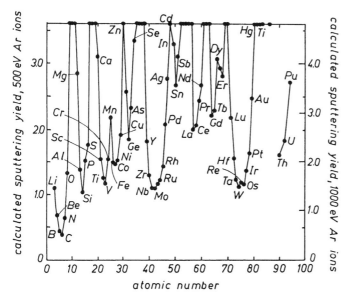

Fig. 5.11. Calculated pure element sputtering yields for argon ions [5.23]

to 16 % of its plateau value. This convention arises from the observation that many profiles resemble error functions and that these points represent the median position $\pm \sigma$ where σ is the standard deviation of the Gaussian generating the error function. A number of separate terms contribute to ΔZ and each of these terms should be added by means of a cross correlation integral with the real profile. In practice, for all but two terms this is equivalent to the quadratic sum approach [5.39]:

$$\Delta Z = \sqrt{\sum_i \Delta Z_i^2} \quad . \tag{5.21}$$

Several terms will be dealt with briefly below so that the reader may appreciate how the instrumental operating conditions may be optimized and how meaningful the measured profile will be.

The principal limitation in many experimental systems is the flatness of the ion beam sputtered crater over the area of analysis. This produces an interface broadening, ΔZ_i, proportional to the depth Z etched:

$$\Delta Z_i = a_i Z \quad , \tag{5.22}$$

where, on the best-arranged AES instruments, a_i may be in the range 0.1 %– 1 %, and on routinely used instruments 5 % or worse may be found. In any experimental arrangement, reducing the analyzed area will reduce a_i to an extent at worst proportional to the analyzed diameter. For AES instruments with electron beam diameters below 1 μm the crater shape should not limit the depth resolution, however, electron stimulated deposition may then become a problem. For AES,

173

electron stimulated deposition (ESD) [5.40] causes a limitation on the minimum area that may be analyzed. As a result of ESD the electron beam will create its own mini-crater within the ion beam crater. If the ion beam sputter rate is $\dot{Z}$ and if the electron beam has a Gaussian profile of characteristic σ_e, the interface blurring arising from ESD may be shown to be ΔZ_e, where

$$\Delta Z_e = \frac{I_e Q 10^{-9}}{34 \dot{Z} \sigma_e^2 e} Z \quad , \tag{5.23}$$

where I_e is the electron beam current, Q is the target atom deposition cross section and e is the electronic charge. It is instructive to put some figures into this relation to determine the effect in practice. For stable oxides Q is typically $10^{-24} \, \text{cm}^{-2}$, for a reasonable signal-to-noise ratio in the peak detection I_e may be greater than $0.2 \, \mu\text{A}$ and a typical sputter rate may be 100 nm in 1–2 h. Thus

$$\Delta Z_e = \frac{2Z}{\sigma_e^2} \quad , \tag{5.24}$$

where σ_e is expressed in micrometers. For a 1 % depth resolution this would limit the electron beam FWHM to 34 μm. For many inorganic compounds the Q values are 10–100 times higher than for the stable oxides [5.39], whereas for metallic layers much smaller Q values are appropriate.

Atomic mixing occurring during depth profiling is the next problem to be considered. The cascade mixing contribution, ΔZ_c, at steady state has been calculated for a range of elements for argon ion sputtering in the energy range 1–10 keV by *Andersen* [5.41]. ΔZ_c increases as the square root of ion energy and, for 1 keV ions incident along the surface normal, is in the range 2–4 nm. Due to the random scattering involved it is unlikely that this value will be greatly reduced for nonnormal incidence in the range of angles commonly suitable for AES or XPS. The development of ΔZ_c with Z is shown by the calculations of *Littmark* and *Hofer* [5.42]. At small depths Z, the random diffusion scattering gives ΔZ_c proportional to the square root of time, and hence

$$\Delta Z_c = a_c \sqrt{Z} \tag{5.25}$$

until Z exceeds the projected range of the primary ion and its recoils. At depths between 1 and 10 times the ΔZ_c value given above, the value of ΔZ_c saturates. The effect of atomic mixing does not give a Gaussian broadening term but a skewed function with a long tail. This, to some extent, distinguishes the effect of atomic mixing from other terms.

Roughening of the surface caused by sputtering and roughness of the outer surface are the last problem which will be discussed here [5.23]. Roughening of the surface by ion sputtering has long been thought to degrade depth profiles. At very small depths, such as the first few atom layers, a slight roughening occurs due to the statistical effect of removing atoms [5.43]. More complex calculations of this effect show that the statistical effect saturates at a blurring term of only 1 nm or so [5.44]. In practice a different type of roughness appears to develop, whose magnitude is given by

$$\Delta Z_r = a_r \sqrt{Z} \quad , \tag{5.26}$$

arising from random events. Empirically it is found that a_r for polycrystalline metals is 0.86 ± 0.22 [5.39], where ΔZ and Z are expressed in nanometers.

In many instances roughness of the outer surface may not affect the depth profiling resolution, however, in other cases it can destroy the chance of a meaningful measurement. The situations in which a good result is obtained are those in which an even layer is grown or deposited on the rough surface. If the thickness of this layer is less than the wavelength of the roughness, the surface roughness may have a minimal effect, especially for near normal incidence sputtering [5.23].

5.3 X-Ray Diffraction

X-ray diffraction is a method of investigating the fine structure of material. The first experiments performed by Laue in 1912 revealed the structure of bulk crystals. Nowadays this method can be used to determine not only the structures of epilayers but also the composition of a ternary compound or the stress in heteroepitaxial films.

Diffraction is due essentially to the existence of certain phase relations. It is well known that two rays are completely in phase whenever their path lengths differ by either zero or a whole number of wavelengths. Differences in the path length of various rays arise quite naturally when we consider how a crystal diffracts x-rays. Figure 5.12 shows the situation where a beam of parallel and monochromatic x-rays of wavelength λ is incident on a crystal at an angle θ_B, the so-called Bragg angle, which is measured between the direction of the incident beam and the crystal plane under consideration. The interaction with the first of these planes will produce a reflected component at the specular reflection angle θ_B, which will be rather weak if the x-rays are deeply penetrating. Since this first plane has a periodic structure, it obviously acts as a two dimensional grating for the x-rays. For the second and all subsequent planes, there will be

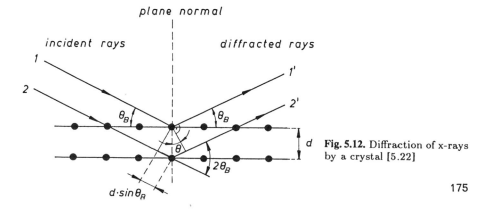

Fig. 5.12. Diffraction of x-rays by a crystal [5.22]

similar components of reflected energy at the specular angle θ_B. It can be seen from Fig. 5.12, that the additional path of ray 2 compared to ray 1 is $2d \sin \theta_B$. Consequently all reflected components can interfere constructively in phase if this distance is a multiple of the wavelength. This condition for efficient specular reflection is called Bragg's law [5.22]

$$2d \sin \theta_B = n\lambda \quad , \tag{5.27}$$

where n is an integer. Since the atomic arrangement is the same in all the planes under consideration, the Bragg diffraction condition depends on the spacing of the planes, but is independent of the atomic arrangement within each plane. Since $\sin \theta_B$ cannot exceed unity, the basic condition $n\lambda < 2d$ must be satisfied to obtain any diffraction. The spacing of planes is of the order of 3 Å or less, which means that λ cannot exceed about 6 Å. On the other hand, if λ is very much smaller than d, the diffraction angles θ_B are too small to be conveniently measured.

Experimentally, Bragg's law can be utilized in two ways. By using x-rays of known wavelength λ and measuring θ_B, one can determine the spacing d of the planes in a crystal. This is structure analysis and is mainly used to investigate the crystalline quality of the epitaxial layers. Alternatively, a crystal with planes of known spacing d can be used at a fixed reflection angle θ to a fixed wavelength λ of the x-ray source. This effect is used to construct filters to obtain perfectly monochromatic x-ray beams.

All the considerations mentioned above can be transformed from real space to reciprocal space as described in Sect. 4.1.1 for electron diffraction. So we will not repeat the theory in this chapter and refer to the literature for more detailed descriptions [5.22, 45, 46].

5.3.1 Diffraction Under Nonideal Conditions

So far we have assumed in our derivation certain ideal conditions, namely, a perfect and infinite crystal and a perfectly parallel and strictly monochromatic x-ray beam. Particularly the model of the infinite crystal cannot be used for thin epitaxial layers and we must consider the effect of finiteness on diffraction, which results in line broadening of the diffraction spectra [5.46].

If the path difference between rays scattered by the first two planes differs only slightly from an integer number of wavelengths, then the plane scattering a ray exactly out of phase with the ray in the first plane will lie deep within the layer. If the layer is so thin that this plane does not exist, then complete cancelation of all the scattered rays will not result. Let us assume a layer of thickness D_{hkl} consisting of $(m + 1)$ planes (Fig. 5.13a). We will take the angle θ as a variable and call the angle which exactly satisfies the Bragg equation θ_B. Then the rays $1', 2', \ldots, M'$ are completely in phase and interfere to form a diffracted beam at maximum amplitude.

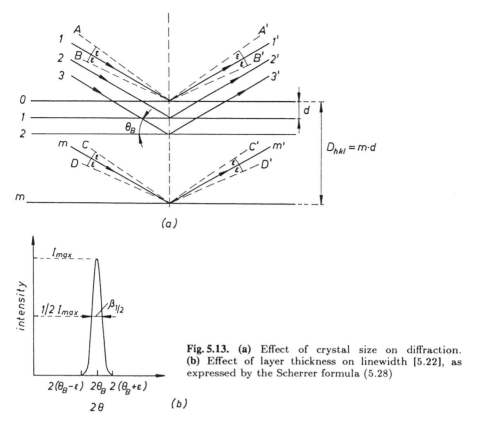

Fig. 5.13. (a) Effect of crystal size on diffraction. (b) Effect of layer thickness on linewidth [5.22], as expressed by the Scherrer formula (5.28)

Let us now consider rays which deviate by a small angle ε from the Bragg angle θ_B, like A and B shown in Fig. 5.13a. Ray A makes a slightly larger angle $(\theta_B + \varepsilon)$, such that ray C' from the mth plane below the surface is $(m + 1)$ wavelengths out of phase with A'. This means that midway in the crystal there is a plane scattering a ray with a phaseshift of $(m + \frac{1}{2})$ wavelengths with respect to ray A' form the surface plane. These rays cancel one another, with the net effect that all rays scattered by the top half of the crystal annul those scattered by the bottom half. The intensity of the diffracted beam at $2(\theta_B + \varepsilon)$ is therefore zero.

The same arguments also hold for the rays D' and B' at an angle $(\theta_B - \varepsilon)$. Consequently the diffracted beam intensity is not zero between the angles $(\theta_B - \varepsilon)$ and $(\theta_B + \varepsilon)$ and has a maximum value at $2\theta_B$. A typical line shape with a width $\beta_{1/2}$ at an intensity equal to half of the maximum intensity is shown in Fig. 5.13b. An exact treatment of the problem gives the following relation between linewidth (in radians) and crystal size:

$$\beta_{1/2} = \frac{0.9\lambda}{D_{hkl} \cos \theta_B} \quad , \tag{5.28}$$

which is known as the Scherrer formula [5.46]. It is evident that the width of the diffraction curve increases as the thickness of the crystal decreases. For example, if a GaAs layer contains only 500 planes the linewidth of the (400) reflection with Cu K_{α_1} radiation ($\theta = 33°$, $a_0 = 5.653$ Å, $\lambda = 1.540$ Å) would be relatively broad, namely 5×10^{-4} rad = 120″.

Of the many kinds of crystal imperfections, the one which the crystal grower always has to face is strain in the epitaxial layers caused by lattice mismatch between the substrate and single-crystalline film. Especially in superlattices which consist of layers with different lattice constants, this effect becomes dominant and changes the diffraction spectrum. The change in shape of a single layer in the superlattice is determined not only by the forces applied to the layer, but also by the fact that each layer retains contact on its two boundaries with the neighboring layers. Because of this interaction between the layers, a single layer in a superlattice structure is not free to deform in the same way as an isolated layer on a thick substrate. The strain in an isolated layer may result in curvature of the substrate plus film and consequently in a nonuniform stress distribution within the film. On the other hand, in a superlattice the forces causing strain are symmetric from the top and the bottom, resulting in a uniform strain throughout the layer.

The effect of both uniform and nonuniform strain on the x-ray diffraction spectrum is illustrated in Fig. 5.14. The situation of an unstrained layer with a lattice constant d_0 is shown in Fig. 5.14a. If the layer is exposed to a given tensile strain parallel to the lattice planes, then the spacing between the lattice planes in the direction perpendicular to the layer surface decreases, $d_\perp < d_0$, whereas the lattice constant within the planes increases, $d_\parallel > d_0$. If the observed x-ray diffraction is symmetric with respect to the surface, the reflection angle θ_B is determined only by the layer spacing $d_\perp$, and the corresponding diffraction line shifts to lower angles but does not change otherwise (Fig. 5.14b). In general both

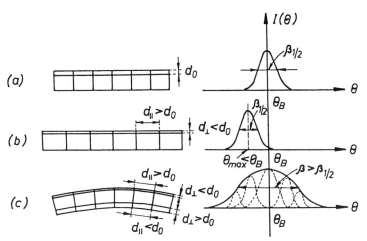

Fig. 5.14. Effect of lattic strain on line position and linewidth: (a) unstrained, (b) uniform strain, (c) nonuniform strain [5.22]

components ($d_{\parallel}$ and $d_{\perp}$) are necessary to characterize the strain within the layer. The mismatch components ($\Delta d_{\perp}/d_0$) and ($\Delta d_{\parallel}/d_0$) can be determined by using x-ray diffraction of lattice planes making an angle ϕ with the crystal surface [5.47]. The x-ray spectrum then contains the diffraction peaks of the layer and the substrate.

The strain in the layer results in a difference $\Delta\phi$ of the inclination angle ϕ and the reflection angle θ is changed by $\Delta\theta$ because of the different lattice spacing in the layer (Fig. 5.15). The angular separation between the diffracted peaks from the substrate and the layer are equal to ($\Delta\theta + \Delta\phi$) or ($\Delta\theta - \Delta\phi$), when the x-ray reflections from the substrate appear at the angle ($\theta + \phi$) or ($\theta - \phi$) (Fig. 5.15).

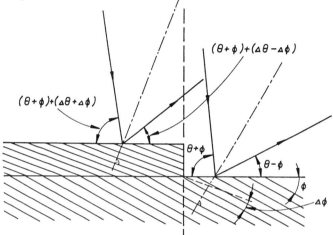

Fig. 5.15. X-ray diffraction of lattice planes making an angle ϕ with the crystal surface [5.47]

The resulting values of $\Delta\theta$ and $\Delta\phi$ can be used to calculate ($\Delta d_{\perp}/d_0$) and ($\Delta d_{\parallel}/d_0$) according to [5.48, 49]

$$\frac{\Delta d_{\perp}}{d_0} = \Delta\phi \tan\phi - \Delta\theta \cot\theta \quad ,$$

$$\frac{\Delta d_{\parallel}}{d_0} = -\Delta\phi \cot\phi - \Delta\theta \cot\theta \quad . \tag{5.29}$$

A difference between ($\Delta d_{\perp}/d_0$) and ($\Delta d_{\parallel}/d_0$) is related to a deformed unit cell. To determine, for example, the composition of the layer from the lattice constant, one wishes to know the lattice constant of the undeformed unit cell of the layer as compared to its substrate. The value ($\Delta d/d_0$)$_{\text{relaxed}}$, when the layer is in the relaxed state, can be calculated from elasticity theory. A general model which describes the unit cell deformation was given by *Hornstra* and *Bartels* [5.47]. The relaxed lattice mismatch can be obtained after taking the anisotropic elastic deformation of the crystal lattice of the layer into account:

$$\left(\frac{\Delta d}{d}\right)_{\text{relaxed}} = \frac{\varepsilon_{\parallel}}{\varepsilon_{\parallel} - \varepsilon_{\perp}} \left[\left(\frac{\Delta d_{\perp}}{d_0}\right) - \left(\frac{\Delta d_{\parallel}}{d_0}\right) \right] + \left(\frac{\Delta d_{\parallel}}{d_0}\right) , \qquad (5.30)$$

where $\varepsilon_{\parallel}$ and $\varepsilon_{\perp}$ are the parallel and perpendicular strains of the layer. The ratio $\varepsilon_{\parallel}/(\varepsilon_{\parallel} - \varepsilon_{\perp})$ depends on the orientation of the crystal lattice and can be expressed in terms of the elastic constants of the material. Examples for strained IV–VI compound superlattices are given by *Fantner* et al. [5.50, 51].

In the case of a single layer on a substrate the layer is bent by the curvature of the substrate, and consequently the strain is nonuniform. On the top side of the plane the spacing $d_{\perp}$ is smaller than on the bottom side, whereas the spacing $d_{\parallel}$ is increased on the top side compared to the bottom. We may consider this layer to be composed of a number of smaller layers, each with substantially constant plane spacing inside but different from the spacing in adjoining layers. These individual layers cause the narrow dotted lines of Fig. 5.14c. But in the experiment the sum of these sharp lines is observed, which is illustrated by the broadened diffraction line shown by the full curve in Fig. 5.14c.

Generally speaking, we can say that the position of the diffraction peak is determined by the lattice constant d_0 and the linewidth is a measure of the variation Δd of the lattice constant within the sample volume tested by the x-rays. This statement is only valid if the monochromaticity of the x-ray source $\Delta\lambda$ and the divergence of the beam $\Delta\theta$ are small enough that the influence of the variation of the lattice constant Δd can be resolved. If this condition is satisfied, the shift of the diffraction peak can be used to calculate the homogeneous change of the lattice constant due to changes in composition or homogeneous stress. On the other hand the line broadening of the diffraction peak can be taken as a quality criterion for the homogeneity of the crystal lattice, and furthermore a certain line shape can be modeled by a distribution of composition or strain within the layer.

5.3.2 High Resolution X-Ray Diffraction

Variations in the x-ray wavelength λ and the angle of incidence θ determine the sensitivity for resolving diffraction peaks due to small variations in the lattice constant. By differentiation of the Bragg equation (5.27) the following relation is obtained:

$$\frac{\Delta d}{d} = \frac{\Delta\lambda}{\lambda} - \frac{\Delta\theta}{\tan\theta} . \qquad (5.31)$$

A sensitivity scale with some characteristic values determining x-ray diffraction is given in Fig. 5.16. Even when the Cu K_{α_2} radiation of the doublet has been removed with a narrow collimator, we still have the problem of the natural linewidth of the Cu K_{α_1} x-ray line. However, the reflection width of a perfect crystal allows the resolution of lattice constant variations down to 10^{-5}. This lower limit is determined by the interference of scattered waves, which is the basis of x-ray diffraction. The lattice perfection can be even better than 10^{-5}

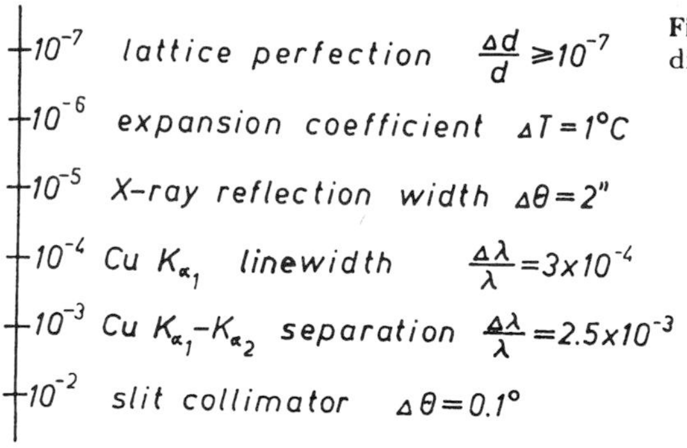

-10^{-7} *lattice perfection* $\frac{\Delta d}{d} \geq 10^{-7}$

-10^{-6} *expansion coefficient* $\Delta T = 1°C$

-10^{-5} *X-ray reflection width* $\Delta \theta = 2''$

-10^{-4} *Cu K_{α_1} linewidth* $\frac{\Delta\lambda}{\lambda} = 3 \times 10^{-4}$

-10^{-3} *Cu $K_{\alpha_1} - K_{\alpha_2}$ separation* $\frac{\Delta\lambda}{\lambda} = 2.5 \times 10^{-3}$

-10^{-2} *slit collimator* $\Delta\theta = 0.1°$

Fig. 5.16. Sensitivity scale of x-ray diffraction [5.52]

and also small temperature variations do not interfere with observations in the indicated range of interest.

A highly parallel and monochromatic incident x-ray beam of sufficient intensity is required for investigating layer structures on semiconductor single crystals. This problem is partly solved by using crystal-collimated radiation in a double-crystal diffractometer [5.53]. However, with such an instrument lattice constants can be measured only on a relative scale using the substrate diffraction peak as a reference.

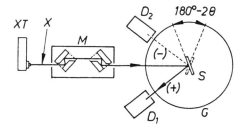

Fig. 5.17. Schematic arrangement of the high-resolution diffractometer. XT: x-ray tube; X: x-ray beam; M: monochromator with fourfold diffraction on germanium crystals; G: goniometer carrying the detectors D_1 and D_2 and the sample S. The two positions of the sample are necessary for measuring absolute lattice constants [5.54]

The x-ray diffractometer developed at Philips Research Laboratories has overcome this disadvantage [5.48]. The main components of the instrument (Fig. 5.17) are the x-ray tube, the monochromator, and a goniometer carrying two detectors and the sample holder with the crystal under investigation. The extremely compact monochromator (about 15 cm long) consists of two U-shaped blocks made from dislocation-free germanium single crystals. The fourfold diffraction at the (440) planes of the germanium crystals results in an x-ray beam with a wavelength spread $\Delta\lambda/\lambda$ of 2.3×10^{-5} and a divergence $\Delta\theta$ of $5'' = 2.4 \times 10^{-5}$ rad. With this arrangement, diffraction measurements can be performed for each lattice plane of any given material in any direction of diffraction, so that absolute measurements of the lattice constant can also be made.

A special feature of the monochromator setup is the possibility of "tuning" (Fig. 5.18). When the two blocks are rotated in opposite directions, the angle θ changes but the emergent beam does not change in direction or position. This enables the pass band of the monochromator to be tuned to the different wavelengths of x-ray tubes with different anode materials.

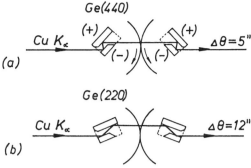

Fig. 5.18a,b. Tuning the monochromator. (a) The (440) reflection of the CuK_{α_1} line has a very small divergence. (b) By rotating the two crystal blocks it is also possible to use the (220) reflection. This reflection has 30 times the intensity, but greater divergence. The monochromator can be tuned to a different wavelength and hence adapted to an x-ray tube with a different anode [5.54]

Another feature is that the intensity of the emergent beam can be increased, although at the expense of the collimation (and hence the resolution). Instead of using the (440) diffraction in the germanium crystals (Fig. 5.18a), diffraction at the (220) planes can also be used (Fig. 5.18b). If the (220) plane diffraction is used, the intensity of the emergent beam is 30 times higher. The divergence, however, then increases from $\Delta\theta = 5''$ for the (440) reflection to $\Delta\theta = 12''$ for the (220) reflection, so that the resolution is rather less [5.54].

5.3.3 X-Ray Diffraction at Multilayers and Superlattices

The periodicity of multilayer structures results in a similar effect in x-ray diffraction, as the periodicity of the crystal lattice is reflected in the Bragg equation. The diffraction from a multilayer structure is therefore modulated and exhibits a well-defined satellite structure. Consequently, the superlattice period Λ can be obtained from the positions of the satellite peaks (θ_{SL}) of order n according to the equation [5.55]

$$\frac{2\sin\theta_n - 2\sin\theta_{SL}}{\lambda} = \pm\frac{n}{\Lambda} \quad . \tag{5.32}$$

The period Λ of the superlattice unit cell is given by

$$\Lambda = \frac{a_W N_W + a_B N_B}{2} \quad , \tag{5.33}$$

where a_W and a_B are the lattice parameters along the growth direction and N_W and N_B are the corresponding numbers of layers. The thickness of the well and the barrier are then given by $t_W = a_W N_W/2$ and $t_B = a_B N_B/2$. However, this description can only be used for superlattices with very small lattice mismatch between the layers and with ideally shaped interfaces, as for example InGaAs/InP

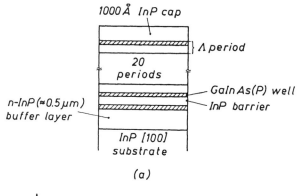

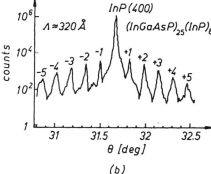

Fig. 5.19. (a) Schematic of a GaInAs(P)/InP multi-quantum-well structure. (b) X-ray diffraction scan of the InP(400) reflection of an encapsulated quaternary InGaAsP/InP multilayer with period $\Lambda = 320\,\text{Å}$ and zero mismatch [5.57]

superlattices [5.55–57]. Figure 5.19 shows a schematic of a multi-quantum-well structure and the obtained x-ray diffraction spectrum.

For a detailed understanding of the x-ray diffraction phenomena of multilayers and superlattices it is necessary to elaborate the dynamical theory for x-ray diffraction. The first theoretical description of the characteristic asymmetric diffraction profile of an absorbing perfect crystal was given by *James* [5.58]. The x-ray diffraction theory for crystals with a strain gradient perpendicular to the surface was developed by *Takagi* [5.59] and *Taupin* [5.60]. They described the diffraction of curved crystals and calculated the contrast of dislocations visible in x-ray topography. This problem became very important starting from the time when crystals were modified by epitaxy, ion implantation and heavy diffusion. Ideal superlattices were described by *Vardanyan* et al. [5.61] with Chebyshev polynomials, whereas *Speriosu* and *Vreeland* [5.62] used a geometric series of the kinematic theory. A general solution for the dynamical reflection of an epitaxial layer of arbitrary thickness can be found in [5.63].

A kinematical step model was introduced by *Segmüller* and *Blakeslee* [5.64] and is now used to analyze x-ray scans with sharp interfaces. Figure 5.20 shows the x-ray diffraction scan of InGaAs/InP multilayers together with the simulation calculated from the step model.

A semi-kinematical approximation based on Taupin's differential equation was used by *Tapfer* and *Ploog* [5.65] to model the observed diffraction curves.

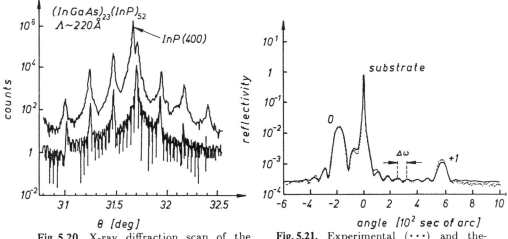

Fig. 5.20. X-ray diffraction scan of the InP(400) reflection of an encapsulated ternary InGaAs/InP multilayer with period $\Lambda = 220\,\text{Å}$. The bottom part shows the simulation of the fit calculated from the step model [5.57]

Fig. 5.21. Experimental ($\cdots$) and theoretical (—) diffraction curve of an AlAs/GaAs superlattice in the vicinity of the (004) reflection [5.65]

This approximation is valid if the thickness of the deformed layer is small compared with the x-ray extinction length. The reflectivity is then given in integral form, which reduces the computation time for a diffraction curve. The excellent agreement between theory and experiment can be seen in Fig. 5.21, where the diffraction spectrum of an AlAs/GaAs superlattice is depicted together with the theoretical curve.

5.4 Photoluminescence

Photoluminescence is one of the most sensitive nondestructive methods of analyzing both intrinsic and extrinsic semiconductor properties. The method is particularly suitable for characterization of the centers responsible for shallow donor and acceptor species which control the electrical properties of the layers, as well as near-band-gap luminescence. It can also be applied to certain deep states (Sect. 6.5.1), provided that the carrier exchange with these centers does not involve so much lattice relaxation that the associated transitions become nonradiative. In most cases, measurements at low temperatures (below 77 K) are necessary to obtain the most complete spectroscopic information, which is usually needed to characterize a given type of transition and to discriminate between species within a given class. There are three major types of impurity-related recombinations: bound excitons (BE), donor acceptor (DA) pairs, or free to bound transitions [in most cases only conduction band to acceptor (CA) can be

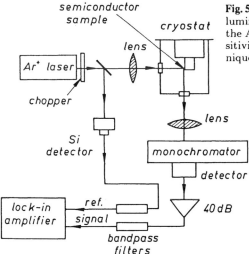

Fig. 5.22. Experimental setup used for photoluminescence measurements. The output of the Ar$^+$ laser is chopped to increase the sensitivity of the apparatus by a lock-in technique

seen]. Photoluminescence has the advantage of the ability to discriminate between species involved in recombinations and can provide simultaneous information on many types of centers. However, compared to elementary techniques of electrical characterization, it is less well suited for the determination of impurity concentrations, particularly for the electrical majority species.

Photoluminescence is in general a convenient technique, requiring in the simplest form a suitable source of optical excitation, a spectrometer and a detector suitable for the emitted light. A schematic drawing of a typical experimental setup is shown in Fig. 5.22. In particular, basic assessment of a semiconductor usually requires measurements at low temperatures, so the sample must be placed in a cryostat. Free suspension, strain-free sample mounting techniques are essential for most detailed studies [5.66]. In this case the samples can float in He, which should be in the superfluid state to avoid additional light scattering by bubbles of the cooling liquid. Low temperatures are necessary for two principal reasons. First, specific information about the centers which promote electrical conductivity, the donors and acceptors, can be obtained only when the electrical carriers are frozen out in these centers. Once the electrical carriers are thermally liberated, the impurities or defects which released them reveal their presence only through some inhibition of carrier mobility. The second advantage of low temperature is the dramatic reduction of spectral broadening due to vibronic processes.

5.4.1 Photoluminescence in Binary Compounds

The major features that appear in the photoluminescence spectrum are shown schematically in Fig. 5.23. In order of decreasing energy the following structures are usually observed: CV, transitions from the conduction band to the valence band; X, free excitons; D^{0}X and A^{0}X, excitons bound to neutral donors or

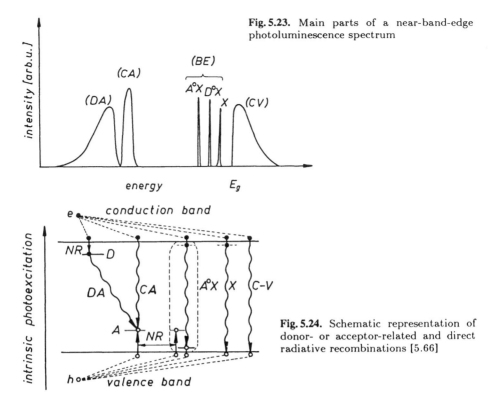

Fig. 5.23. Main parts of a near-band-edge photoluminescence spectrum

Fig. 5.24. Schematic representation of donor- or acceptor-related and direct radiative recombinations [5.66]

acceptors, respectively; CA, transition from conduction band to an acceptor; DA, transition from donor to acceptor. All these recombination processes are represented schematically in Fig. 5.24 [5.66]. In the following we will discuss each type of transition separately and point out what can be learned from the structure concerned.

The transitions from the conduction band to the valence band are usually observed at temperatures well above 100 K. The line shape is structureless and given by an effective density of states term ($\sqrt{h\nu - E_g}$) multiplied by a Boltzmann filling factor:

$$I_{\mathrm{CV}} \propto \sqrt{h\nu - E_g}\, \exp\left(-\frac{h\nu - E_g}{k_B T}\right) \quad ; \tag{5.34}$$

E_g is the energy gap of the semiconductor, $h\nu$ is the energy of the luminescence light and $k_B T$ has its usual meaning. It can be seen from (5.34) that the low-energy edge of the structure gives the energy gap of the semiconductor, and from the slope of the high-energy part, the temperature of the carriers can be obtained. At temperatures above 10 K the electron temperature coincides with the lattice temperature, so the actual sample temperature can also be evaluated from the CV transitions [5.67].

Recombination of electrons and holes through free (X) and bound excitons (BEs) are the first processes to appear in the near-gap luminescence spectra as the donor or acceptor concentrations are increased from very low levels, and particularly for slightly compensated material. A prominent type of recombination process involves the annihilation of an exciton bound to either of these species in their neutral charge state. The sensitivity of these sharp photoluminescence components to local strains in the sample may be exploited to reveal macroscopic built-in stress [5.68]. Consequently, the linewidth of the (A^0X and D^0X) luminescence can be taken as a quality criterion for the epitaxial layer, as is widely accepted for CdTe/GaAs heterostructures [5.69].

Macroscopic stress may be detected from its contribution to the spectral linewidth when stress effects dominate in the broadening, rather than electric field processes which are frequently important in dilutely doped and compensated layers [5.70]. The detailed properties of these states were first extensively studied in CdS [5.71]. In this case Zeeman spectroscopy was essential to distinguish between spectra involving donor and acceptor species [5.72].

Satellite spectra "two-electron" and "two-hole" transitions may be considered, when the A^0X and D^0X states are inadequately resolved. In this cases the following transitions are observed:

$$D^0X \to D^0(2p) + h\nu_{em} \quad \text{or} \quad A^0X \to A^0(2p) + h\nu_{em} \quad ,$$

where $D^0(2p)$ and $A^0(2p)$ are excited donors and acceptors and $h\nu_{em}$ is the energy of the emitted photons. These lines appear below the principal bound exciton transitions separated by energies which carry almost the entire chemical shift of the impurity ground state. These satellite transitions are the most suitable for providing information on the donor and acceptor centers, including their basic binding energies derived from the analysis of the relative energies of several excited states [5.73]. This then allows a given transition to be assigned to a given chemical dopant.

The recombination process from the conduction band to an acceptor state (CA) is a typical example of a so-called free-to-bound transition. It can be seen from Fig. 5.24 that the low-energy threshold of this CA transition can be described as in (5.34), but E_g must be replaced by $E_g - E_A$:

$$h\nu_{min}(CA) = E_g - E_A \quad , \tag{5.35}$$

where E_A is the position of the energy of the acceptor above the valence band edge. The CA band is mainly broadened by the kinetic energy of the electrons before the recombination. So, the study of the line shape as a function of temperature and excitation energy provides detailed information on energy relaxation processes for the hot electrons. However, from an analytical point of view, we are interested in the fact that these CA transitions provide direct spectroscopic information on the acceptor species present in the semiconductor layer. It is advantageous to record spectra at a temperature at which donors are already ionized, so that the DA transition is fully quenched in favor of the CA transition and just a single band appears for each acceptor species. On the other hand, the temperature and the excitation rate must be sufficiently low that the CA luminescence peaks remain as narrow as possible [5.74]. The CA luminescence is often used in conjunction with the "two-hole" satellite transition technique to distinguish the shallow acceptors in GaAs or InP [5.75]. Figure 5.25 shows the energy position of the CA tansitions and the "two-hole" satellites for different acceptors in GaAs. For comparison the corresponding A^0X transitions are marked on the right side of Fig. 5.25. It is evident that it is difficult to distinguish between the different acceptor species by A^0X transition, whereas the CA transitions together with the "two-hole" satellites exclude any misinterpretation.

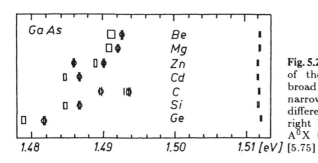

Fig. 5.25. Schematic representation of the energy positions for the broad CA transitions (□) and the narrow "two-hole" satellites (⧫) for different acceptors in GaAs. On the right hand side the corresponding A^0X transitions (|||) are indicated

The DA recombination processes compete strongly with the A^0X and D^0X transitions, especially when the composition ratio is high and particularly when the concentration of donor and acceptor species is increased. A detailed description of these processes was given by *Dean* [5.76]. In this chapter we have to restrict ourself to some basic features. The electrostatic interaction within the ionized donor-acceptor pair after the transition, as illustrated in Fig. 5.24, is responsible for the Coulomb term in the expression for the transition energy [5.66]:

$$h\nu_{em}(DA) = E_g - (E_A + E_D) + \frac{e^2}{\varepsilon r} - E_{VdW} \quad , \tag{5.36}$$

where e is the elementary charge, r the separation of the donor-acceptor pair and E_{VdW} a correction term due to polarization interaction in the initial state of the transition. To first order, the term $e^2/\varepsilon r$ in (5.36) is responsible for the spectral dispersion into a very large number of discrete zero-phonon lines. Each line is associated with a different discrete value of r allowed by the lattice structure which defines the relative position of the donor and acceptor sites. But this discrete line structure is in most cases smeared out if lattice relaxation accompanies the electronic transition, which is mostly true for II–VI compounds when deep acceptors are involved [5.77]. The discrete DA lines may also be lost if E_D is so small that only recombinations at relatively distant donor-acceptor pairs give bound states according to (5.36). Such a case occurs in even well-refined semiconductor crystals of GaAs and InP, where E_A is modest but E_D is very small.

188

Table 5.3. Near-band-gap emission features of undoped MBE GaAs at 1.6 K [5.75, 80–82]

Energy [eV]	Assignment
1.5156	Free exciton (X)
1.5144	Excited states of (D^0, X)
1.5140	Exciton bound to neutral donor (D^0, X)
1.5134	Exciton bound to ionized donor (D^+, X)
1.5124	Exciton (doublet) bound to neutral carbon
1.5122	acceptor (A^0, X)
1.5110	Exciton bound to neutral Ga-site defect
1.5095	15 sharp lines (d, X) attributed to excitons bound to neutral
1.5049	acceptor-like point defects
1.4939	Two-hole transition (doublet) of exciton bound to neutral C
1.4937	acceptor $(C^0, x, 2h)$
1.4935	Conduction band to neutral C acceptor (e, C^0)
1.4915	Conduction band to neutral Be acceptor (e, Be^0)
1.4900	Donor to C acceptor (D^0, C^0)
1.4850	Conduction band to neutral Si acceptor (e, Si^0_{As})
1.4830	Donor to Si acceptor (D^0, Si^0_{As})
1.4790	Conduction band to neutral Ge acceptor (e, Ge^0_{As})
1.4770	Donor to Ge acceptor (D^0, Ge^0_{As})
~ 1.44	Si atom + Ga vacancy complex (?)
1.4065	Conduction band to neutral Mn acceptor (e, Mn^0_{Ga})
1.4046	Donor to Mn acceptor (D^0, Mn^0_{Ga})

In principle, a study of the ratio of the DA and the CA luminescence intensity can be used to determine the compensation rate in the semiconductor, i.e., the ratio of donors to acceptors, N_D/N_A. This method is not very accurate but has been successful in the case of C acceptors in n-type GaAs [5.78].

To increase the resolution, excitation spectroscopy is used. In this case the spectrum of excitation for a given luminescence line is studied. That means, a selective absorption is measured to investigate the line structures. This method takes advantage of the fine tuning of a dye laser, and the resolution of the spectroscopy is governed by the linewidth of the excitation light, which can be as small as 0.01 meV. If many dopants are present in a material, which causes overlap in a normal luminescence spectrum, excitation spectroscopy is useful to resolve the structure [5.79].

A summary of the near-band-gap emission features of undoped MBE-grown GaAs is given in Table 5.3 [5.75, 80–82]. A large variety of high resolution photoluminescence spectra for doped and undoped GaAs layers can be found in the review article of *Ilegems* [5.83].

5.4.2 Photoluminescence in Ternary and Quaternary Compounds

In photoluminescence of ternary and quaternary compounds, in general two lines are observed. One is due to bound excitons (BEs), and the other represents the sum of CA transitions and DA emissions. Compared to binary compounds, additional broadening is usually observed because of random fluctuations of the alloy composition or strain effects which can be temperature dependent. To associate the lines with the corresponding transitions the following dependence of the line intensity on excitation intensities can be used: The CA transition intensity is linear with the excitation intensity, while the BE line is superlinear, since two particles are involved in this process. At higher intensity of excitation also a slight shift of the CA line towards higher energies can be observed due to the saturation of the donor-acceptor pair emission while the position of the BE line stays unchanged.

Provided that the identity of the transition is well established, the composition of the alloy can be determined from the line position of the bound excitons. Quite recently this method has been used to solve a problem with $Al_xGa_{1-x}As$ mixed crystals involving the accuracy of the Al content determination and thus the band-gap evolution with composition [5.84]. Until recently the value of $dE_g^\Gamma(x)/dx$ =1.247 eV for the change of the direct gap of $Al_xGa_{1-x}As$ with composition, reported by *Dingle* et al. [5.85], was generally accepted [5.86]. In several recent papers this value has been strongly contested. Independent groups reported linear slopes ranging from 1.34 eV [5.87] to as much as 1.455 eV [5.88]. The value of the slope was even found to be slightly decreasing with temperature increase [5.89]. The key reason for the revision of the well-accepted value of 1.25 eV came from a critique of the exact determination of the Al content. It was pointed out by *Kuech* et al. [5.88] that the sputter Auger electron spectroscopy used by *Dingle* et al. [5.85] leads to Al enrichment of the surface layer and thus to overestimation of the Al content. Quite similar problems are met with the use of the electron microprobe technique. *Kuech* et al. [5.88] used a nondestructive nuclear profiling to determine the x value. The other authors relied on either precise lattice constant measurements [5.87, 89] or very careful electron microprobe analysis [5.90].

The main difference between emission processes in high-purity binary and ternary semiconductors is the occurrence of an additional emission broadening mechanism in the latter due to statistical fluctuations of the composition [5.91]. They lead to a fluctuation of the band gap, and thus to a broadening of all electron-hole recombination processes. The magnitude of the fluctuation and the resulting emission linewidth increases with the localization of the quasi-particles involved in the recombination. The linewidth of the BE emission is governed by the effective exciton volume, while for CA emission it is the hole bound to the acceptor whose effective radius governs the linewidth. For BE emission the variation of the energy gap composition (i.e., dE_g/dx) counts. In the latter case only the fluctuations of the absolute energy of the valence band (i.e., dE_v/dx) do. This is so because the mean free path of the electron recombining with a hole bound at an acceptor is much larger than the Bohr radius of the hole.

The composition-fluctuation-limited linewidth Γ equals

$$\Gamma = 2.36Q\frac{dE_g}{dx}\sqrt{\frac{x(1-x)}{KV}} \quad , \tag{5.37}$$

where KV is the effective number of cations whose fluctuations result in a broadening (K is the number of cations in a unit volume, $2.21 \times 10^{22}\,\mathrm{cm}^{-3}$ for AlGaAs), V is the effective volume of the bound quasi-particle (exciton for the BE transition and acceptor for the CA transition), $Q = 1$ for BE, and $Q = Q_{\mathrm{vb}}$ [$Q_{\mathrm{vb}} = (dE_{\mathrm{vb}}/dx)/(dE_{\mathrm{g}}/dx)$ is a valence-band fractional percentage of the band-gap discontinuity between the two binary compounds forming the ternary alloy: in our case GaAs and AlAs]. The effective volume definition is somewhat arbitrary, but it is proportional to a_{B}^3, with a_{B} being the Bohr radius of the bound quasi-particle. In the effective mass approximation a_{B} is inversely proportional to the effective mass m^* of the bound quasi-particle. For the exciton, $1/m^* = 1/m_{\mathrm{e}}^* + 1/m_{\mathrm{hh}}^*$ (m_{e}^* and m_{hh}^* are the electron and heavy-hole effective masses, respectively). For the shallow acceptor, m_{hh}^* is a reasonable approximation for m^*. It is clear then from (5.37) that the ratio of the Gaussian half-width of the CA and BE transitions are simply given by [5.92]

$$\frac{\Gamma(\mathrm{CA})}{\Gamma(\mathrm{BE})} = Q_{\mathrm{vb}}\left(1 + \frac{m_{\mathrm{hh}}^*}{m_{\mathrm{e}}^*}\right)^{3/2} . \qquad (5.38)$$

Since effective masses are rather well known for most semiconductors [5.93] this ratio yields directly a value of the fractional valence-band offset equal to Q_{vb}, provided dE_{vb}/dx is a constant value.

This method was recently used by *Raczyńska* et al. [5.94] for epitaxial layers of $\mathrm{Al}_x\mathrm{Ga}_{1-x}\mathrm{As}$, which showed a composition-fluctuation-limited linewidth, to determine the band-offset partition between the conduction and valence bands of $\mathrm{GaAs/Al}_x\mathrm{Ga}_{1-x}\mathrm{As}$ heterostructures. The result is 0.38 for Q_{vb}, in excellent agreement with the currently accepted partition of the band offset between conduction and valence bands in AlGaAs [5.95].

5.4.3 Photoluminescence of Quantum Well Structures and Superlattices

The basic element of a superlattice is the single quantum well (SQW). It is created in compositional superlattices from heterostructures of constituent semiconductors, and in doping superlattices by differently doped neighboring layers of the host semiconductor [5.96]. Sometimes the term superlattice is used to refer to a periodic structure which can be considered as a set of independent quantum wells [5.96, 97]. In such a structure the barrier layers are thick enough to prevent any coupling between the wave functions describing the energy states of the carriers confined to the individual quantum wells. In fact, such periodic structures do not exhibit properties typical of superlattices, but they show rather the properties of a single quantum well in multiplication. Therefore, such structures are called multi-quantum-well (MQW) structures. They can be treated as superlattices with infinitely thick barriers. As a matter of fact, the first direct experimental observation of bound-electron and bound-hole states of rectangular potential wells was reported for MQW $\mathrm{GaAs/Al}_x\mathrm{Ga}_{1-x}\mathrm{As}$ structures consisting of 50 GaAs layers separated by $\mathrm{Al}_x\mathrm{Ga}_{1-x}\mathrm{As}$ layers thicker than 25 nm [5.97].

The simplest quantum well which appears in superlattices is the rectangular, or square-shaped, potential well. If a particle is completely confined to a layer of thickness L_z by an infinite potential well, then the energies of the stationary bound states resulting from the solution of the time-independent Schrödinger equation are equal to

$$E = E_{ni} + \frac{h^2}{8\pi^2 m}(k_x^2 + k_y^2) \quad , \quad \text{where} \tag{5.39}$$

$$E_{ni} = \frac{h^2}{8m_i}\left(\frac{n}{L_z}\right)^2 \quad , \quad n = 1, 2, 3, \ldots, \infty \quad ; \tag{5.40}$$

k_x, k_y are here the wave vectors of the particle in the directions parallel to the well walls. The motion of the particle in the z direction is quantized, while in the x and y directions the particle moves classically. One can see from the last equation that the particles confined to the quantum well exhibit what is known as the quantum size effect, i.e., the dependence of the energies E_n on the width L_z of the well.

In the case of a GaAs/Al$_x$Ga$_{1-x}$As MWQ structure, this problem must be solved for electrons and two sets of holes, the heavy (hh) and light holes (lh). For each of these types of carriers the Schrödinger equation has been solved separately with the particle mass m_i being replaced by the relevant effective mass m^* ($m_{hh}^* = 0.45m_0$, $m_{lh}^* = 0.08m_0$) [5.97]. The theoretically predicted behavior has been confirmed experimentally [5.96,97], which shows that the simple effective mass approximation is quite appropriate to describe the particle in finite square wells.

A GaAs/AlGaAs MQW structure was used by *Mendez* et al. [5.98] to study the influence of the electric field in a Schottky barrier on the photoluminescence of the MQW structures. Six identical GaAs quantum wells (width = 35 Å) in an AlGaAs medium were formed as illustrated schematically in Fig. 5.26. With an increasing applied field perpendicular to the quantum wells, the luminescence intensity decreases and becomes completely quenched at an average field of a

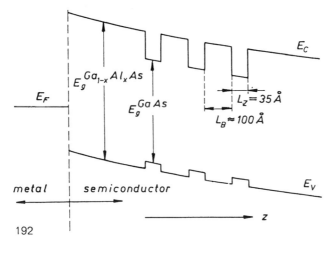

Fig. 5.26. Schematic energy band diagram of a multi-Schottky junction [5.98]

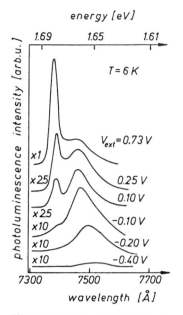

1.69   1.65   1.61

T = 6 K

$V_{ext} = 0.73$ V

x1

x25   0.25 V
      0.10 V

x25
x10   −0.10 V

x10   −0.20 V
x10   −0.40 V

7300   7500   7700

wavelength [Å]

photoluminescence intensity [arb.u.]

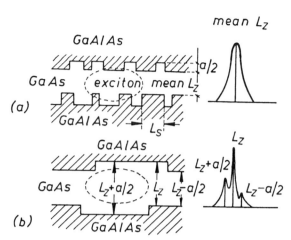

Fig. 5.27. Photoluminescence intensity for various applied voltages versus emission wavelength for GaAs quantum well with $L_z = 35$ Å [5.98]

mean L_Z

GaAlAs

GaAs   (exciton)   mean L_z

(a)

GaAlAs L_s $a/2$

GaAlAs

GaAs $L_z + a/2$ L_z $L_z - a/2$ $L_z + a/2$ $L_z - a/2$

(b)   GaAlAs

L_z

Fig. 5.28. (a) Effect of the interface disorder on the luminescence linewidth when the lateral size L_s of the growth islands is smaller than the exciton Bohr diameter. The linewidth corresponds to the confinement energy fluctuations experienced by the exciton. (b) When L_s is larger than the exciton Bohr diameter, the exciton experiences only one confinement energy corresponding to $L_z - a/2$, L_z, or $L_z + a/2$ and discrete peaks are observed in luminescence [5.100]

few tens of kV cm^{-1}, (applied voltage, −0.4 V), as shown in Fig. 5.27. This is accompanied by a shift to lower energies of the peak positions. The observation is interpreted as caused by the field, which indicates a separation of electrons and holes in the quantum well with the modification of the quantum wells, resulting in luminescence quenching.

Photoluminescence of superlattices is very sensitive to interface disorder. The model introduced by *Weisbuch* et al. [5.99] is shown in Fig. 5.28. The upper part (a) corresponds to superlattices grown without growth interruption. In such a case, growth islands of one monolayer thickness will be present at each interface, so that the well width will show three different values: the normal width L_z and $L_z + a/2$ or $L_z - a/2$. Part (b) shows the situation where a superlattice is grown with growth interruption. Similar smoothing of both interfaces is assumed.

The energy levels E_n of an effective mass carrier of mass m^* in a square well potential of a width L_z with infinitely high barriers are simply given by (5.39). An uncertainty of the well width ΔL_z, caused as discussed below, will then cause an uncertainty ΔE_1 of the $n = 1$ energy level:

$$\Delta E_1 = \Delta L_z \frac{h^2}{4m^* L_z^3} \ . \tag{5.41}$$

We assume that the positions of the heterointerfaces in the z direction vary statistically along x and y. Consequently the distribution of well width in the probe volume is statistical (Gaussian). Thus the $n = 1$ energy level will also not be sharp, but will show a Gaussian distribution according to (5.41). The radiative recombination in GaAs quantum wells of reasonable quality of a width $L_z < 11$ nm is of excitonic origin at temperatures up to room temperature [5.101]. The photon energy of the $n = 1$ heavy hole exciton recombination is

$$h\nu = E_g + E_e + E_h - E_{X\ (2D)} \qquad (5.42)$$

where E_g is the three-dimensional gap, E_e and E_h are the energy shifts of the $n = 1$ electron and heavy hole energy levels according to (5.39), and $E_{X(2D)}$ is the binding energy of the two-dimensional exciton. Since this binding energy is typically of the order of 1 % of the band gap, its broadening due to a distribution of well width can be neglected in a first approximation. If we assume that the exciton density of states is occupied over an energy range much smaller than the magnitude of the energy uncertainties described by (5.41), this implies that at low excitation intensity and low temperatures, the line shape of the exciton recombination will have a Gaussian character. The half-width of the recombination line is related to the half-width of the L_z distribution via

$$\Delta L_z = \Delta h\nu \frac{\mu L_z^3}{h^2 \pi^2} \quad , \qquad (5.43)$$

where μ is the reduced mass of the exciton. Thus, more generally, the luminescence intensity distribution is directly proportional to the distribution function of quantum well widths. It does not matter whether this distribution is continuous or has discrete values. An evaluation of the line shape will distinguish qualitatively between both cases and will lead to a determination of the parameters of the distribution.

Thus three discrete energy levels are expected for the exciton in a structure shown in Fig. 5.28. At low temperatures the excitons will thermalize to the lowest energy level, generated by the quantum well island having the largest width (L_s), leaving the other energy levels, caused by well islands of a smaller width, unoccupied. With increasing temperature the energy levels which stem from narrower wells are increasingly thermally occupied. The relative size of the sum of surfaces of the three different types of quantum well islands constituting one "quantum well" as a function of growth interruption time can be derived from the temperature dependence of the respective luminescence lines. Thus the existence/nonexistence of high energy lines in the interrupted growth/noninterrupted growth sample and their respective temperature dependences are readily understood. The observation of thermalization also allows a derivation of an upper limit for the size of the islands. Their mean size must be smaller than the diffusion length of the exciton, otherwise they would not thermalize.

This model was used by *Bimberg* et al. [5.102] to interpret the luminescence data obtained from superlattices grown with and without growth interruption. A

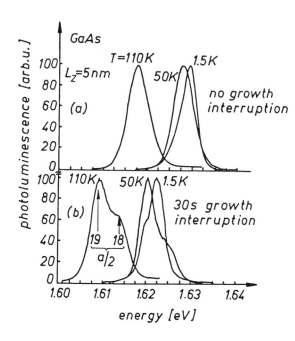

Fig. 5.29. Temperature dependence of the luminescence of the 5 nm GaAs/AlGaAs quantum well structure grown (a) without interruption of the growth and (b) with 30 s interruption of the growth in the range 1.5–110 K [5.102]

typical result for the appearance of line splitting in photoluminescence spectra due to interface smoothing is shown in Fig. 5.29.

Direct images of growth islands differing by 2.8 Å in height at GaAs/AlGaAs heterointerfaces were observed by scanning cathodoluminescence [5.103]. By this new method atomic-scale structure information is expressed in terms of spectral properties.

5.5 Electrical Characterization

Electrical characterization can give considerable information about the purity of epitaxial layers. Such information is important for the evaluation and control of growth parameters used to prepare high quality epitaxial layers for device applications. Included under the term "electrical characterization procedures" are Hall coefficient and resistivity measurements, and deep level transient spectroscopy (DLTS). The analysis of Hall coefficient and resistivity data versus temperature provides information about the concentration of impurities, while the spatial distribution of the impurities can be investigated by DLTS.

5.5.1 Determination of Carrier Concentration and Mobility

The experimental techniques used for the measurement of the Hall coefficient and resistivity of semiconductors as well as the basic theory are well known and described in detail in [5.13].

To determine the donor and acceptor concentrations from the temperature variation of the carrier concentration, a theoretical expression must be fitted to the experimental data. The usual model for nondegenerate n-type material with a donor concentration N_D, acceptor concentration N_A and the concentration of free electrons n_0 is given by the solution of

$$\frac{n_0(n_0 + N_A) - n_i^2}{N_D - N_A - n_0 - (n_i^2/n_0)} = \frac{N_c}{g_1} \exp(-E_D/k_B T) \tag{5.44}$$

where $N_c = 2(2\pi m_D^* k_B T/h^2)^{3/2}$, n_i is the intrinsic carrier concentration ($n_i^2 = n_0 p_0$), m_D^* is the conduction band density-of-states effective mass, g_1 is the degeneracy of the ground state and E_D is the energy of the ground state below the conduction band.

In particular temperature ranges, it is possible to approximate (5.44) by a simpler relation:

1. At very low temperatures, where $n_0 \ll N_A$, $N_D - N_A$ and $p_0 = n_i^2/n_0 \simeq 0$,

$$n_0 \simeq \frac{N_c}{g_1} \frac{(N_D - N_A)}{N_A} \exp(-E_D/k_B T) \quad . \tag{5.45}$$

2. At slightly higher temperatures or for low values of N_A, where $n_0 \gg N_A$ and $p_0 \simeq 0$,

$$n_0 \simeq \sqrt{\frac{N_c}{g_1}(N_D - N_A)\exp(-E_D/k_B T)} \quad . \tag{5.46}$$

3. At still higher temperatures, where $E_D \ll k_B T$ but $n_0 \gg n_i$,

$$n_0 \simeq N_D - N_A = \text{const.} \tag{5.47}$$

By plotting the experimental carrier concentration versus $1/T$ on a semilog graph, simple estimates can be made of E_D and $(N_D - N_A)/N_A$ from the slope and intercept of the straight line fitted to the data in the appropriate temperature range using (5.45–47).

The analysis of the temperature variation of the Hall coefficient is not a reliable method for determining the donor and acceptor concentrations in samples which do not have sufficient carrier freeze-out. One method that has frequently been used to estimate N_D and N_A is to calculate these concentrations from the Brooks-Herring [5.104] mobility formula for ionized impurity scattering using the experimental data of mobility and carrier concentration measured around liquid nitrogen temperature. The mobility limit for ionized impurity scattering is given by

$$\mu_1 = \frac{3.28 \times 10^{15}\sqrt{m^*/m}\varepsilon_0^2 T^{3/2}}{(2N_A + n)[\ln(b+1) - b/(b+1)]} \quad \left[\frac{\text{cm}^2}{\text{V s}}\right] \quad , \tag{5.48}$$

where

$$b = \frac{1.29 \times 10^{14}(m/m^*)\varepsilon_0 T^2}{n^*} \tag{5.49}$$

and n^* is an effective screening density:

$$n^* = n + (n + N_A)\frac{N_D - N_A - n}{N_D} \quad [\text{cm}^{-3}] . \tag{5.50}$$

The rest of the terms have their usual meanings. For temperatures and/or samples in which donor deionization occurs, it is not possible to solve for N_D and N_A explicitly, and N_D and N_A must be adjusted separately to obtain agreement between the experimental and calculated mobilities. Where donor deionization does not occur, $n = N_D - N_A$ and so $n^* = n$. Thus the acceptor concentration N_A can be calculated directly from the measured mobility and carrier concentration.

For polar binary compounds the important electron scattering mechanisms are by optical phonons through polar interactions [5.105] (μ_{PO}), polar scattering due to piezo-electrically active acoustic phonons [5.106] (μ_{PE}), deformation potential scattering by acoustic phonons [5.107] (μ_{DP}), and scattering by ionized impurities [5.104] (μ_I). Each of these separate scattering mechanisms exhibits a unique dependence on temperature: μ_I varies as $T^{3/2}$, μ_{PO} as T^{-2}, and μ_{PE} as $T^{-1/2}$. Assuming that the different scattering mechanisms can be treated independently of each other, the local mobility can be calculated by a simple inverse sum

$$\frac{1}{\mu_H} = \sum_i \frac{1}{r_{H_i}\mu_i} , \tag{5.51}$$

where μ_H describes the overall Hall mobility, μ_i the component of the drift mobility, and r_{H_i} the corresponding Hall factor. The results of such a calculation together with the experimental data for a GaAs epilayer are shown in Fig. 5.30 [5.108].

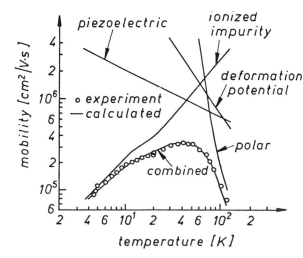

Fig. 5.30. Experimental temperature variation of the mobility for a GaAs layer with $N_D = 4.8 \times 10^{13}\,\text{cm}^{-3}$ and $N_A = 2.13 \times 10^{13}\,\text{cm}^{-3}$ [5.108]

Errors may be introduced in the determination of free carrier concentrations by Hall measurements because of carrier depletion in the epitaxial surface layer and at the epitaxial layer–substrate interface. These errors lead to underestimation of $(N_D - N_A)$ and become important in thin low-doped layers. Assuming that the Fermi level of the free surface is pinned by an amount ϕ_s below the conduction band because of surface states [5.109], the width of the surface depletion layer l_s is given in the abrupt depletion approximation by [5.110]

$$l_s \sim \sqrt{\frac{2\varepsilon V_{bs}}{q(N_D - N_A)}} \quad , \quad \text{where} \tag{5.52}$$

$$V_{bs} \sim \phi_s - \left[\frac{E_g}{2q} - \frac{k_B T}{q} \ln\left(\frac{n}{n_i}\right)\right]$$

represents the surface band bending, E_g the semiconductor bandgap, n_i the intrinsic carrier density, and the other quantities have their usual meanings. Similarly, when layers are grown on semi-insulating substrates, band bending at the layer–substrate interface occurs as a result of the diffusion of free carriers from the epitaxial layer into the substrate, where they are trapped in deep levels. Neglecting the band bending in the substrate, one obtains for the width of the depletion layer in the epitaxial material

$$l_i \sim \sqrt{\frac{2\varepsilon V_{bi}}{q(N_D - N_A)}} \quad . \tag{5.53}$$

The built-in potential is given by

$$V_{bi} \sim \phi_i - \left[\frac{E_g}{2q} - \frac{k_B T}{q} \ln\left(\frac{n}{n_i}\right)\right] \quad ,$$

where ϕ_i represents the depth below the conduction band of the deep trap levels in the semi-insulating substrate.

The effective electrical thickness of the epitaxial layer l_e, taking into account these depletion effects, is equal to

$$l_e = l_m - l_s - l_i \quad , \tag{5.54}$$

where l_m is the metallurgical layer thickness.

Calculations presented by *Chandra* et al. [5.110] for the case of GaAs layers with a net doping level around 10^{15} cm^{-3} show that the depletion widths l_s and l_i are of the order of 1 μm. Values of $(N_D - N_A)$ evaluated without depletion layer corrections can be considerably lower than the actual concentrations present in the layers.

A high value of carrier mobility in an epitaxial layer is widely accepted as a quality criterion, because it allows the conclusion that no scattering centers are present to limit the mobility. To overcome this unavoidable problem of impurity incorporation during doping, the so-called modulation doped structures

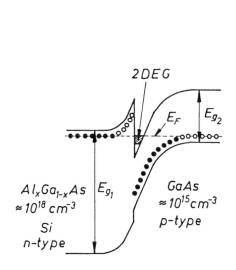

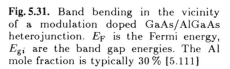

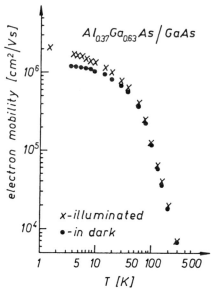

Fig. 5.31. Band bending in the vicinity of a modulation doped GaAs/AlGaAs heterojunction. E_F is the Fermi energy, E_{g_i} are the band gap energies. The Al mole fraction is typically 30 % [5.111]

Fig. 5.32. Temperature dependence of electron mobility for a two-dimensional electron gas in heterostructures with a 333 Å thick undoped spacer at the interface [5.112]

were developed. Figure 5.31 shows the band diagram of an $Al_xGa_{1-x}As/GaAs$ heterostructure with the typical features of a modulation doped structure [5.111]. In a multilayer structure, selective doping of the wide gap (E_{g_1}) material only results in the diffusion of free carriers into the narrow gap (E_{g_2}) material with higher carrier affinity. The transfer of the carriers to the narrow gap material results in a changed depletion layer in the wide gap material near the interface. This space charge region is the origin of a strong electric field causing a band bending in the narrow gap material and the formation of a nearly triangular potential well confining the two-dimensional electron gas (2DEG).

The basic difference in the low field transport of electrons in the two-dimensional channel of the modulation doped heterojunction and carriers in the bulk material is the reduction or elimination of Coulombic scattering by ionized impurities. Ionized impurity scattering is the dominant process limiting the mobility at low temperatures, as explained above, so the mobility enhancement in the 2DEG is most pronounced at low temperatures.

The experimental results obtained by *Weimann* and *Schlapp* [5.112] in AlGaAs/GaAs heterostructures are shown in Fig. 5.32. In the high temperature region the mobility is limited by the polar optical scattering mechanism. However, in the low temperature region there is no limitation due to ionized impurity scattering as obtained in high purity GaAs (compare Fig. 5.30).

The improvement of the low temperature mobility of the two-dimensional electron gas is quite evident, however, there are still limiting scattering mech-

199

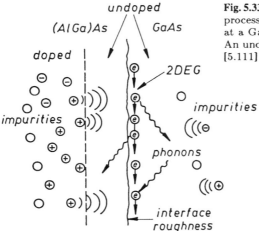

undoped

(AlGa)As / GaAs

doped

2DEG

impurities

impurities

phonons

interface
roughness

Fig. 5.33. Some of the remaining scattering processes of a two-dimensional electron gas at a GaAs/AlGaAs heterojunction interface. An undoped AlGaAs spacer layer is included [5.111]

anisms, which can be taken as a test for the interface quality. Following the model presented by *Störmer* [5.111] we list briefly the major remaining scattering mechanisms, which are shown schematically in Fig. 5.33.

1. Bulk phonons, optical as well as acoustic
2. Interface phonons
3. Alloy scattering due to penetration of carriers into the barrier
4. (Remote) ionized impurities in AlGaAs
5. Ionized impurities in GaAs from unintentional doping
6. Neutral impurities in GaAs from unintentional doping
7. Interface roughness

Consequently, the high mobility limit at low temperatures can be taken as a quality criterion for the interface. Based on this argument, *Ploog* [5.113] showed the improvement of the growth process of AlGaAs/GaAs short-period superlattices.

5.5.2 Deep Level Transient Spectroscopy

In a metal-semiconductor rectifying contact the application of a reverse bias creates a space charge region in the semiconductor. The width of the depletion layer is a function of the reverse bias, giving rise to a voltage-dependent capacitance [5.3]. Electronic transitions between localized defects and the conduction band alter the space-charge density and can be detected as a change in the device capacitance.

Deep level transient spectroscopy (DLTS) utilizes the trapping of majority carriers in a selected region of the depletion layer, and the capacitance is monitored during the thermal reemission of the trapped charge. So the rate at which the capacitance relaxes strongly depends on the temperature. The essential feature of this method is the implementation of a rate window for monitoring a given relaxation time as a function of temperature. The DLTS method and its extension to DDLTS is well established by now and has been reviewed in [5.4, 114, 115]. Here we will concentrate only on the relations which are necessary to evaluate the data.

The concentration of filled trap centers $N_T^-(t)$ is determined by the following basic rate equation:

$$\frac{dN_T^-(t)}{dt} = c_n N_T^0 - e_n N_T^- - c_p N_T^- + e_p N_T^0 \quad , \tag{5.55}$$

where N_T^0 is the concentration of neutral traps, c_n and c_p are the capture rates, and e_n and e_p are the emission rates for electrons and holes, respectively.

The filling process is governed by the following solution of the rate equation (5.55):

$$N_T^0(t) = N_T[1 - \exp(-c_n t)] \quad , \tag{5.56}$$

where N_T is the total trap concentration in the material. The capture rate c_n is given by

$$c_n = \frac{1}{t_n} = \sigma_n \langle v_n \rangle n + e_p + e_n \quad ; \tag{5.57}$$

t_n is the characteristic capture time for electrons, σ_n the capture cross section, $\langle v_n \rangle$ the thermal velocity of electrons, and n the free-carrier concentration. The electron capture cross section σ_n is often temperature dependent, and over some temperature range may be approximated by [5.116]

$$\sigma_n(T) = \sigma_\infty \exp(-E_B/k_B T) \quad , \tag{5.58}$$

where σ_∞ is the limiting value of σ_n at $T \to \infty$, and E_B is the activation energy of the capture cross section.

During the thermally activated emission of electrons, the second part of the rate equation becomes dominant and can be calculated by the principle of detailed balance:

$$e_n = \sigma_n \langle v_n \rangle N_v \exp\left(-\frac{E_T}{k_B T}\right) = \frac{1}{\tau} \quad ; \tag{5.59}$$

N_v stands for the effective density of states in the valence band, E_T is the energy separation between the defect level and the valence band, and τ is the characteristic emission time.

If, however, an exponential variation of the capture cross section according to (5.58) is assumed, one obtains

$$e_n = \sigma_\infty \langle v_n \rangle N_v \exp[-(E_T + E_B)/k_B T] \quad . \tag{5.60}$$

Since a Schottky diode is a majority carrier device, in n-type material only defect levels in the upper half of the band can be detected.

The trapped electrons are thermally emitted at a characteristic rate e_n, as described in (5.59). Therefore also the capacitance transient decays exponentially and, as the temperature is increased, the electron emission rate from each defect level increases exponentially. The DLTS technique is based on the idea of a rate window, measuring the capacitance decay at the times t_1 and t_2 after the filling pulse. A maximum value in DLTS signal $C(t_1) - C(t_2)$ is obtained at

$$\frac{1}{e_n} = \tau_{max} = \frac{t_2 - t_1}{\ln(t_2/t_1)} \quad , \tag{5.61}$$

and only those defects which are emitting electrons with a $\tau \sim \tau_{max}$ are observed. With fixed t_1 and t_2 a series of peaks is obtained in a DLTS spectrum corresponding to various defect levels. By changing the observation times t_1 and t_2, another τ_{max} is fixed and the maximum of the DLTS signal is reached at another temperature. Then (5.59) can be used to obtain the defect energy level from the variation of e_n with T. However, if we take into account an exponential temperature dependence of the capture cross section, (5.58), we have to correct this value for the activation energy E_B of the capture cross section, (5.60).

The electron capture cross section σ_n of electron traps is measured with a majority carrier filling pulse of constant height. The pulse width t_i is varied to change the time during which electrons may be captured. The capture process is governed by (5.56) and it follows straightforwardly that the characteristic filling time $t_n = 1/c_n$ is

$$t_n = t_i \left[\ln\left(\frac{N_T}{N_T - N(t_i)}\right) \right]^{-1} . \tag{5.62}$$

Since $N(t_i)$ is proportional to the peak height in the DLTS spectrum, (5.62) can be used to analyze the peak height as a function of t_i for the characteristic filling time. With known carrier concentration and thermal velocity the capture cross section is calculated from (5.57).

The double correlation DLTS (DDLTS) proposed by *Lefevre* and *Schulz* [5.115] can be used to define a narrow spatial window in the depletion layer of the semiconductor. For the transient capacitance analysis in DDLTS two pulses are used, as shown in Fig. 5.34. The two pulses have different amplitudes, V_p and V_p' respectively. The correlation signal $\Delta^2 C$ is obtained by the weight function depicted in the lower part of Fig. 5.34, i.e., by relating the single correlation signal ΔC after the two pulses. Using the same delay times t_1 and t_2,

$$\begin{aligned} \Delta^2 C &= \Delta C(t_1) - \Delta C(t_2) \\ &= [C'(t_1) - C(t_1)] - [C'(t_2) - C(t_2)] \quad . \end{aligned} \tag{5.63}$$

The physical meaning of the correlation technique is obvious from Fig. 5.35. During the two pulses, trap levels in the space charge layer are filled to depths x_{fp} and x_{fp}', respectively. Only those traps located in the observation window $x_{fp} - x_{fp}'$ contribute to the double correlation signal $\Delta^2 C$. These traps have a definite emission time constant because they are located in a well-defined field region. The time constant measured is also independent of the position of the Fermi level because the effect of traps located in the vicinity of the Fermi level

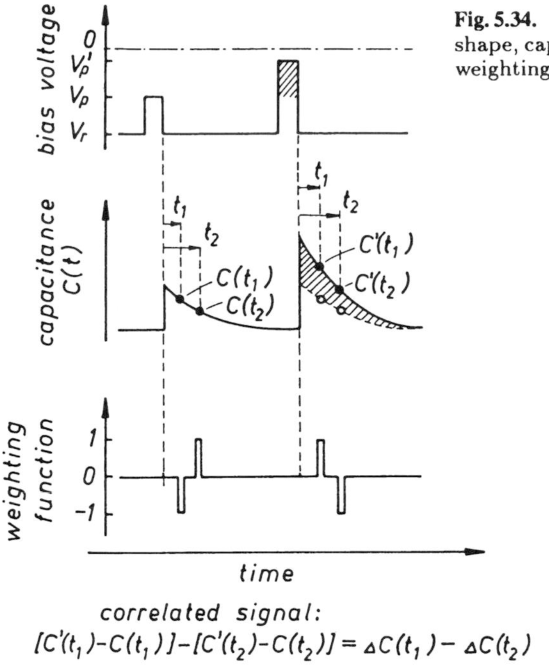

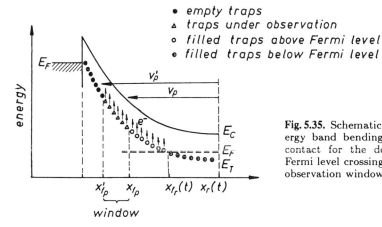

Fig. 5.34. Schematic drawing of the pulse shape, capacitance transient, and correlation weighting function as used in DDLTS [5.115]

correlated signal:

$$[C'(t_1)-C(t_1)]-[C'(t_2)-C(t_2)] = \Delta C(t_1) - \Delta C(t_2)$$

- • empty traps
- ▲ traps under observation
- ○ filled traps above Fermi level
- ◉ filled traps below Fermi level

window

Fig. 5.35. Schematic drawing of energy band bending in a Schottky contact for the definition of the Fermi level crossing points and the observation window [5.115]

is subtracted. Contact noise is also reduced by setting the observation window. The measurement sensitivity is therefore increased. Profiles of the deep level distribution can be determined because the observation window can be positioned in the space charge layer by varying the steady-state reverse bias or the pulse height. Since in most cases the time constant of the traps is field dependent, the pulse height used has to be kept constant and only the reverse bias varied. In this case, the local field in the observation window is constant to a first-order approximation for a large range of variation of the reverse bias. The magnitude of the double correlation signal $\Delta^2 C(V_r)$ is a measure of the deep level profile for a given energy when the measurement temperature is set to the peak signal in the temperature scan corresponding to the level.

For deep level densities that are much smaller than the shallow level, the concentration of the deep level N_T is given by [5.115]

$$N_T(x_{fp}) = 2\exp\left(\frac{t_1}{\tau}\right)(\varepsilon A)^2 n_0(x_r(t_1))\frac{\Delta C(t_1)}{C_r^3(t_1)}\frac{1}{x_{fp}^2 - x_{fp}'^2} \quad , \qquad (5.64)$$

where $n_0(x_r)$ is the shallow doping profile, A the constant area, ε the permittivity and C_r the space charge capacitance. The Fermi level crossing point x_{fp} can be determined by

$$E_f - E_T = \frac{q^2}{\varepsilon}\int_{x_{fp}}^{x_p}(x_p - z)n_0(z)dz \quad , \qquad (5.65)$$

where x_p is the depth of the space charge region when the voltage pulse V_p is applied.

The defect structure at the interface of CdTe/PbTe heterostructures grown by hot-wall epitaxy has been characterized by DDLTS measurements [5.117]. Figure 5.36 shows the defect concentration profiles for differnt traps obtained from samples grown with different substrate temperatures. The sample grown with a substrate temperature of 500° C has a fairly constant defect concentration with layer depth. In contrast the sample grown at 400° C showed an increase in defect concentration near the interface. With lower substrate temperature the growth rate is increased and therefore a larger number of defects are incorporated, especially in the region near the interface.

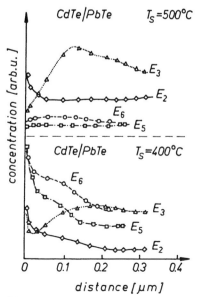

Fig. 5.36. Concentration profiles obtained by DDLTS measurements for CdTe/PbTe heterostructures grown with different substrate temperatures T_s by hot-wall epitaxy [5.117]

Traps in n-GaAs grown by MBE have been reported by several researchers [5.118–122]. Recently the production of the most common electron trap (EL2) by rapid thermal processes was investigated by DLTS measurements [5.123]. It was found that the spatial distribution of EL2 corresponds to that of thermal stress introduced by the rapid thermal processing.

A clear relation between growth conditions and deep levels in GaAs was found by *Xu* et al. [5.124]. Samples were grown under As- or Ga-stabilized conditions as well as in the transition region between the corresponding surface reconstructions. As shown in Fig. 5.37 the layers exhibit different sets of DLTS peaks depending on the surface structure during growth. A minimum concentration was measured in samples grown in the transition region, therefore the origin is related to the surface stoichiometry.

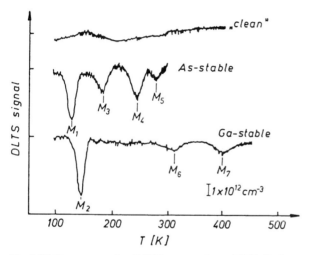

Fig. 5.37. Representative DLTS spectra from MBE GaAs grown under different conditions, i.e., different surface structures. The sample giving the "clean" spectrum is grown in the transition range. The magnitude of the concentration is indicated by the bar. Sample carrier concentrations were 10^{15} cm^{-3} and the rate window used was 20.7 ms [5.124]

5.6 Sophisticated Characterization Methods

There are a great variety of characterization methods which are based on very expensive equipment and require specialized staff to run the complicated apparatus. Also, the interpretation of the experimental data is not always straightforward, so that a special group of experts in the field is necessary to work mainly on these analysis problems. Therefore, crystal growers cannot use these sophisticated characterization methods as a routine tool, but must rely on collaboration with specially equipped and trained groups. However, results of such investigations give very detailed insight into the problems the crystal grower has to deal with, and all information obtained is of enormous help in improving the crystalline and electrical quality of the grown epilayers.

5.6.1 Transmission Electron Microscopy

Transmission electron microscopy (TEM) is based on the diffraction of electrons when they pass through a very thin sample. In the simplest view, electron diffraction from a crystalline lattice can be described as a kinematic scattering process that meets the wave reinforcement and interference conditions given in the Bragg equation (Sect. 5.3). In many cases samples must be thinned to a thickness of a few thousand angstroms by chemical etching or ion milling.

The patterns are formed by diffraction of an electron beam (50–200 keV) transmitted through the thin sample. As in the case of x-ray diffraction, the electron diffraction patterns are spot patterns from single crystalline films, ring-shaped patterns from randomly oriented microcrystalline films and superimposed ring and spot patterns from large grain polycrystalline films. Figure 5.38a shows as an example a TEM micrograph of square-shaped islands of PbTe grown by hot-wall epitaxy on (001) oriented KCl substrates [5.125]. As can be seen from the diffraction patterns in Fig. 5.38b, the islands were all oriented in the (001) direction. For comparison a polycrystalline PbTe film grown on BaF$_2$ is shown

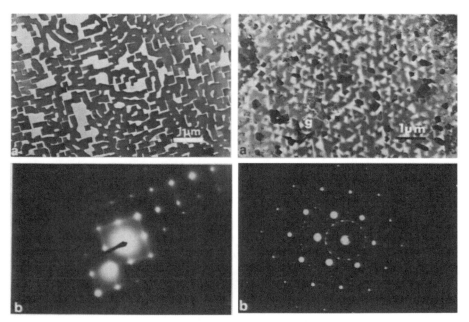

Fig. 5.38. (a) TEM micrograph of PbTe islands grown on (001) surface of KCl. (b) Diffraction patterns of the PbTe islands near the [001] pole [5.125]

Fig. 5.39. (a) TEM micrograph of a PbTe film grown on BaF$_2$ (111) cleavage plane. (b) Diffraction pattern showing the [111] pole of the matrix islands and randomly oriented ⟨001⟩ islands [5.125]

in Fig. 5.39. The dark spots in the TEM micrograph result from the misoriented grains in the epitaxial (111) oriented matrix of the PbTe film. The diffraction pattern shows the (111) pole originating from the matrix and the ring-shaped pattern from the randomly oriented (001) islands. In the vicinity of cleavage steps of BaF_2 substrates the (001) islands are not randomly oriented but show a preferential orientation parallel to the cleavage step. This can be seen from Fig. 5.40a where the diffraction pattern of this area is shown. Due to the three common type $\langle 110 \rangle$ directions in the matrix, the three equivalent $\langle 001 \rangle$ diffractions are visible, reflecting the threefold symmetry of the $\langle 111 \rangle$ cleavage plane. Figure 5.40b gives the indices of this diffraction pattern. Matrix reflections (black points) and the three variants of $\langle 001 \rangle$ islands are indicated by squares, each of them having a common (220) reflection with the matrix.

Transmission electron microscopy as an analytical technique for structural determination differs from the other structural techniques (channeling, LEED, x-ray diffraction) in that the sample preparation is generally destructive during the thinning process. Consequently, TEM analysis is usually viewed as the last step

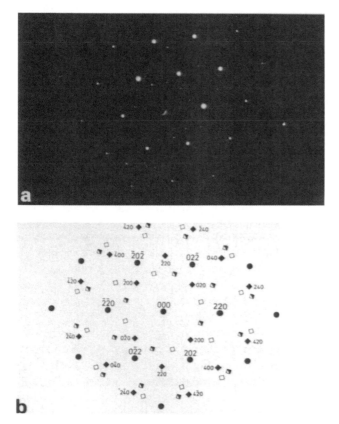

Fig. 5.40. (a) Electron diffraction of a PbTe film on BaF_2 in the vicinity of a cleavage step of the substrate. **(b)** Indices of the diffraction pattern of **(a)** [5.125]

in the history of a sample rather than as one part of a succession of measurements. In this regard, one can take the sample preparation a step beyond the conventional thinning technique and prepare cross-sectional TEM samples [5.126]. In this case, one cuts the sample into millimeter-thick slices and mounts them sideways in a holder for polishing to about 50 μm. Then they are ion milled to a thickness of about 500–1000 Å, as shown schematically in Fig. 5.41. Sample preparation involves care because different materials erode in the ion miller at different rates and the final sample is fragile. However, the finished cross section allows direct examination of the thin film reaction zone edge-on, rather than a top view through the sample.

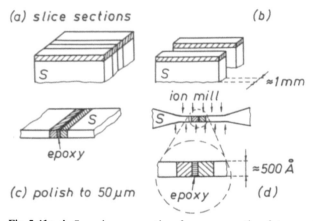

Fig. 5.41a–d. Sample preparation for cross sectional transmission electron spectroscopy. (a, b) slicing sections of about 1 mm thick; (c) using epoxy to hold two sections together and then polishing to about 50 μm thickness; (d) ion milling to produce the final sample thickness (S : substrate) [5.126]

Cross-sectional TEM samples show the planarity of the interface and the structure of the various reacted layers. During recent years it has become possible to resolve the crystalline layers separately by high resolution electron microscopy (HREM) [5.127–129]. As an example we show the results published by *Tsai* and *Lee* [5.130]. They used HREM to investigate the effect of post-annealing on the defect structure at the GaAs/Si interface. Two types of misfit interfacial dislocations were found, as shown in Fig. 5.42 with the illustrative Burgers circuits. The indicated Burgers vectors indicate the difference in atomic bonding at the interface between the two types of dislocations. In Fig. 5.42b it can be seen that the misfit dislocation is formed at a surface step. A sketch showing the formation of a distortion at a surface step is given in Fig. 5.43. The atomic structure was drawn to match the HREM image. The extra atomic plane is marked in Fig. 5.43. The Burgers vectors of this type of dislocation at steps are always along a $\langle 110 \rangle$ direction lined up with the step edge. The other type of dislocation shown in Fig. 5.42a was always found in the flat region of the interface. It is thus clear that the substrate surface quality affects the type of defects formed.

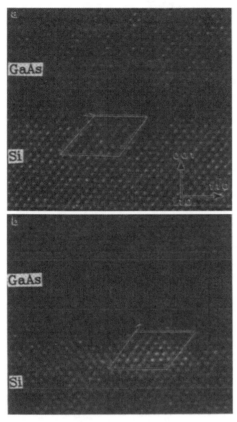

Fig. 5.42a,b. High resolution images of GaAs/Si interfaces showing two types of dislocations. The projected Burgers vectors are indicated on the Burgers circuits [5.130]

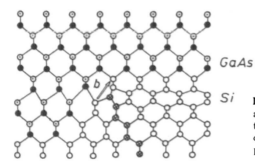

GaAs

Si

Fig. 5.43. Formation of a dislocation at
a surface step. The sketch was drawn
to match the region inside the Burgers
circuit in Fig. 5.42b. The extra atomic
plane in the Si lattice is marked ⊗ [5.130]

5.6.2 Rutherford Backscattering and Channeling

Rutherford backscattering spectrometry (RBS) is based on the analysis of energy loss dE/dx of light ions (H^+, He^+) at high energies (0.5–5 MeV) during their passage through the solid. The energy lost in penetration is directly proportional to the thickness of material traversed so that a depth scale can be assigned directly and qualitatively to the energy spectra of detected particles. The yield of backscattered particles is proportional to the scattering cross section, so the composition depth profile can be found from knowledge of energy loss and cross section. The arrangements of atoms in single crystals determine the magnitude of incident ion–target atom interactions. The influence of the crystal lattice on the trajectories of ions penetrating into the crystal is known as channeling – a term that visualizes the atomic rows and planes as guides that steer energetic ions along the "channels" between rows and planes. The steering action is effective and can lead to a hundredfold reduction in the yield of backscattered particles. By this channeling effect the depth resolution of the ion scattering technique can be improved, as well as its sensitivity to light impurities.

The use of Rutherford backscattering and channeling measurements in the analysis of superlattices started with the investigation of GaSb/InAs superlattices [5.131, 132]. These studies were followed by investigations performed on GaSb/AlSb [5.133] and Ge/GaAs [5.134] superlattices. Recently, detailed compositional and structural analyses have been carried out to analyze the mismatch strain in heteroepitaxial CdTe on $Cd_{0.96}Zn_{0.04}Te$ [5.135] or ZnSe on GaAs [5.136]. Through the combined use of RBS and channeling, depth profiles of strain, composition, and crystalline quality were able to be determined.

The principles, technique, and various applications are summarized in a monograph by *Feldman* et al. [5.137]. Here we will discuss the problem only briefly using as an example the investigations on the GaSb/AlSb superlattice [5.133]. This selection of the GaSb/AlSb system has the advantage of sharing the same anion, Sb, which enables the separation of the row offset model from the strain study. Periodic structures of GaSb/AlSb (30 nm/30 nm) with 10 periods were grown epitaxially on (100) GaSb substrates by MBE. Channeling measurements and analysis were made using a 2 MeV Van de Graaff accelerator. Figure 5.44 shows a series of energy spectra of 1.76 MeV $^4He^+$ ions backscattered from a superlattice sample under various experimental conditions. The best channeled spectrum, i.e., the lowest curve in this figure, is obtained with the ion beam aligned with the [100] growth direction (0° with respect to the surface normal) of the GaSb/AlSb superlattice. The next lowest spectrum, plotted as solid dots, is an aligned spectrum for [110] axial channeling (45° tilted from normal). A very broad energy window is set for the alignment to ensure that the alignment is based on a sampling of scattering from several layers inside the superlattice. The higher dechanneling along the [110] direction compared with that along the [100] growth direction for this superlattice sample is consistent with all the earlier measurements [5.131, 132] on superlattices. The random spectrum was obtained by tilting the surface normal of the sample to 45° away from the beam direction with a 10° rotation away from the (110) plane. The energy window was set to 15 keV (three channels) which corresponds to a depth in the superlattice between the second and the third layer. Since the energy scale of this figure can be translated to a depth scale, a change in the energy window corresponds to probing different depths of the sample. In Fig. 5.44 the energy positions are indicated which correspond to the individual GaSb and AlSb layers in the Sb part of the spectrum (1.4–1.6 MeV) and to those of the overlapping counts from Ga (below 1.4 MeV). Each layer corresponds to alternately a GaSb (1,3,5,...) and an AlSb (2,4,6,...) layer, with a layer thickness of 30 nm.

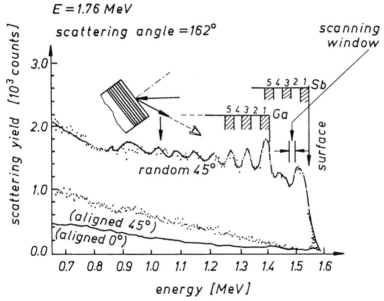

Fig. 5.44. Energy spectra of 1.76 MeV He⁺ ions backscattered from (100) GaSb/AlSb superlattices, [100] aligned, [110] aligned and random. The depth scale, based on Sb and Ga signals, is marked in units of layers (30 nm/layer) [5.133]

Figure 5.45 shows the backscattering yield normalized to the random level (around 45° tilt) plotted as a function of the tilt angle. Two window settings are given. The first one is set between 1.535 and 1.550 MeV, corresponding to the Sb signal of the surface layer of GaSb (plotted as solid circles), and the second window is set between 1.490 and 1.505 MeV, which corresponds to the Sb signal of the second layer of the superlattice. The center of an angular yield curve is defined by the average of the two angular positions corresponding to the half-height on either side of the angular scan. It is believed that the center corresponds to an angle of incidence giving the least dechanneling – "best" channeling – for a given layer. One can see from Fig. 5.45 that the two angular yield curves for two different depths do not have a common center. The first layer is centered at 45.09°, and the second layer at 44.92°. This is direct evidence that the [110] axis of the second layer is not in line with that of the first layer. A small angular difference of 0.17° is observed. This difference is presumably due to the lattice strain caused by intrinsic lattice mismatch between GaSb and AlSb in the superlattices.

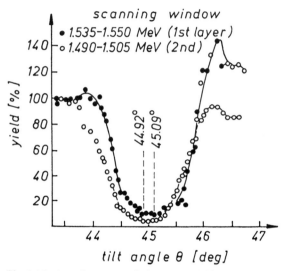

Fig. 5.45. Angular scan of the RBS yield for two different energy (depth) windows. (●) first layer; (○) second layer. The center position of the angular scan changes from layer to layer, indicating the variation of the [110] direction [5.133]

These observations of the strain can be elaborated as shown in Fig. 5.46. The top part of this figure illustrates a model of a strained superlattice, and its effect on the channeling along the [1$\bar{1}$0] or [110] direction. It has been shown that lattice mismatched layers can be grown with essentially no misfit defect if they are sufficiently thin and the mismatch is accommodated totally by uniform lattice strain [5.138]. The resulting strains in the superlattices consist of both hydrostatic and [100] uniaxial components which alter the lattice constants. For the discussed superlattice sample, the following relations are valid:

$$a_0(\text{AlSb})_\perp > a_0(\text{AlSb}) > a_{0\|} > a_0(\text{GaSb}) > a_0(\text{GaSb})_\perp \quad, \tag{5.66}$$

where $a_0(\text{AlSb})$ and $a_0(\text{GaSb})$ are bulk lattice constants, $a_{0\|}$ is the lattice constant in the planes parallel to the interfaces, and $a_{0\perp}$ are lattice constants for GaSb or AlSb perpendicular to the interface. Lattice strain causes the [1$\bar{1}$0] and [110] channeling directions to oscillate between an angle greater than 45° and an angle less than 45° degrees. That has been observed experimentally as presented in the lower part of Fig. 5.46. In this figure the angular position of the minimum yield,

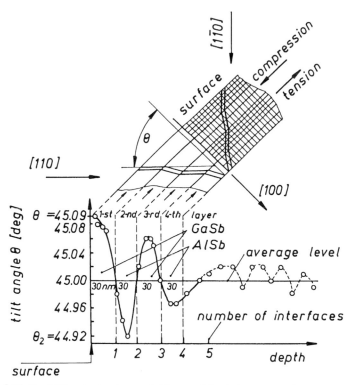

Fig. 5.46. *(top)* The strained superlattice and its influence on the channeling. *(bottom)* The angular position of the minimum yield, defined in Fig. 5.45, as a function of depth. The oscillation can be explained by a kinked [110] channel [5.133]

as defined in Fig. 5.45, is plotted as a function of depth of the superlattice. The window is three channels wide, equivalent to a 15 keV energy interval, and is the same as the energy resolution of the backscattering system. The oscillation of the angular position of the minimum is direct evidence of the alternating tensile and compressive nature of the strain. The damping of the oscillations is due to the fact that ion channeling in a given layer is always influenced by the previous history of the ion trajectory. However, the channeling experiment indicates clearly a $0.17° \pm 0.03°$ "kink" between the layers of GaSb/AlSb (Fig. 5.45).

Experimental results as described above can be modeled by computer simulations. Recently computer programs have been developed to simulate random and channeling backscattering spectra with high speed [5.139]. The limiting physical conditions for a standard analysis procedure of backscattering spectra are discussed by *Rauhala* [5.140].

Results from RBS experiments can also be compared with the analysis of x-ray rocking curves, as was done for a GaSb/AlSb superlattice [5.140]. These data allow the following conclusions to be drawn [5.141]:

213

1. X-ray rocking curve analysis is a very powerful method for characterizing the strain in superlattices as a function of depth, since it determines the geometry of the distortion (parallel, perpendicular, negative, or positive strain). The strain determined by this method is in excellent agreement with that calculated from the elastic constants and lattice misfit of the layer materials.

2. Backscattering and channeling measurements, combined with computer simulation, also allow a determination of the strain by providing a value of the associated "kink angle". The strain determined in this way is in reasonable agreement with the x-ray and elastic misfit values.

3. He ion-beam bombardment at high doses relieves a large portion of the strain of the AlSb/GaSb strained superlattice. This limits the beam dose that can be used in ion backscattering measurements without perturbing the specimen. On the other hand this ion-beam perturbation might be utilized to modulate the strain in superlattices in a controllable manner.

6. Fundamentals of the MBE Growth Process

The MBE growth process occurs on the substrate crystal surface, i.e., in the third zone of the deposition chamber arrangement (Sect. 1.1.2). Many different experiments using, for example, modulated molecular beam mass spectrometry [6.1–3] or RHEED pattern intensity oscillations [6.3–13] have been devoted to study the growth mechanism in MBE. These experiments dealt mainly with the MBE growth process of GaAs and related compounds, however, some fundamental rules creating a physical picture of the MBE growth process in general may be concluded from the wealth of data which became available [6.3, 14, 15].

The purpose of this chapter is to present in a consistent form a physical picture of the MBE process, taking into consideration the atomistic nature of the crystallization phenomena occurring in MBE growth. This presentation will be based on the concept of a near surface transition layer which is believed to appear in all modifications of the MBE growth technique.

6.1 General View of the MBE Growth Process

As a crystal growth process, MBE is, by definition, a dynamic phenomenon. A consideration of thermodynamics as well as kinetics is therefore essential. While thermodynamics embodies the essence of the behavior of the physical system at equilibrium, the kinetics controls the ability of the system to move towards the equilibrium state within the given conditions. It is important to specify the notion of equilibrium in relation to the physical system involved in the MBE growth process, and then relate the measured effects to the appropriate state of the system [6.3].

6.1.1 Equilibrium States in MBE

It seems to be reasonable [6.3, 14, 15] to distinguish three different states of equilibrium for the MBE growth system. These are: (i) the equilibrium state of the entire system, i.e., the substrate, the growing film, and the gaseous phase; (ii) the partial equilibrium between the substrate and the growing film; and (iii) the spatially localized equilibrium between the substrate and the growing film. For brevity we will refer to these, following *Madhukar* [6.14], as overall system equilibrium, or global equilibrium (i), partial equilibrium (ii), and local equilibrium (iii) (Fig. 6.1).

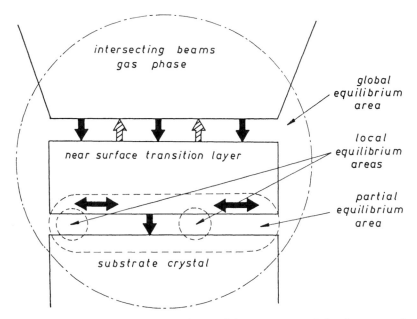

Fig. 6.1. Schematic illustration of the spatial arrangement of the three areas in the MBE growth zone where, according to *Madhukar* [6.14], different kinds of thermodynamical equilibrium may be defined

It is evident that, for growth to occur, a driving force for attachment of vapor phase species to the solid surface at a rate higher than the rate of loss due to reevaporation must exist, implying a definite departure from the global equilibrium. One should also consider the temperature of each part of the system, since equilibrium implies an isothermal state. Since in MBE the species arriving at the surface are at different temperatures defined by the temperatures of the effusion cells, which in turn are different from the substrate temperature, no isothermal state for the whole system exists. This could be ignored if all incident atoms and molecules were thermally accommodated (Sect. 1.1.2), i.e., acquired the temperature of the substrate before being involved in any other process. Unfortunately, on the basis of the existing experimental data, no unequivocal statement can be made, so far, about the closeness of approach to equilibrium of the whole system involved in the MBE process, in relation to an isothermal state of the system [6.3].

Assuming for the MBE growth a far-from-equilibrium state of the whole system, one should take into account that the quality of the growing film, and especially its structural perfection, will depend significantly on whether partial, or at least local, thermodynamic equilibrium is established on a time scale of relevance to the system. If local thermodynamic equilibrium is established, one may use the usual notions of absolute rate theory [6.16] and the resulting rate expressions for kinetic processes, such as atom migration and evaporation, when describing the MBE growth process [6.14]. However, whether the combined influence of individual kinetic processes is able to move the growing film towards its partial equilibrium state [equilibrium (ii)] and whether or not it is able to reach this state on some relevant time scale of changes in the growth conditions depends upon the growth conditions and the specific nature of the system involved in the MBE growth. It is entirely possible that under certain growth conditions, local thermodynamic equilibrium (iii) may remain operative throughout growth, but the kinetics may not be fast enough to ensure partial equilibrium (ii) during the growth. The resulting microstructural and chemical properties of a film grown under local equilibrium conditions (iii) may be quite different from those which are exhibited by a film grown under conditions of partial equilibrium (ii), for the same degree of departure from global equilibrium (i) [6.14].

When turning to the partial equilibrium between the growing film and the substrate, i.e., when considering only the condensed phase part of the entire system involved in MBE growth (Fig. 6.1), it is often useful to split this problem further [6.3] and deal firstly with the dissociation reactions of molecular species adsorbed at the surface (see the precursor state introduced into the GaAs growth mechanism by *Foxon* and *Joyce* [6.1–3]) and subsequently with the rearrangement of atomic species so formed to produce the crystalline film.

According to *Heckingbottom* et al. [6.17] any species that will be finally incorporated into the crystal lattice of the growing film dwell in the surface layer, where the growth process occurs, for a definite time, making a large number of site changes before becoming incorporated (for typical MBE growth conditions of GaAs, the time is equal to about 1s and the number of site changes is about 10^6 [6.3, 17]). These authors have also pointed out that there exists extremely hindered kinetics as soon as the species become incorporated into the grown crystal. Otherwise the abrupt geometries that can be produced with MBE would be rapidly degraded by interdiffusion. Crucially, however, in the surface layer where the crystal forms, there must be sufficiently facile kinetics to allow the growth of highy perfect crystals.

6.1.2 The Transition Layer Concept

The presented considerations lead to the conclusion that the entire MBE system directly involved in the growth process consists of three different parts. The crystalline solid phase is one extreme, the gaseous phase is the second extreme, and a transition layer in between, where a gradual transition from a gas to a crystal occurs, creates the third part of the system (Fig. 6.1).

The crystalline phase, namely, the substrate or the overgrown epilayer, exhibits short-range order as well as long-range order in the spatial distribution of the constituent atoms. The gaseous phase created by the beams intersecting each other near the substrate surface is characterized by complete lack of order in the spatial distribution of the constituent atoms or molecules.

The transition layer, where all processes leading to epitaxy occur, is obviously the most important part of the growth system. Its geometrical form and the processes occurring there depend strongly on the growth conditions chosen. Consequently, if one wishes to grow a definite layer structure with MBE, one has to be able to adjust properly the properties (structural and compositional) of the transition layer [6.18]. We will hereafter call this layer the near-surface transition layer (NSTL).

The presented picture of the MBE growth system involving a specific NSTL is consistent with the general theory of epitaxy [6.19, 20]. This picture will be used when the specific MBE growth processes are discussed in the following sections. Keeping it in mind, we will consider first the role of the substrate surface in the growth process and then the phenomena occurring in the transition layer near this surface. It seems to be reasonable to proceed in this sequence because the growth process is realized at the interface between the substrate surface and that region of the NSTL which is nearest to this surface. In doing this we will try to be general enough to ensure the validity of the considerations presented below for different material systems which may be grown with MBE.

6.2 Relations Between Substrate and Epilayer

The surface of the substrate crystal plays a crucial role in the MBE growth process, because it influences directly the arrangement of the atomic species of the growing film through interactions between the outermost atomic layer of the surface and the adsorbed constituent atoms of the film.

Generally speaking, epitaxy, and so MBE, is a growth process of a solid film on a crystalline substrate in which the atoms of the growing film mimic the arrangement of the atoms of the substrate [6.21]. Consequently, the epitaxially grown layer should exhibit the same crystal structure and the same orientation as the substrate. This is true for epitaxial layers and structures of many practically important material systems, such as GaAs/AlGaAs or CdTe/HgCdTe. However, there are also many important cases [6.22] where the epitaxial layer has either a different orientation than the substrate crystal, e.g., CdTe (111) layers grown on GaAs (100) substrates [6.23, 24], or a totally different crystal structure, e.g., sphalerite Si (111) on hexagonal Al_2O_3 [6.25], zincblende CdTe (111) on hexagonal Al_2O_3 [6.26], or rocksalt PbTe (100) on zincblende GaAs (100) [6.27]. Moreover, it has also been demonstrated, with so-called grapho-epitaxy [6.28, 29], that single crystalline films may be grown on noncrystalline, suitably prepared substrates.

Three factors are important from the point of view of the mutual relation between the substrate and the epilayer. Lattice constant matching or lattice mismatch is the first, crystallographic orientation of the substrate is the second, and its surface geometry, or surface reconstruction, is the third. The main problem related to these three factors will be presented and briefly discussed in the following sections.

6.2.1 Critical Thickness for the Formation of Misfit Dislocations

The most frequent case of MBE growth is heteroepitaxy, namely, the epitaxial growth of a layer with a chemical composition and sometimes structural parameters different from those of the substrate. The heteroepitaxy crucial problems are related to lattice mismatch. When lattice mismatch occurs, it is usually accommodated by structural defects in the layer [6.30] or by strain connected with a relevant interfacial potential energy [6.31–33].

Following *van der Merwe* [6.32], we will use hereafter the term "misfit" to refer to the disregistry of the equilibrium interfacial atomic arrangements of the substrate and the overgrown epilayer. This disregistry results from differences in atomic spacings or lattice symmetries, which are characteristic of each of the two crystals in the absence of interfacial interaction between them.

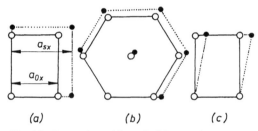

(a)   (b)   (c)

Fig. 6.2. Examples of interfacial atomic arrangements in "substrate-epilayer" systems of different crystallographic symmetry [6.32]. Black dots represent the substrate atoms while empty circles stand for overlayer atoms

In the simplest case, where both interfacial arrangements have rectangular symmetry (Fig. 6.2a), and the epilayer is fairly thin in comparison to the thickness of the substrate crystal, the misfit may be quantitatively defined as [6.33]

$$f_i = \frac{a_{si} - a_{oi}}{a_{oi}} \quad , \quad i = x, y \quad , \tag{6.1}$$

where a is the lattice constant (the natural lateral atomic spacing) and s and o designate the substrate and the overgrown layer, respectively. The convenient features of this definition are as follows. If a film is strained so that the lattices of film and substrate are in register at the interface, then the misfit strain defined by

$$e_i = \frac{a_{oi}^{str} - a_{oi}}{a_{oi}} \quad , \quad i = x, y \tag{6.2}$$

will be equal to f_i. In (6.2) a_{oi}^{str} stays for the lateral atomic spacing in the strained overgrown layer. If, however, the misfit is shared between dislocations and strain, then

$$f_i = e_i + d_i \quad , \quad i = x, y \quad , \tag{6.3}$$

where d_i is the part of the misfit accommodated by dislocations. A positive value for f implies that the misfit strain is tensile and that the misfit dislocations are positive Taylor dislocations [6.34] i.e., extra atomic planes lie in the overgrowth (Fig. 6.3).

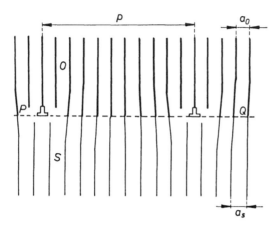

Fig. 6.3. A pure misfit dislocation geometry of edge type at interface PQ of crystals O (overgrowth) and S (substrate), with lattice spacings a_o and a_s, respectively [6.32]

The above definition does not allow for reconstruction [6.35] or unevenness of the substrate surface. Therefore, for very thin epilayers (1–2 monolayers thick), for which the equilibrium interatomic spacing may differ from that of the bulk crystal by as much as 5 % [6.36], *Frank* and *van der Merwe* defined misfit by [6.37]

$$f_i = \frac{a_{oi} - a_{si}}{a_{si}} \quad , \quad i = x, y \quad . \tag{6.4}$$

By this definition misfit is related to the fixed substrate interatomic spacing. However, when the overlayer is thick enough (if its thickness becomes comparable to the thickness of the substrate), both crystals are usually treated on the same footing by defining the misfit as [6.38, 39]

$$f_i = 2 \frac{a_{si} - a_{oi}}{a_{si} + a_{oi}} \quad , \quad i = x, y \quad . \tag{6.5}$$

It is well known that if the misfit between a substrate and a growing layer is sufficiently small, the first atomic layers which are deposited will be strained to match the substrate and a coherent (pefectly matched) epilayer will be formed. For such a state of the system "epilayer–substrate" the term "pseudomorphism" has been introduced by *Finch* and *Quarrell* [6.40]. However, as the layer thickness increases, the homogeneous strain energy E_H becomes so large that a thickness is reached when it is energetically favorable for misfit dislocations to be introduced.

The overall strain will then be reduced but at the same time the dislocation energy E_D will increase from zero to a value determined by the misfits f_1 and f_2. These misfits are defined by (6.5) in which, however, a_{oi} is replaced by a_{oi}^{str}, the lattice constant of the overgrowth strained at the interface.

Subscripts 1 and 2 stand for the two directions in the plane of the interface (not always perpendicular to each other, see, e.g., Fig. 6.2b, c) for which different strains occur. The existence of this critical thickness was first indicated in the theoretical study by *Frank* and *van der Merwe* [6.37], then treated theoretically by others [6.31–34, 41], and confirmed by various experimental observations [6.33, 39, 42–48]. Discussing the problem of the critical thickness, we will follow the considerations presented by *Ball* and *van der Merwe* [6.49]. Consequently, the system "overgrown epilayer–substrate crystal" will be called hereafter a "bicrystal".

The basic assumption of the theory is that the configuration of the epitaxial system is the one of minimum energy. For a particular bicrystal consisting of a semi-infinite substrate A and an epitaxial layer B of thickness t, the interfacial energy per unit area E_I will be [6.49]

$$E_I = E_H + E_D = \frac{\mu t}{1 - \nu}(e_1^2 + 2\nu e_1 e_2 + e_2^2)$$
$$+ \frac{\mu b}{4\pi(1 - \nu)} \sum_{i=1}^{2}\left[\frac{|e_i + f_{0,i}|}{\cos \gamma_i \sin \beta_i}(1 - \nu \cos^2 \beta_i)\ln\left(\frac{\varrho R_i}{b_i}\right)\right] \quad , \qquad (6.6)$$

where μ and ν are the interfacial shear modulus and Poisson's ratio, respectively, e_i are the strains in the overgrowth defined by (6.2) in which the rectangular coordinates x, y are replaced by the directions 1 and 2, respectively, b is the magnitude of the Burgers vector [6.30, 50] characterizing the dislocation at the interface, $f_{0,i}$ is the natural misfit between the layer and the substrate [given by (6.5) with x, y replaced by 1, 2], β and γ are the angles between the Burgers vector and the dislocation line, and between the glide plane of the dislocation and the interface, respectively, R_i stands for the cut-off radius of the dislocation which defines the outermost boundary of the dislocation's strain field, and ϱ is a numerical factor used to take the core energy of the dislocation into account (usually $\varrho = 4$ [6.49]). In many cases, such as the (001) interface of fcc, diamond, or sphalerite lattices, the misfits and lattice parameters will be identical with respect to the two perpendicular interfacial directions [110] and [1$\bar{1}$0] in which misfit dislocations are observed to lie [6.51]. For such a situation, (6.6) can be simplified, because also the homogeneous strains e_1 and e_2 will be the same $e_1 = e_2 = e$. It becomes

$$E_I = 2\mu t e^2 \frac{1 + \nu}{1 - \nu} + \mu b \frac{(|e + f_0|)(1 - \nu \cos^2 \beta)}{2\pi(1 - \nu)\cos \gamma \sin \beta}\ln\left(\frac{\varrho R}{b}\right) \quad . \qquad (6.7)$$

One should notice that the homogeneous strain energy E_H [first term on the right hand side of (6.7)] is zero at zero strain (if $e = 0$), while the dislocation energy E_D (the second term) falls to zero at

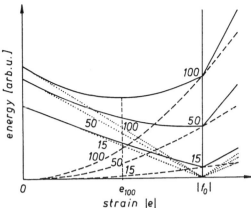

$$f = e + f_0 = 0 \quad , \tag{6.8}$$

which is the condition of pseudomorphism in the "layer–substrate" system, or of coherency of the layer with the substrate [6.52].

The actual configuration of the considered system will be given by the strain for which $E_I = E_I(e)$ is a minimum [6.49]. It can be seen from Fig. 6.4 that for film thicknesses of 15 and 50 times a_s this minimizing strain is $|e| = |f_0|$, corresponding to a coherent epilayer which is free of misfit dislocations. However, in the case of a film of thickness 100 times a_s the minimizing strain e_{100} is somewhat smaller, indicating that misfit dislocations will be present and spaced at intervals of [6.32]

$$p_{100} = \frac{a_s}{|f_0 + e_{100}|} \quad . \tag{6.9}$$

It will be noticed in the curves in Fig. 6.4 that when the thickness t is small the curve of E_I has a negative slope at $|e| = |f_0|_-$, and this is the condition for a coherent film with no misfit dislocations. As the thickness increases, the slope of E_I at $|e| = |f_0|_-$ increases to zero and then becomes positive. The thickness t_c at which the slope is zero is the critical thickness, above which misfit dislocations will be introduced. The criterion for the critical thickness is

$$0 = \frac{\partial E_I}{\partial |e|} \quad \text{evaluated at} \quad |e| = |f_0|_- \quad , \tag{6.10}$$

which gives the relation

$$t_c = \frac{b(1 - \nu \cos^2 \beta)}{8\pi |f_0|(1 + \nu) \sin \beta \cos \gamma} \ln\left(\frac{\varrho t_c}{b}\right) \tag{6.11}$$

from which t_c may be calculated for given natural misfit f_0. Since the introduction of the first dislocations is being considered the cut-off radius R is equal to the thickness of the film.

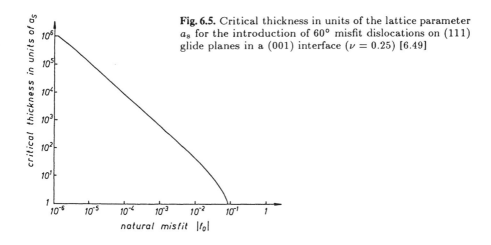

Fig. 6.5. Critical thickness in units of the lattice parameter a_s for the introduction of $60°$ misfit dislocations on (111) glide planes in a (001) interface ($\nu = 0.25$) [6.49]

Figure 6.5 shows a plot of $t_c(|f_0|)$ for the case of $\nu = 0.25$ where the misfit dislocations are of the $60°$ type lying on (111) glide planes in an (001) interface. Again, ϱ is taken as 4.

The (001) interface of diamond and sphalerite bicrystals has eight different possible misfit dislocation slip systems of the $60°$ type which will produce interfacial dislocations, and all of these are equivalent in the relief of misfit. The assumption was made that the misfit is relieved simultaneously in both of the interfacial directions 1 and 2 since the critical thicknesses for the $60°$ dislocations along these two directions will be the same.

In the case of interfaces of lower symmetry the dislocation slip systems are generally not equivalent in the relief of misfit and each system will have its own critical thickness. The dislocations which are in fact introduced will according to the minimum energy principle be those with the lowest critical thickness. The critical thickness for each system may be calculated from (6.10) with $e = e_1$ and E_I given by (6.6), and the interfacial direction 1 is perpendicular to the interfacial dislocations being considered.

As an example consider the (133) interface of the fcc structure. It is well known that the glissile dislocations in the structure generally lie along one of the $\langle 110 \rangle$ directions and are of the $60°$ or pure screw type. The screw dislocations can be ignored as they make no contribution to the relief of misfit. Only one of the $\langle 110 \rangle$ directions, namely $[01\bar{1}]$, lies in the interface and the misfit dislocations with this direction will be on either the (111) or the ($\bar{1}$11) glide plane. On each of these glide planes there are two $60°$ dislocations which will have the same critical thickness. For the critical thicknesses of the dislocations on each glide plane one obtains

$$t_c = \frac{b_1(1 - \nu \cos^2 \beta_1)}{8\pi(|f_{0,1}| + \nu|f_{0,2}|)\sin\beta_1 \cos\gamma_1} \ln\left(\frac{\varrho t_c}{b}\right) , \qquad (6.12)$$

which reduces to (6.11) if $|f_{0,1}| = |f_{0,2}|$, which is always the case for bicrystals of the same structure in parallel orientation.

223

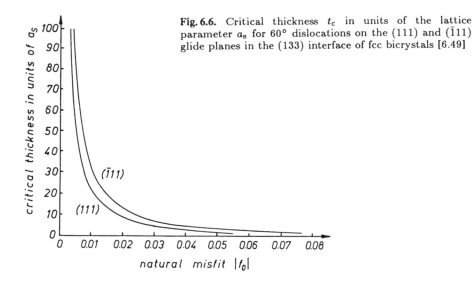

Fig. 6.6. Critical thickness t_c in units of the lattice parameter a_s for 60° dislocations on the (111) and ($\bar{1}$11) glide planes in the (133) interface of fcc bicrystals [6.49]

In the particular case of the (133) interface, the (111) glide plane is at an angle of $\gamma_1 = 22°$ and the ($\bar{1}$11) at an angle of $\gamma_1 = 48.5°$. For 60° dislocations $\beta_1 = 60°$. Figure 6.6 shows the curves of the critical thicknesses as they depend on the natural misfit for 60° dislocations on the two glide planes. It will be seen that the critical thickness for the introduction of misfit dislocations on the (111) glide plane is considerably lower than that for the dislocations on the ($\bar{1}$11) glide plane [6.49].

The comparison between the predicted critical thickness and experimental observations has been discussed by *Matthews* [6.33]. In metals the predicted and experimental values of t_c agree fairly well, but after coherency is lost the observed strains appear to be significantly larger than the theoretical predictions. In some materials, however, the thickness to which coherency persists is very much larger than the predicted value. For example, germanium films on gallium arsenide are coherent to a thickness of approximately 2 μm, while the theoretical result is approximately 3000 Å [6.53].

The explanation for this discrepancy is not known, but the mechanism by which the misfit dislocations are introduced must be of significance. For example, it has been observed [6.49] that in (133) interfaces of III–V compounds the misfit dislocations are on the ($\bar{1}$11) glide plane rather than the (111) glide plane, which has the lower critical thickness. The explanation probably lies in the fact that the resolved shear stress along the ($\bar{1}$11) glide is larger and the misfit dislocations are generated more readily on this plane, even though they are less efficient in the relief of misfit.

The calculated values of critical thicknesses, however, do serve as a useful indication of the lower limit of the thickness at which misfit dislocations are introduced. There is no case recorded, so far, of misfit dislocations being introduced at thicknesses below the critical thickness.

It is also possible to change the critical thickness of an epitaxial layer by growing one or more subsequent layers [6.51, 54]. For example, if an epilayer of B were deposited on substrate A, and the layer thickness t exceeded the critical thickness t_c it might be possible in theory to restore coherency by the deposition of a layer of the same material as the substrate. It has been shown [6.54] that there is a second critical thickness of the layer B, above which it is not possible to restore coherency, no matter how thick the top layer is grown.

It has also been shown [6.55] that a direct correspondence exists between the lattice constant of coherently strained multilayer structures and the lattice constant of a single coherently strained layer of equal thickness with a misfit equal to the misfit between the constituent layers spatially averaged over a single period of the mulitlayer. In particular, this correspondence allows one to predict the critical thickness of coherently strained multilayers if single-layer critical thicknesses are known. This has a special meaning for the MBE growth of strained-layer superlattice structures [6.56], and for device technology, which will be discussed in more detail in the next chapter.

6.2.2 Role of the Crystallographic Orientation of the Substrate

Many experimental results indicate that the crystallographic orientation of the substrate plays an important role in MBE. When III–V compounds are grown with MBE on III–V substrates, the substrate orientation influences considerably the incorporation process of dopants. This concerns the intentionally introduced dopants like Si in GaAs and AlGaAs/GaAs heterostructures [6.57–60], as well as the unintentionally incorporated contaminants like C in GaAs [6.59, 61].

The orientation of the substrate surface influences also the electrical and optical properties of GaAs [6.60] and the AlGaAs/GaAs heterostructures [6.59]. This results from the difference in the band structure for different substrate orientations [6.62], and from the influence of the substrate orientation on the electrical compensation of the dopants introduced into the heterostructure [6.58, 63]. These dependences are especially important for the so-called modulation doped configuration [6.64] of GaAs/AlGaAs heterostructures [6.57, 62, 65]. Drastic improvement in electrical and optical quality of the epitaxial layers grown on the GaAs(111)B surface has been achieved by slightly misorienting the substrates [2° off towards the (100) orientation]. Such misorientation introduces surface steps [6.62] (Fig. 6.7), and thus changes the growth mechanism. The effect of substrate misorientation on the surface morphology and photoluminescence spectra of III–V layers on III–V substrates has been further demonstrated by growing AlGaAs layers on lenticular (planoconvex shaped) GaAs substrates with orientations close to (100) [6.66]. It has been found that the smoothest areas, which simultaneously exhibited the narrowest photoluminescence lines (only band-to-band emission was present) were centered 6° off (100) towards the (111)A face, i.e., where growth occurred on monatomic steps terminated by Ga atoms.

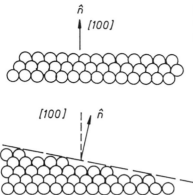

Fig. 6.7. Schematic illustration of how surface steps occur *(lower panel)*, by slightly misorienting the (100) surface *(upper panel)* of a substrate crystal

A strong influence of the substrate orientation on the growth parameters has also been demonstrated in MBE of narrow gap II–VI compounds (e.g., $Hg_{1-x}M_xTe$; M =Cd, Mn, Zn) [6.67, 68]. In this case the substrate orientation considerably influences the surface condensation coefficients [Ref. 6.69, pp. 1–28] of Hg and Cd atoms. During the MBE growth of these compounds on CdTe(111) surfaces at 185° C, about one order of magnitude less Hg is required for the growth when MBE is performed on the (111)B face (the outermost atomic layer is formed by Te atoms) instead of the (111)A face (the outermost atomic layer is formed by Cd atoms). The amount of Hg necessary for growth on the CdTe (100) surface lies in between. The dependence of the surface condensation coefficient on substrate orientation is less dramatic for Cd than for Hg, however, a similar tendency is observed.

As the next example of the influence of substrate orientation on MBE growth, the experimental data concerning polar-on-nonpolar MBE may be considered [6.70–72]. When a polar semiconductor (a compound like GaAs) is grown on a nonpolar semiconductor substrate of similar crystal structure (an elemental semiconductor like Si or Ge), new physical phenomena arise, which are not present in the heteroepitaxial growth of one III–V compound upon another. These are (i) antiphase disorder on the compound side of the interface [6.70], (ii) lack of electrical neutrality at the interface [6.73], and (iii) cross-doping [6.72, 74]. All these phenomena are sensitive to substrate orientation, as has been shown by growing GaAs on differently oriented Si substrates [6.70–72], or by investigating the surface structure of GaAs films grown on Ge surfaces [6.75].

The presented experimental data may be understood by analyzing the geometry of chemical bonds on the substrate surface for different crystallographic orientations of crystals with the same bulk structure. Figure 6.8 shows schematically the geometry of surface bonds for some selected orientations of GaAs, which is the substrate material most frequently used for III–V MBE. Similar schematic illustrations may be constructed for other semiconducting crystals, showing the geometry of surface bonds when the orientation is changed [6.76].

226

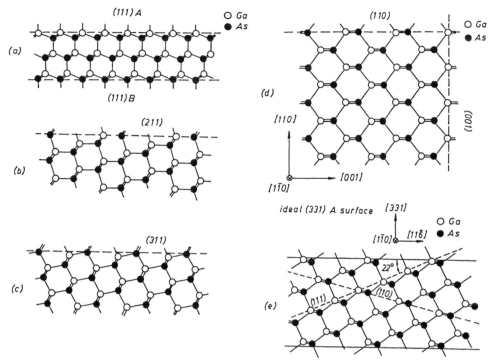

Fig. 6.8a–e. Surface bonds for differently oriented GaAs substrate crystals. (a) (111) surface viewed along the [01$\bar{1}$] direction, showing the A(Ga) and the B(As) faces [6.57]. (b) (211) surface viewed along the [01$\bar{1}$] direction [6.58]. (c) (311) surface viewed along the [01$\bar{1}$] direction [6.61]. (d) A cross section of the crystal lattice showing the planes (110) and (001) [6.73]. (e) The (331) surface viewed along the [110] direction [6.59]

As an example of how the surface bond geometry may help to explain the orientational dependences of the MBE process, we will discuss in more detail the problems of intentional and unintentional doping in GaAs with group IV elements. We will follow the considerations of *Wang* et al. [6.57, 61].

The starting point for this discussion is that the GaAs(111) orientation exhibits the (111)A and (111)B surfaces, as shown in Fig. 6.8a. One would expect Si doping atoms to bond to surface Ga atoms, taking up sites in the As sublattice and acting as a p-type dopant on the (111)A surface. This has been confirmed experimentally by electron diffraction observations by *Cho* et al. [6.77] and by Hall measurements by *Wang* et al. [6.57]. For the GaAs(111)B surface, *Wang* et al. found Si to be n-type, as has already been reported by *Cho* [6.77]. Typical room-temperature Hall measurements of (111)A homoepitaxial layers showed $p = 1.3 \times 10^{18} \, \text{cm}^{-3}$ at a mobility of $100 \, \text{cm}^2 \, \text{V}^{-1} \text{s}^{-1}$, while on (111)$B$, $n = 2.9 \times 10^{17} \, \text{cm}^{-3}$ at a mobility of $2200 \, \text{cm}^2 \, \text{V}^{-1} \text{s}^{-1}$ has been obtained. These mobilities indicate compensation of Si, which is most likely due to reconstruction into a somewhat nonpolar (110)-like surface [6.78]. Surface polarity on

GaAs exists not only on the (111) orientation, but in principle on (211), (311), etc. *Wang* et al. [6.57] have also grown bulk GaAs, bulk $Al_{0.26}Ga_{0.74}As$ and $Al_{0.26}Ga_{0.74}As/GaAs$ heterostructures on (211), (311), (511), (711), and (911) surfaces. All these substrates were cut from the same boule, and the A and B polarities were noted, although for some high index planes, the A, B polarities no longer yield distinguishable doping results. It was found that Si predominantly occupies As sites and acts as a p-type dopant on $(111)A$, $(211)A$, and $(311)A$ orientations, and incorporates on Ga sites as an n-type dopant on $(111)B$, $(211)B$, and $(311)B$. For (511) and higher index planes, Si always behaves as an n-type dopant.

These results can be explained based on the surface bonding of each orientation (Fig. 6.8). Recall that the (211) surface consists of twice as many single dangling bond sites as there are double dangling bond sites. Therefore, the (111)-like character outweighs the (100)-like character. The (311) surface consists of equal densities of single and double dangling bond sites. Therefore, (111) and (100) components are equally weighted. For (511) orientation and higher index planes, the (100)-like double dangling bonds outnumber the (111)-like single dangling bonds. This explains the observation that Si behaves as a p-type dopant only on $(111)A$, $(211)A$, and $(311)A$ surfaces.

The surface morphologies of all these epitaxial layers are excellent, as can be best illustrated by the typical two-dimensional carrier mobilities (Table 6.1) in modulation doped $Al_{0.26}Ga_{0.74}As/GaAs$ heterostructures grown on these high index planes (Sect. 5.5.1). The two-dimensional nature of the hole gas on $(311)A$ orientation has been confirmed by magnetoresistivity measurements at low temperatures [6.57, 79, 80].

The polar (311) surface (Fig. 6.8c), on which there is one single dangling bond site and one double dangling bond site in each surface unit cell, provides a tool for studying incorporation of amphoteric impurities. Since bonding between

Table 6.1. Characteristics of modulation doped $Al_{0.26}Ga_{0.74}As/GaAs$ heterostructures on high-index planes [6.57]

Orientation	Sheet carrier density $[10^{11}$ cm$^{-2}]$	4.2 K mobility $(10^5$ cm^2 V^{-1}s$^{-1})$
$(211)A$	p 1.2	0.63
$(211)B$	n 5.1	1.9
$(311)A$	p 1.7	0.88
$(311)B$	n 4.3	4.1
$(511)A$	n 4.9	6.4
$(511)B$	n 5.2	5.9
$(711)A$	n 4.5	1.7
$(711)B$	n 5.2	2.1
$(911)A$	n 5.2	6.1
$(911)B$	n 5.6	3.6

the adatom at the single dangling bond site and the substrate atom is weaker than the bonding between the adatom at the double dangling bond site and the substrate atom, one would expect that the limiting process in GaAs crystal growth by MBE is the incorporation of adatoms on the single dangling bond site. Therefore, the single dangling bond site will be more readily available and the group IV impurity will bond to the Ga atoms, taking up As lattice sites on the $(311)A$ surface and bond to As atoms, taking up Ga lattice sites on $(311)B$. This is the case of Si [as already discussed, an exclusively n-type dopant on the (100) orientation] which behaves as an n-type dopant on the $(311)B$ orientation and p-type on $(311)A$. If carbon were amphoteric, then it would also behave consistently as n-type on $(311)B$ and p-type on $(311)A$.

For comparison studies [6.61], normally three substrates, with (100), $(311)A$, and $(311)B$ orientations, were loaded side by side on a molybdenum substrate holder. MBE growth of GaAs employed the typical growth conditions used for (100) orientation, i.e., a substrate temperature of $580°$–$600°$ C under As_4-rich conditions. Epitaxial GaAs $7\,\mu m$ thick was grown at a rate of $1\,\mu m/h$. Capacitance–voltage, van der Pauw Hall measurements, (Sect. 5.5.1), and low-temperature photoluminescence (PL) (Sect. 5.4.3) were used to characterize the samples. The major electric results of six growth runs are as follows [6.61].

1. All the GaAs epilayers grown on (100)-oriented substrates showed p-type conductivity with a room-temperature hole concentration of $(1–3) \times 10^{14}$ cm^{-3}, and 77 K mobilities ranging from 6000 to $8500\,cm^2\ V^{-1}s^{-1}$. These results are typical of good undoped GaAs grown by MBE.
2. All the GaAs grown on $(311)A$ orientation showed p-type conductivity with a room-temperature hole concentration of 6×10^{14}–1×10^{15} cm^{-3}, i.e., always 3–5 times higher than was obtained for (100)-oriented substrates. Most importantly, on $(311)A$ the liquid-nitrogen hole mobility was always 10 %–40 % higher than for undoped GaAs films grown side by side on (100)-oriented substrates. This indicates that all the unintentionally doped p-type GaAs layers grown on the (100) orientation were heavily compensated, since they showed lower hole concentrations and lower mobilities than were observed for $(311)A$.
3. Of the six runs grown on the $(311)B$ orientation, two showed n-type conductivity. One showed a room-temperature electron concentration of $7 \times 10^{13}\ cm^{-3}$ and mobility of $8200\,cm^2\ V^{-1}s^{-1}$, and a liquid-nitrogen mobility of $1.3 \times 10^5\ cm^2\,V^{-1}s^{-1}$. Another showed a room-temperature electron concentration of $3 \times 10^{13}\ cm^{-3}$ with an electron mobility of $8050\,cm^2\ V^{-1}s^{-1}$, and a liquid-nitrogen mobility of $1.2 \times 10^5\ cm^2\ V^{-1} s^{-1}$. The other four $(311)B$ GaAs layers were of high resistivity.

The fact that all p-type (100) GaAs layers have a lower hole concentration (together with a lower hole mobility) than p-type simultaneously grown $(311)A$ GaAs films provides conclusive evidence that the normally undoped p-type GaAs (100) grown by MBE is heavily compensated. This raises the question of what the compensating donors are. Previously, sulfur and silicon have been identified by photothermal ionization spectroscopy [6.81–83] as major residual donors in MBE-grown GaAs. Since the growth of GaAs on the (311)B orientation does not consistently provide n-type films, it seems unlikely that the residual donor is carbon, because carbon-related species are always present in the MBE growth system.

If carbon is not the predominant residual donor in the (311)B n-type GaAs films, then it means that it is extremely difficult to force carbon to take up the Ga site to become a donor. Two arguments may explain this: (1) the strain associated with a very small impurity like carbon occupying a large Ga lattice site (but the strain effect [6.84] is on the order of phonon energy, which may be too small to be the driving force), and (2) the driving force of electronic origin due to carbon's strongly electronegative nature would appear to be more important.

It is therefore more likely that sulfur, which is contained in the As source, is the residual donor. Fluctuation of the As beam pressure and minor growth temperature differences from run to run could vary sulfur incorporation, leading to varying doping results.

Similar considerations, based on the surface bonding structure which is a characteristic feature of definite crystallographic orientations of the substrate crystal, may be used in order to explain the other experimental results presented at the beginning of this section. Instead of doing this, we will emphasize here that a classical thermodynamic approach, neglecting the atomistic properties of the substrate surface, is not well adapted to understanding the MBE growth, even though such an approach can provide very useful information [6.68].

6.2.3 Role of the Substrate Surface Reconstruction

The significance of surface reconstruction (Sect. 4.1.3) for the growth process and the resulting surface morphology, stoichiometry and other properties of the MBE grown films has remained largely unexamined [6.85]. In part this is a consequence of the poor accessibility of the dynamic growth front for in situ, real time, probing. Even the UHV environment typical for MBE, which permits use of external coherent beams such as electrons, ions or light for diagnostic purposes during the growth, has not enabled many experimental data to be obtained on this subject. Indirect glimpses of the role of reconstruction on a static or growing GaAs(100) surface are to be found in the differences between materials and heterostructures grown on As-stabilized (2 × 4) versus Ga-stabilized (4 × 2) GaAs(100) surfaces [6.4, 86]. On the other hand, the reconstruction of the GaAs(100) surface itself, as well as the MBE conditions required for creation of a definite reconstruction of this surface, has been studied for a long time [6.3]. This is a consequence of the fact that the (100) orientation is the one most frequently used for epitaxy of GaAs, as it has two sets of orthogonal (110) cleavage planes, which is important for device technology [6.87]. Reports have also appeared of the presence of long-range ordering in the Al distribution in $Al_x Ga_{1-x} As$ films

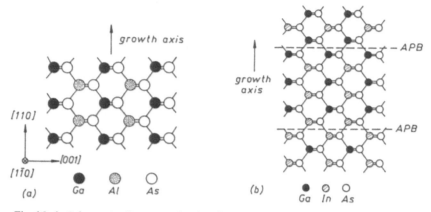

Fig. 6.9a,b. Schematic diagrams of ordered III_1–III_2–V ternary compounds grown by MBE on III–V (110) substrates. **(a)** The long-range order in $Al_{0.5}Ga_{0.5}As$ grown on GaAs(110) [6.88, 89]. **(b)** The ordered $In_{0.5}Ga_{0.5}As$ structure with two antiphase boundaries (APB) grown on InP(110) [6.90]

grown under certain conditions on GaAs (100) and (110) surfaces [6.88, 89]. For the (100) growth an in-plane ordering with a short-range (near atomic) periodicity and for (110) growth similar ordering in the growth direction as well have been found. Thus, the growth along the [110] direction produces much stronger ordering than the growth along the [100] direction (Fig. 6.9a). A similar long-range ordering effect has also been found for $In_{0.5}Ga_{0.5}As$ grown on InP substrates along the [110] direction [6.90]. However, in this case a high density of antiphase boundaries in the ordered ternary compound phase has been found (Fig. 6.9b). Earlier, only a quasi-periodic ordering in the [110] growth direction for growth on GaAs (110) with a periodicity of $\sim 10\,nm$ had been reported [6.91].

For these long-range ordering in III–V ternary compounds a charge transfer dipole stabilization model was proposed [6.88]. The physics of this model is as follows. The configurational-dependent electrostatic energy of a ternary III–III–V compound is due to the interaction of two charged species III_1 and III_2 occupying one of the face-centered-cubic (fcc) sublattices in the zinc blende structure. It is plausible that in the ground state, the energy of the system is minimized by having a completely segregated monolayer superlattice structure as shown in Fig. 6.9. This means that the electrostatic interaction due to the difference in charge transfer between III_1–V and III_2–V causes a "dipole" interaction which stabilizes the ordered structure [6.88]. Taking this model into consideration one may wonder whether the origin of the long-range ordering phenomenon occurring in MBE growth may not be directly related to the surface reconstruction. This point of view may be supported by the experiments concerning ordering effects in $GaAs_{1-x}Sb_x$ grown on GaAs(100) and on InP(100) substrates. The first substrate is lattice mismatched while the second one is lattice matched to the ternary compound [6.92]. These experiments have revealed a unique evolution of the ordering with the growth temperature, showing the formation of a short-range order at lower temperatures and a long-range order at higher temperatures. A distinct anisotropy of the formation of ordered structures between [011] and [0$\bar{1}$1] axes was observed for all epilayers investigated. The analysis of the structure conditions, as well as of the evolution and anisotropy of the ordering effect has led the authors of [6.92] to the conclusion that these effects have been caused by the surface atomic structure of the growing epilayers.

In order to explain the role of surface reconstruction in MBE growth we will present here the model and the relevant considerations of *Ogale* et al. [6.85]. We will begin with the recognition that surface reconstruction suggests the presence of a periodic array of local potentials and charge distributions associated with inequivalent sites. Consequently, such surface kinetic rates as migration, desorption, and dissociative molecular reaction would vary periodically from one such site to another. In addition, the rates for two different species would naturally be different. It should thus be expected that the nature of surface reconstruction during growth and its restoration towards bulk epitaxial sites over some lateral and vertical (i.e., the growth direction) length scale would play a significant role in determining the outcome for the grown material. Ogale et al. have employed Monte Carlo computer simulations based upon the configuration dependent reactive incorporation (CDRI) model of the MBE growth process [6.15, 93–96]. The details of the physical ingredients and basis of the model, as well as the simulation procedures and reliability have been previously established [6.15]. These previous simulations, however, were for unreconstructed surfaces. Information on the specific positions of atoms at and near the surface for the As(2 × 4) reconstruction is still unsatisfactory although the papers published by *Chadi* [6.97], *Frankel* et al. [6.98], *Larsen* and *Chadi* [6.99], and *Farrell* et al. [6.100] have changed the situation considerably. The two-unit periodicity has been shown to be along the direction of the dangling orbitals of the surface atoms. Proposed structural models of the surface imply the four-unit periodicity to be similar to the two-unit behavior with a difference only in the nature of the buckling of the surface bonds. Consequently, *Ogale* et al. [6.85] considered in the simulations a surface with a two-unit periodicity in this direction as well, i.e., a (2 × 2) reconstructed GaAs (100) surface. This was sufficient for examining the possible role of growth kinetics modulated by surface reconstruction.

They labeled the two inequivalent sites favorable (f) and unfavorable (u), based on the inequivalent kinetic rates for migration from, adsorption on, and desorption from such sites. Dynamic recovery from the surface reconstruction in the growth direction was assumed, in these first simulations, to occur over a length scale of two monolayers at the growth front. For GaAs homoepitaxy, the activation energy for Ga hopping to (from) a f-site was reduced (increased) by a factor of four with respect to the rates employed for an unreconstructed surface. The As$_2$ and Ga desorption rates were made inequivalent for the f and u sites in the same proportion while leaving unchanged the desorption rates for As$_2$ pairs for which the separation between As atoms is not significantly influenced by the reconstruction.

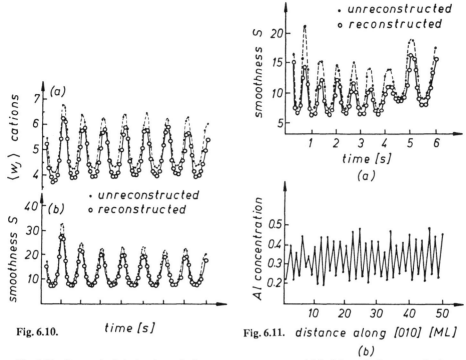

Fig. 6.10.   time [s]

Fig. 6.11.   distance along [010] [ML]

(b)

Fig. 6.10. Dynamical behavior of the average terrace width **(a)** and the growth front smoothness parameter **(b)** for GaAs(100) homoepitaxy on reconstructed surfaces [6.85]

Fig. 6.11. (a) The smoothness parameter S during growth of 6ML of $Al_{0.33}Ga_{0.67}As(100)$ followed by GaAs on reconstructed and unreconstructed As-stabilized GaAs(100) substrate. **(b)** The Al concentration in successive planes along the [010] direction in the $Al_{0.33}Ga_{0.67}As(100)$ alloy [6.85]

The results of the performed simulations are presented in Fig. 6.10 for GaAs (100) homoepitaxy and in Fig. 6.11 for $Al_{0.33}Ga_{0.67}As(100)$ heteroepitaxy [6.85]. Figure 6.10a shows the time dependence of the average terrace width at the growth front in one of the two principal directions of the (2×2) surface unit cell. The average terrace width for the reconstructed surface is seen to be smaller than for the unreconstructed surface. This is due to the preferential filling of the f-sites arising from the longer residence time of cations at such sites encountered during the migration process. While average terrace width provides an insight into surface smoothness in the lateral direction, a measure of smoothness in the growth direction is provided by the mean square fluctuation in the growth front step height profile, σ_h^2. This quantity is defined by [6.101]

$$\sigma_h^2 = \frac{1}{N} \sum_{i,j} (h_{ij} - \langle h \rangle)^2 \quad , \tag{6.13}$$

where $h_{i,j}$ is the height to which the film has grown over the substrate site

(i, j), $\langle h \rangle$ is the average height of the film, and N is the number of lattice sites in one layer. Following *Madhukar* et al. [6.15, 101, 102] an overall measure of the growth front smoothness (S) has been employed in [6.85]. This smoothness parameter is defined by

$$S = \frac{\langle W_I \rangle \langle W_J \rangle}{\sigma_h^2} \quad , \tag{6.14}$$

in which $\langle W_{I,J} \rangle$ are the growth front average terrace widths in the two reconstruction directions.

Figure 6.10b shows the behavior of S for both reconstructed and unreconstructed surfaces. While the damped oscillatory nature of the growth front smoothness arising from an essentially layer-by-layer incorporation of material is maintained, the degree of roughness is seen to be enhanced by the reconstruction-induced introduction of inequivalent kinetic rates within the surface unit cell. Note that the oscillation period coincides with the delivery and incorporation of a monolayer's worth of Ga since, for the experimental conditions most commonly employed for GaAs, the desorption rate of Ga in the simulations is essentially zero.

The role of surface reconstruction modulated surface kinetics is manifested even more significantly in the atomistic nature of lateral and vertical chemical distribution at heterointerfaces. In Fig. 6.11a the behavior of the smoothness parameter is shown for growth of 6 ML of $Al_{0.33}Ga_{0.67}As$ on a static GaAs surface followed by deposition of GaAs without interruption of growth. The GaAs growth rate (1 ML/s) and kinetic rates are the same as for Figs. 6.10a and b, whereas the Al kinetic rates are suitably chosen to incorporate its greater reactivity and slower surface migration [6.102, 103]. Once again, one notices the reduced degree of surface smoothness for the reconstructed case and a lower smoothness for $Al_{0.33}Ga_{0.67}As$ compared to GaAs. The true significance of the surface reconstruction is however revealed in Fig. 6.11b. Plotted is the concentration of Al atoms in successive planes along the [010] direction. One notes a remarkable long-range, in-plane ordering of Al in what was expected to be a random alloy. No such ordering is found for growth on an unreconstructed surface. Rather, only random fluctuations with an amplitude $< 1\%$ are found. The long-range ordering has been observed experimentally for $Al_x Ga_{1-x}As$ alloys [6.88, 89]. The observation has caused speculation and controversy as to the origin of this effect. The presented simulation results provide one possible explanation in terms of surface reconstruction dependent kinetic rates during epitaxial growth. An important feature of this proposed explanation is that it predicts that the degree of occurrence of the phenomenon is tied to the kinetics of growth that depend on the chosen growth conditions (i.e., substrate temperature, group V pressure, and growth rate) as well as to the nature of the surface reconstruction on the starting substrate surface and during growth [6.85, 96].

6.3 The Near-Surface Transition Layer

The concept of a transition layer located between the substrate surface and the gaseous environment of the intersecting beams is based on experimentally well-documented phenomena, namely, on chemisorption and physisorption, as well as surface migration and particle clustering (Sect. 1.1.2). The reason for the transition layer to be formed near the substrate surface is the existence of interactions between this surface and the particles of the impinging beams. The solid surface, as seen by an approaching particle (atom or molecule), is not neutral and thus causes some distance-dependent interaction described by a relevant interaction energy. The parameters describing this energy depend on the definite surface-particle system, and in many cases they are hard to determine exactly. They are, however, of importance, because the interaction energy describes the geometry, and to a large extent also the processes characteristic of the transition layer, and consequently for the growth process in MBE.

6.3.1 Physical and Chemical Adsorption

We first present some definitions and explanations concerning the adsorbed state of particles interacting with a solid surface (in the case of MBE this is the substrate surface).

Surface adsorption is a state in which the particle entering, or dwelling in, the near-surface region interacts with the surface. This interaction is described, in terms of thermodynamics, by the heat of adsorption, which is defined according to the first law of thermodynamics by

$$\Delta H = \Delta E + P(V_g - V_s) \quad , \tag{6.15}$$

where ΔE is the change in the internal energy of the system "gas particles – adsorbed particles" and the second term in the sum is the work connected with the adsorption process: P is the pressure in the gas phase and V_g and V_s are the volumes of the gas particles in the gas phase and in the surface adsorbed phase, respectively. It has become conventional to distinguish two kinds of adsorption, namely, physical adsorption (called hereafter physisorption) and chemical adsorption (called hereafter chemisorption) [6.104–106]. The indication of the type of adsorption is given by the value of the heat of adsorption. In the case of physisorption the particle–solid surface interaction is weaker than in chemisorption, therefore, also the adsorption heat for physisorption ΔH_p is smaller than ΔH_c the adsorption heat for chemisorption. Typical values are in the range [6.104] $\Delta H_p \leq 25\,\text{kJ/mol} = 0.26\,\text{eV/particle}$, and $25\,\text{kJ/mol} \leq \Delta H_c \leq 10^2 - 10^3$ kJ/mol or $0.26\,\text{eV/particle} \leq \Delta H_c \leq 1\text{–}10\,\text{eV/particle}$, although the value of $0.38\,\text{eV/molecule}$ has been found for the desorption energy of As_4, molecules from the GaAs(100) surface and has been attributed as a value corresponding to a physisorbed state [6.107].

The nature and the strength of the bonding forces responsible for adsorption depend on the interaction between the surface and the particle (atom or molecule) entering the near-surface region. If the particle becomes stretched or bent but retains its idenity, the van der Waals forces bind it to the surface, and one should speak of physisorption. If, on the other hand, the particle loses its identity, chemical bonds (e.g., ionic or covalent) attract it to the solid surface. This is the case of chemisorption.

Physisorption and chemisorption are additive, i.e., the van der Waals interaction is still operative in the case of chemisorption and may add a significant contribution to the total binding energy of the particle which is adsorbed on the solid surface [6.105]. As a result of the nature of the binding forces, in physisorption the adsorbate determines the interaction with the solid surface, while only little, or even no dependence on surface orientation is observed. But in chemisorption the surface plays the dominant role. Consequently, for different substrate orientations different bonds may be created with the same adsorbate (Sect. 6.2.2).

Physisorption and chemisorption appeared in the MBE formalism at its very beginning, as a result of studies on interaction of As_2 and As_4 molecules with the GaAs(100) surface [6.107–119]. The results of these studies have been presented in a schematical form illustrating the kinetics of the MBE growth of GaAs at temperatures below 600 K from Ga and As_2, and also from Ga and As_4 beams [6.111] (Fig. 6.12a,b). These diagrams can be found in many reviews, e.g. in [Ref. 6.112, Chap. 2; Ref. 6.113, Chap. 3] or [6.1–3, 114].

In the growth from dimeric arsenic and gallium, As_2 molecules are first adsorbed into a mobile, weakly bound precursor state (Fig. 6.12a). Dissociation of adsorbed As_2 molecules can occur only when the molecules encounter paired Ga lattice sites while migrating on the surface. In the absence of free surface Ga adatoms, As_2 molecules have a measurable surface lifetime but no permanent condensation occurs. The sticking coefficient of As_2 molecules is thus a function of the arrival rate of Ga atoms I_{Ga}, being unity only for a monolayer of Ga atoms or for $2I_{Ga} > I_{As}$. From these kinetic processes it follows that stoichiometric GaAs can be grown for relative fluxes (beam intensities) of As_2 and Ga larger than or equal to unity.

In growth from tetrameric arsenic and gallium, As_4 molecules are adsorbed into a mobile precursor state and in the absence of a surface Ga population have a sticking coefficient equal to zero, but a measurable surface lifetime. The temperature dependence of this lifetime reveals a desorption energy of the As_4 molecules equal to about 0.4 eV. With a coincident Ga flux, As_4 exhibits a temperature-independent sticking coefficient (in the temperature range between 450 and 600 K) which is, however, a function of I_{Ga}, the Ga beam intensity. The desorption of As is a second-order process with respect to its adsorption when $I_{Ga} \gg I_{As_4}$, and is the prevailing process when $I_{Ga} \ll I_{As_4}$. The sticking coefficient of As_4 is always less than or equal to 0.5, even when $I_{Ga} \gg I_{As_4}$. However, when $I_{Ga} \ll I_{As_4}$ one As atom sticks for every Ga atom supplied.

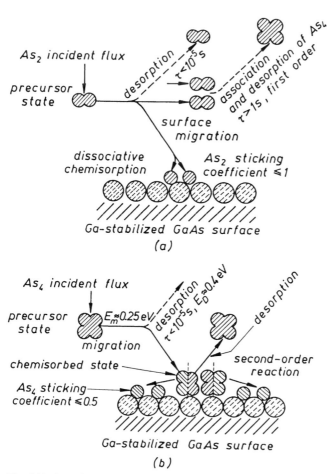

Fig. 6.12a,b. Schematic illustration of the MBE growth models of GaAs on the Ga-stabilized GaAs(100) surface at temperatures below 600 K. (a) Growth from Ga and As₂ beams. (b) Growth from Ga and As₄ beams [6.111]

These results can be explained by a process of dissociative chemisorption, with a pairwise interaction of As_4 molecules adsorbed on adjacent Ga lattice sites (Fig. 6.12b). When the As_4 surface population is small compared to the number of Ga sites, the rate-limiting step is the encounter probability between As_4 molecules, leading to second-order kinetics. As the As_4 surface population is increased, there is an increasing probability that an arriving molecule will find adjacent sites occupied, and the desorption rate becomes proportional to the number of molecules being supplied (a first-order process).

The presented growth models describe well the real physical situation when the growth temperature of GaAs is lower than or equal to 600 K. During the growth of GaAs at growth temperatures above 600 K an additional process occurs on the substrate surface. Molecules of As_2 are lost by desorption from the GaAs surface at a rate which increases with increasing temperature. This desorption process creates a Ga adatom population on the surface, and in the absence of an impinging arsenic flux results in a Ga-stabilized structure being established. In extreme cases free gallium may be produced on the substrate surface. As_4 lost in this way from the surface may be replaced by dissociation of incident As_2 or As_4 beams to maintain an As-stabilized structure provided that the arsenic supply rate exceeds the rate of loss of As_2 by thermal desorption. The flux ratio I_{As_4} to I_{Ga} required to maintain a specific surface structure during growth will therefore depend upon the substrate temperature and the absolute magnitude of the fluxes. The growth rate of MBE-crystallized GaAs films is entirely controlled by the flux density of the Ga beam impinging on the substrate surface. This implies that nearly all the Ga atoms incident on the surface get incorporated into the growing epitaxial layer.

The growth model established from kinetic measurements is not unique to GaAs but is also valid for AlAs, InP, a number of other III–V compounds, and, with minor modifications, ternary III–III–V alloys [Ref. 6.113, Chap. 18].

Following the described models, based on experiments performed with modulated molecular beams [6.108–110] and static RHEED [6.3], computer simulations based on Monte Carlo methods have been performed [6.93–95, 115–121], which have also provided a great deal of insight into the atomistic nature of the MBE growth process. By a combination of these experimental and theoretical techniques a conceptual picture of the MBE growth process of a semiconducting compound has emerged [6.119, 121]. In this picture a monatomic cation and molecular anion are involved as shown in Table 6.2. The key results which have led to this picture are as follows [6.118].

1. The cation surface kinetics is critical in controlling the growth-front (and interface) quality. Substrate temperature and, to a lesser extent, anion overpressure control the cation kinetics in MBE growth.

2. For a fixed growth rate, a given semiconductor can grow by the layer-by-layer mode above a certain temperature where the cation hopping rate is 10^4 hops/s for monolayer growth.

Table 6.2. Scheme of MBE of compound semiconductors [6.121]

Cation	Anion
Impingement from vapor	Impingement
↓	↓
Chemisorption	Physisorption
↓	↓
Surface evaporation	Surface migration/evaporation
↓	↓
Surface migration (in-plane; interplane)	Dissociative chemisorption
	↓
	Surface migration/evaporation
↓	↓
Incorporation	Incorporation

3. Since different semiconductors have different bond strengths and hence different activation barriers for cation hopping, the ideal temperatures for growth are different.

4. Due to the observations noted in point 3, the quality of the normal and inverted interfaces can be quite different. The term normal is used for the interface produced by the sequence in which the lower melting temperature component is grown first. The term inverted interface is used for the reverse sequence.

5. The surface roughness of a growing structure gets worse as the film thickness grows due to statistical fluctuations of the impinging cation flux.

The results listed above have been very important in understanding a variety of observations on MBE-grown heterostructures. According to these studies certain generic conclusions can be drawn regarding heterostructures grown by the MBE technique. In general, if one is interested in heterostructures grown from AC-BC, (A, B cations, C anions), one finds that (a) the ideal growth conditions (for layer-by-layer growth) are different for material AC than for material BC, and (b) the ideal growth conditions for the normal and inverted structures may be quite different. The differences will increase as the bond energy difference between A-C and B-C bonds increases.

As shown in Fig. 6.13, there are at least three anion incorporation mechanisms in MBE of GaAs. The simulation procedure described in [6.119, 120] allows one to investigate the influence of the growth conditions on the occurrence of the individual mechanisms, and thus to investigate how the NSTL is

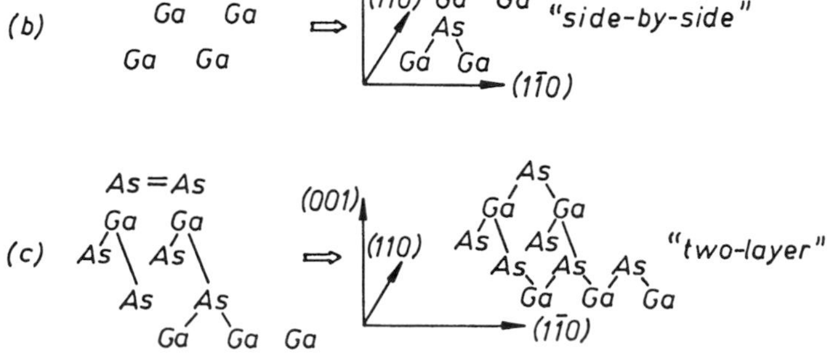

Fig. 6.13a–c. Schematic illustration of the anion incorporation mechanism according to the atomistic model of GaAs MBE [6.120]. On the left the As$_2$ molecules are shown in their precursor state near the substrate surface

Table 6.3. Percentage of anions incorporated by the three different modes shown schematically in Fig. 6.13 for various flux ratios. Note that some do not total 100 due to vacancies remaining in the layers [6.120]

In-line	Side-by-side	Two-layer	Flux ratio
57	30	13	100
46	35	19	20
40	37	23	10
37	36	27	7
32	35	31	5
30	34	34	4
26	31	39	3
17	22	47	2

changing with the growth conditions chosen. Table 6.3 shows the percentage of anions incorporated by the three different modes for various As_2/Ga flux ratios.

The program has chosen a site at random and first looks for possible "two-layer" deposition. If no possible configurations are present it then branches to look for in-line or side-by-side deposition with a branching ratio of one-half. The two-layer sites are scanned first because without doing this pits quickly develop (the surface roughens rapidly). Although the two-layer deposition routine may be suspected to be a weak link, there is a very interesting result in Table 6.3. Note that given the branching ratio of one-half one would expect the in-line mode to prevail over the side-by-side mode, since it requires a configuration of only three cations as opposed to four for side-by-side. This is what is seen for the higher flux ratios. However, as the flux ratio is decreased several things happen: (1) the growth front gets rougher (possibly changing to a three-dimensional-type growth), (2) the two-layer mode of anion deposition now dominates, and (3), very curiously, the side-by-side mode has beaten the in-line mode, which was totally unexpected and remains at the moment unexplained.

More details concerning the results gained by Monte Carlo simulation on the influence of the growth conditions on the incorporation mechanism may be found in [6.15].

Although the experiments which have led to this understanding have been done mainly on GaAs, it is reasonable to believe that the same picture is valid for other III–V and also II–VI compounds [6.119]. The difference between III–V and II–VI compounds is that in the former the cation incorporation rate is close to unity, while in the latter the anions are usually more easily incorporated into the crystal lattice than the cations (see e.g., the MBE growth peculiarities of Hg-containing II–VI compounds [6.67, 68, 122–124], and the experimental results concerning CdTe and $Cd_{1-x}Mn_xTe$ UHV ALE [6.125–131]).

6.3.2 Spatial Arrangement of the Near-Surface Transition Layer

The presented considerations justify the conclusion that in the near-surface transition layer (NSTL) appearing during the MBE growth at least two different areas may be distinguished. The first is the area of chemisorption, which is located nearest to the substrate surface, and the second is the area of physisorption. In both of these areas lateral migration and other surface kinetic effects occur. The question which is still open concerns how far the NSTL extends into the gaseous phase of the intersecting beams in the UHV chamber.

In order to estimate this extent one has first to define the interface between the substrate crystal, or the already grown epilayer, and the chemisorption area of the NSTL. The simplest way of doing this is just to state that those atoms which are incorporated into the solid, and thus bound there with the number of bonds characteristic for the surface of the substrate or the grown epilayer, belong to the crystalline phase. It is evident that these atoms have lost their feasibility for surface migration [6.17]. Conversely, all atoms which have been chemisorbed, but are bound to the solid surface through number of bonds less than the full complement, belong to the NSTL. These atoms are of course able to migrate over the crystalline surface, or may fairly easily desorb into the gaseous phase of the vacuum chamber.

In a two-dimensional growth mode, most frequently realized in a MBE process, the chemisorbed atoms form a growth front one or two monolayers thick [6.120], as shown in Fig. 6.13. A chemisorption region of a similar thickness has also been found in the experiments concerning UHV ALE of CdTe [6.128–131] (see also [6.132]). This two-monolayer-thick chemisorption region is not at odds with the handbook knowledge, according to which chemisorption is usually accompanied by dissociation of the molecule into atoms and one-monolayer coverage is not exceeded, but in some cases models for the chemisorption with nondissociative reactions and extensions over more than one monolayer were proposed [6.104].

Taking into account all the above-mentioned facts, we will consider hereafter the one to two monolayer thick area of the NSTL ($\leq 10\,\text{Å}$) as the chemisorption region of the transition layer.

Besides the chemisorbed atoms, the physisorbed molecules are located in the NSTL. These molecules dwell in the so-called precursor state, for which high migration, easy thermal desorption, but also a fairly low potential barrier for dissociative chemisorption are typical (Fig. 1.4).

The concept of a precursor state in adsorption was initially postulated by *Taylor* and *Langmuir* in 1933 [6.133] on the basis of experimental results concerning the behavior of Cs adsorbed on W. Following this work, precursors for adsorption were investigated experimentally and theoretically [6.134–137] and became a commonly accepted phenomenon. The main conclusions resulting from these investigations may be summarized as follows: (i) A precursor state for chemisorption may occur when the sticking probability of the adparticle is close

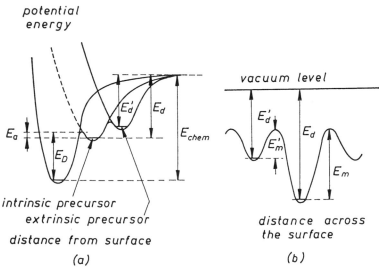

Fig. 6.14. One-dimensional interaction energy (**a**), corresponding to adsorption of a nondissociated molecule, involving two types of precursor states: intrinsic precursor (on empty chemisorption sites) and extrinsic precursor (on filled chemisorption sites). Changes of the interaction energy across the surface (**b**) are shown relative to the vacuum level corresponding to desorption [6.136]

to one, $s \sim 1$, independent of the surface coverage, i.e., also for $\theta > 1$. (ii) In the precursor state the molecule is not dissociated and is attracted to the surface by relatively weak forces, physical in nature. This is accompanied by fairly intensive surface migration. (iii) A distinction has to be made between precursor states which exist over empty surface sites, and states which exist over sites which are filled with chemisorbed species; the former may be described as intrinsic precursor states and the latter as extrinsic precursor states. (iv) The minima in the one-dimensional interaction energy curve determining the equilibrium positions of the adparticles relative to the solid surface occur at distances which increase in the following sequence: r_{min} (dissociative chemisorption) $< r_{min}$ (intrinsic precursor physisorption) $< r_{min}$ (extrinsic precursor physisorption) (Fig. 6.14).

More quantitative data concerning the position of the physisorbed adparticles relative to the solid surface may be deduced from the quantum mechanical mean field theory of multilayer physisorption [6.138, 139], developed for inert atom physisorption on graphite.

So far, no experimental data and no unequivocal theoretical estimates concerning the adlayer position in MBE growth have been published. Therefore, it is reasonable to assume, taking into account the presented ideas on physisorbed precursor states, that the spatial extent of the physisorbed region of the NSTL occurring in MBE may reach 2–3 monolayers toward the UHV environment, in addition to the layers being chemisorbed on the substrate surface. This means that, depending on the particular MBE growth conditions, the whole NSTL may extend toward the UHV environment as far as 1–2 nm, i.e., the thickness of 3–5 monolayers of the adsorbate, counting from the interface between the crystalline phase of the epilayer and the NSTL.

6.4 Growth Interruption and Pulsed Beam Deposition

According to the single scattering model (kinetic in the diffraction sense) of the RHEED intensity oscillations, changes in intensity of the specular beam are directly related to the changes in the surface roughness on an atomic scale, see Fig. 1.9. The oscillations are observed when growth is initiated on a smooth surface, which is usually obtained after annealing of a buffer layer. At first, a maximum reflectivity is obtained, as the surface is smooth. After initiation of the growth the surface becomes relatively rough, giving rise to diffuse scattering of the reflected electron beam. When the surface coverage is half a monolayer the minimum reflectivity is obtained. After this, the surface smoothness recovers and the reflectivity increases. Finally, when the first monolayer is complete, the maximum intensity is again obtained. This "optical analogue" model based on the kinematic theory of RHEED is useful in most cases for explaining the oscillatory behavior of the intensity of the RHEED pattern. It also explains the recovery effect, that is, the surface smoothing effect during growth interruption [6.140].

6.4.1 Recovery Effect During Growth Interruption

Figure 6.15 shows RHEED intensity oscillations obtained during the heteroepitaxial growth of $Al_xGa_{1-x}As/GaAs$. A 20 s growth interruption was made at

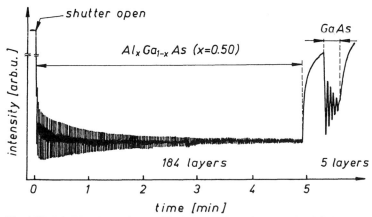

Fig. 6.15. RHEED intensity oscillations during the growth of $(Al_xGa_{1-x}As)_{184}(GaAs)_5$ heterostructure; [100] azimuth. Note that by 20 s growth interruption at the heterointerface, oscillations can be recovered [6.13]

the heterointerface, i.e., the Al and Ga source shutters were closed while leaving the As source shutter open. During the growth interruption period, the specular beam intensity increased rapidly. This process is attributed to the recovery of the surface smoothness, which in turn is caused by the surface migration or the sublimation of atoms adsorbed to the crystalline surface of the grown epilayers [6.13]. It is thus connected with the substantial changes in the NSTL configuration. Increased surface smoothness means increased incorporation of chemisorbed adatoms of the NSTL into the crystalline phase of the grown layer. Sublimation of adatoms into the UHV environment means, however, partial or complete disappearance of the NSTL.

This interpretation requires a more extended explanation, because the recovery effect is not simple (Sect. 4.1.5). This effect can be separated into fast and slow processes [6.4–6, 11] which may be ascribed to different surface phenomena. The initial fast response may be a consequence of the ability of those group III atoms which are essentially isolated at the instant of growth interruption to migrate rapidly towards steps, where they become incorporated into the crystal lattice, whereas the slower subsequent response may be a manifestation of collective migration behavior leading to rearrangement of the step configuration [6.4, 5].

Joyce et al. [6.141] argued for an interpretation based on the assumption that all effects which occur during the growth interruption period are in the plane of the interface of the monolayer level. Thus, they associated the fast process with a change of the surface reconstruction and the slow change with a rearrangement of the terraces [6.11].

Guided by the insight provided by computer simulations of the MBE growth based on the CDRI model (configuration dependent reactive incorporation model [6.93–95]), *Lee* et al. [6.6, 142] have suggested that the long-time recovery may not be a manifestation of cooperative movement of collections of Ga atoms on the surface, but that it is rather evidence of the opportunity to migrate provided to individual or pairs of Ga atoms by the As_2 desorption or adsorption processes.

Let us now call to mind the definition of the NSTL as an area, near to the crystalline surface, in which atomic as well as molecular species may dwell in chemisorbed and physisorbed states which allow fast migration along the direction parallel to the crystal surface. It is also important to remember that the atomic species, after being incorporated into the crystal lattice of the grown epilayer, become bound chemically with a full complement of bonds, thus losing the ability to migrate [6.17]. Keeping this definition in mind, one may interpret the recovery effect as being simply a rearrangement process of the NSTL which finally leads to smoothing of the crystalline surface of the grown epilayer by aiding the incorporation of the movable chemisorbed atoms into the crystal lattice. This point of view may be supported by the observation of the beneficial influence which the deposition of a small amount ($\sim 0.02\,\mathrm{ML}$) of Ga has on the recovery of the $Al_x Ga_{1-x} As$ epilayer upon MBE growth interruption [6.143]. The interpretation of the recovery effect using the NSTL concept is also consistent with the experimental fact that the RHEED pattern intensity is sensitive only to the shape of the crystalline surface, and consequently does not directly "feel" the migration processes of the movable NSTL adparticles.

The smoothing effect caused by growth interruption has been studied in detail by *Tanaka* et al. [6.144, 145] and by *Bimberg* et al. [6.146–149]. They used photoluminescence (Sect. 5.4.3) to investigate the heterointerfaces of MBE-grown GaAs-AlAs and GaAs-Al$_x$Ga$_{1-x}$As quantum well structures when no growth interruption was introduced, and when the growth process was interrupted before a heterointerface started to be grown [6.150–153]. *Tanaka* et al. [6.154] have also demonstrated the interface smoothing related to growth interruption using high resolution transmission electron microscopy (see Sect. 5.6.1 for a description of the method).

6.4.2 Growth of Superlattice Structures by Phase-Locked Epitaxy

The recovery effect has been used by *Sakamoto* et al. [6.155, 156] to develop a special MBE growth technique named phase-locked epitaxy (PLE). This technique has already been described in Sect. 1.3.2 and illustrated in Figs. 1.10 and 1.11, therefore we will describe it here only in order to show how it may be understood by using the concept of the NSTL.

The RHEED intensity oscillations are analyzed by a computer and the source cell shutters are driven at a predetermined phase of the oscillations. The oscillations thus obtained during the growth of a (GaAs)$_3$(AlAs)$_3$ trilayer superlattice structure are shown in Fig. 6.16 [6.13]. In this PLE growth, the Al source shutter was opened first and three monolayers of AlAs were grown. According to the Stranski-Krastanov growth mode (Fig. 1.5b), the growth front becomes rough and islands of the grown crystalline phase are formed. Taking into consideration the thermodynamic equilibrium scheme shown in Fig. 6.1, one may conclude

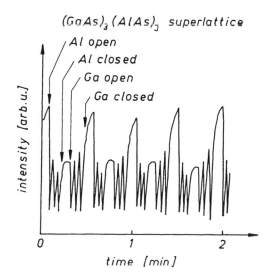

Fig. 6.16. RHEED intensity oscillations during computer-controlled PLE of a trilayer (GaAs)$_3$(AlAs)$_3$ superlattice structure [6.13]

that at this moment only local equilibrium has been reached in the MBE system, namely, in those parts of the crystal–NSTL interface where the islands have been formed.

After growing three monolayers of AlAs the Al source shutter was closed, causing an interruption in the growth. In the time of growth interruption (6 s) the system "crystalline solid–NSTL" approaches a partial equilibrium, due to the kinetic effects occurring in the NSTL. Consequently, a smoothing of the grown AlAs epilayer appears. One should keep in mind that the As source shutter is still open during the growth of superlattice layers, also when the Ga and Al cell shutters are closed. Thus, a NSTL exists near the crystalline surface during the whole time of fabrication of the superlattice structure.

The RHEED specular beam intensity recovers its former value during the time of growth interruption. After it had recovered, the Ga source shutter was opened in order to grow three monolayers of GaAs. As may be seen in Fig. 6.16, this caused the RHEED intensity oscillations to start to increase their amplitude, tending to reach the value characteristic for GaAs (this is the intensity observed before starting the growth of AlAs). This tendency does not mean, however, that the Stranski-Karastanov growth mode does not occur. The evidence for the existence of this growth mode is the fast increase of the RHEED specular beam observed after closing the Ga cell shutter, when the growth of three monolayers of GaAs has been terminated. The MBE process was thereby interrupted again for a short time, allowig partial equilibrium between the GaAs epilayer and the NSTL to be reached, and the surface of the 3-monolayer thick GaAs epilayer to be smoothed. The described procedure is repeated until the desired number of superlattice layers have been grown. Figure 6.17 shows schematic cross-sectional

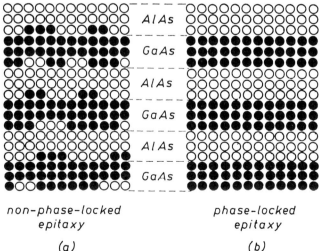

non–phase–locked epitaxy

phase–locked epitaxy

(a)

(b)

Fig. 6.17a,b. Simplified models of the cross-sectional view of $(GaAs)_3(AlAs)_3$ trilayer superlattice structures grown by two methods. **(a)** Conventional non-PLE method (constant-time shutter controlling method) **(b)** PLE method [6.13]

views of trilayer superlattice structures according to a simplified rigid-sphere model. The first structure (Fig. 6.17a), which exhibits fairly rough interfaces, was grown by a conventional MBE technique, whereas the second one (Fig. 6.17b), with ideal, smooth interfaces, was obtained by the PLE technique [6.13].

6.4.3 UHV Atomic Layer Epitaxy

The interrelation between the kinetic processes in the NSTL and the epitaxial growth in a UHV environment may be seen more clearly when considering atomic layer epitaxy (ALE). As already mentioned (Sect. 1.3.3 and [6.132]), ALE may be defined as a crystallization technique for compound films, which is based on chemical reactions at the heated surface of a solid substrate, to which the reactants are transported alternately as pulses of neutral molecules or atoms, either as chopped beams in high vacuum, or as switched streams of vapor, possibly on a carrier gas. A characteristic delay time is introduced in ALE growth between the sequential pulses of the reactant species. During this time, hereafter called dead time, the deposition process is interrupted to enable a free reevaporation of those atoms of the deposited species which are in excess to the first chemisorbed monolayer, and therefore only weakly bound to the substrate surface.

Three different growth modes may be distinguished for ALE [6.131]. The first, relying on sequential surface exchange reactions between compound reactants containing the constituent elements of the grown film, is realized under near-atmospheric pressure conditions in a gas-flow-type reactor, similar to the reactors used in Chemical vapor deposition (CVD) [6.132, 157–162]. In this growth mode, which may be called the "CVD-like mode of ALE", the separation of reaction steps is accomplished by a gas flow over the substrate. The reactants are alternately injected into the growth chamber together with an inert carrier gas, and following each reaction are purged away by the carrier gas stream.

The second growth mode, relying on sequential surface exchange reactions between compound reactants, is realized under high vacuum conditions in a modified MBE growth chamber provided with a special gas admittance system [6.163–165]. The growth conditions are here very similar to those which are found in gas source molecular beam epitaxy (see Sect. 1.3.1 and [6.166–173]). The reactant species are admitted into the high-vacuum growth chamber and impinge with straight trajectories onto the heated substrate surface in the form of molecular beams, without any carrier gas. The beam nature of the reactants flow allows the utilization of mechanical shutters to reduce transient flow effects during cycling in the ALE growth. This ALE mode may be called gas source high vacuum ALE or molecular layer epitaxy (MLE).

The third growth mode of ALE is realized under ultrahigh vacuum conditions in standard MBE growth chambers [6.125–130, 174–179]. It relies on direct surface reaction between the heated substrate and the constituent elements of the growing film which are delivered to the substrate as elemental atomic or molecular beams, generated thermally in effusion cells. This growth mode is the MBE-like mode of ALE, and will be called hereafter UHV ALE.

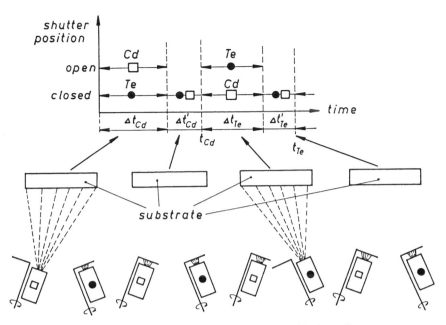

Fig. 6.18. Deposition timing scheme of ALE growth of CdTe [6.128]

Because the CVD-like mode of ALE is performed in near-atmospheric pressure gas flow reactors, we will concentrate our considerations only on the last two growth modes of ALE. We will begin with UHV ALE, taking as an example the growth of CdTe on CdTe (111) substrates [6.174].

The deposition timing scheme for this growth procedure is shown in Fig. 6.18, and the key conceptual steps associated with the growth by UHV ALE are illustrated in Fig. 6.19, following a model proposed by *Herman* et al. [6.130]. This model is indeed for the growth of CdTe epilayers on CdTe(111) substrates, however, it may be considered as quite general for UHV ALE because CdTe and the polar (111) surface may be treated as nearly ideal for studying the mechanism of this growth process [6.132]. The model includes (i) the existence of transition layers of both Cd and Te species intermediate between the crystalline (well-ordered) substrate or epilayer phase and the gaseous, nonordered phase created by the impinging beams (or the residual gas of the UHV chamber in the dead time of the ALE timing scheme), and (ii) partial reevaporation of the first chemisorbed monolayer of the deposited constituent elements [6.129]. The transition layers create reaction zones 3–5 monolayers thick near the substrate surface, where the atoms or molecules are weakly bound to the solid surface, so that they may readily migrate over the surface, become included into the crystal lattice of the growing epilayer, or thermally desorb into the UHV environment [6.130].

The activation energies for reevaporation of this near-surface region were measured to be 1.5 eV for Te on CdTe(111)*A* and 0.5 eV for Cd on CdTe(111)*B*

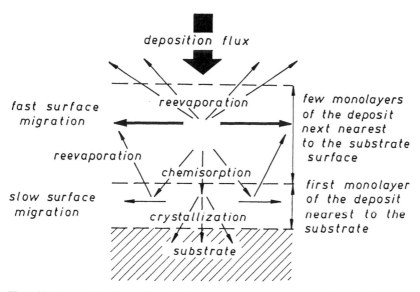

deposition flux

fast surface
migration

reevaporation

few monolayers
of the deposit
next nearest
to the substrate
surface

reevaporation

chemisorption

slow surface
migration

crystallization

first monolayer
of the deposit
nearest to the
substrate

substrate

Fig. 6.19. Schematic illustration of the key conceptual steps associated with the growth of CdTe overlayers by UHV ALE [6.130, 132]. Here "substrate" means the single crystalline bulk substrate wafer and the already-grown epilayer

[6.129]. Because the activation energies for surface migration are only about one-third of the bond energies [6.180], i.e., 0.5 eV and 0.17 eV for Te and Cd, respectively, migration of the adsorbed atoms and molecules should, indeed, take place effectively in the near-surface region.

It should be emphasized that in the transition layers relevant for UHV ALE (these transition layers are created only by one of the constituent elements of the compound to be grown), two different regions may be distinguished, as in the NSTL occurring in MBE (this transition layer is created by all constituent elements of the compound to be grown). Adparticles belonging to different regions of the ALE-relevant transition layer are bound to the crystalline surface with different bonds. The first monolayer, chemisorbed with covalent bonds, creates the first region, nearest to the substrate (Fig. 6.19). Atoms of this monolayer are much more strongly attached to the surface than the adparticles of the other monolayers of the deposited element. Those monolayers next nearest to the substrate create the second region of the transition layer. The adparticles of this region are attracted by van der Waals forces (this corresponds, for example, to physisorption of anion molecules), or by relatively weak chemical interactions (this corresponds to multilayer chemisorption by bonds characteristic of the bulk phase of the pure constituent element of the compound to be grown).

The principal technological parameter in UHV ALE is the substrate temperature during the growth process. Its value should be high enough to break all bonds attracting the adparticles belonging to the second region of the ALE-relevant transition layer, and thus to cause the thermal desorption of these adparticles during the dead time shown in the deposition timing scheme in Fig. 6.18. However, this temperature should be low enough to preserve the chemical bonds of the first monolayer, and thus to cause the growth of the compound as "layer of the first element by layer of the next element" [6.159, 174].

A characteristic feature of UHV ALE is the possibility of partial desorption of the first chemisorbed monolayer if the dead time is too long. This desorption, indicated on the scheme shown in Fig. 6.19, is a consequence of the UHV environment in which the growth process is performed (see, for example, the Langmuir adsorption isotherms relating the surface coverage of a solid to the gas pressure at a given temperature [6.104]).

The model of UHV ALE presented here is based on isothermal desorption experiments [desorption of Cd and Te species from CdTe(111) and from GaAs(100) substrate wafers] performed using Auger electron spectroscopy [6.129] and quadrupole mass spectrometry [6.131]. It has also been verified by studies of surface morphology of CdTe epilayers grown by UHV ALE on CdTe(111)B substrates with different growth conditions [6.128]. These conditions are specified by the numerical data given in Table 6.4. The results of the investigations are shown in Fig. 6.20.

In conclusion, one may consider that UHV ALE is not a self-regulatory process, in contradiction to what is still believed for the CVD-mode of ALE [6.162]. This means that timing and dosing in the deposition cycle play a crucial role in this process. Thus, UHV ALE has to be precisely designed when smooth and flat layers in an atomic scale have to be grown. The smoothest surface morphology is obtained in UHV ALE when the growth proceeds in an exactly two-dimensional layer-by-layer mode. However, this happens only if the deposited atoms of the constituent elements build up exactly one complete monolayer by the end of the respective dead times in the deposition cycle. If this so-called "1 ML coverage criterion" is not satisfied during the UHV ALE process, a three-dimensional growth may occur, causing surface roughness [6.128, 178].

An understanding of the nature of the NSTL occurring in sequential deposition processes of the constituent elements of the grown epilayers seems to be fundamental for handling UHV ALE.

6.4.4 Migration Enhanced Epitaxy

A slight modification of UHV ALE, in which no dead-time in the deposition timing was introduced, has been demonstrated for low-temperature growth of GaAs and GaAs–AlAs quantum well structures by *Horikoshi* et al. [6.181]. They named this modification migration enhanced epitaxy (MEE).

In MEE growth, interruption of the arsenic supply to the growing surface during the Ga or Al supply period is essential to the surface migration enhancement of the surface adatoms. Therefore, Ga or Al atoms and arsenic were alternately supplied to the GaAs (001) surface to obtain metal-stabilized surfaces periodically. The typical switching behavior of As_4 and Ga beam intensity mea-

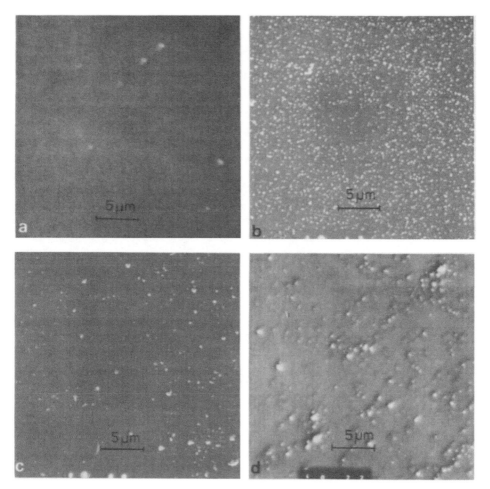

Fig. 6.20a–d. Scanning electron micrographs showing surface morphology of CdTe(111) overlayers grown by UHV ALE on CdTe(111)B substrates under the growth conditions listed in Table 6.4: (a)–(d) correspond to experiments $M1$–$M4$, respectively [6.128]. It may be clearly seen that the smoothest surface is obtained when the 1ML coverage criterion is fulfilled (a). When the growth is performed with a 3ML coverage at the end of the dead time, three-dimensional nucleation may occur, leading to the roughest surface (d)

sured by an ionization gauge placed at the substrate holder position is shown in Fig. 6.21. Bearing in mind the response time of the ionization gauge amplifier, the observed response indicates that the beam intensities change very quickly following the shutter operation. The growth process was monitored by observing RHEED specular beam intensity oscillations, where the 12 keV electron beam was used in the [100] azimuth.

It is believed that in MEE, growth proceeds in a layer-by-layer manner. To substantiate this effect, Horikoshi et al. recorded the RHEED specular beam intensity during the growth of GaAs at 580° C. Figure 6.22 compares the MEE

Table 6.4. Deposition parameters used in ALE of CdTe layers grown on CdTe(111)B substrates (according to the timing scheme shown in Fig. 6.18) [6.128]

Experiment		$M1$	$M2$	$M3$	$M4$
T_s	[K]	553	553	553	553
Flux of Cd, Φ_{Cd}	[nm/s]	0.11	0.16	0.16	0.40
Δt (Cd)	[s]	2.5	2.3	2.5	2.2
$\Delta t'$ (Cd)	[s]	0.5	0	1.4	0.3
Flux of Te$_2$, Φ_{Te}	[nm/s]	0.12	0.12	0.12	0.40
Δt (Te)	[s]	2.8	2.9	3	2.4
$\Delta t'$ (Te)	[s]	1.0	0	1.2	0.4
$\theta_{Cd}(t(Cd))$	[ML]	0.94	1.32	1.32	3.14
$\theta_{Te}(t(Te))$	[ML]	1.01	1.07	1.08	2.95
$\Delta\theta$	[ML]	-0.07	0.25	0.24	0.19
Surface morphology		Fig. 6.22a	Fig. 6.20b	Fig. 6.20c	Fig. 6.20d

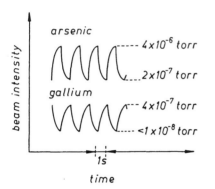

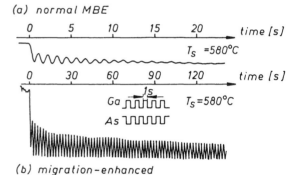

Fig. 6.21. Beam intensity change for Ga and As$_4$ species caused by shutter operation characteristic of MEE [6.181]. Ga response was measured under As-free conditions

Fig. 6.22. RHEED specular beam intensity oscillation during the growth of GaAs for normal MBE growth **(a)**, and for MEE growth **(b)** [6.181]

results with those of normal MBE growth. In both cases, they used the beam flux intensities of Ga and As$_4$ of $J_{Ga} = 6 \times 10^{14}\,\mathrm{cm}^{-2}\mathrm{s}^{-1}$ and $J_{As_4} = 4 \times 10^{14}\,\mathrm{cm}^{-2}\mathrm{s}^{-1}$. In normal MBE growth, the RHEED oscillations almost completely disappear after about 20 periods because the surface flatness on the atomic scale deteriorates. In contrast, the oscillations continue during the entire layer growth process when the MEE method is applied. This result indicates that a better surface flatness is maintained during MEE growth even after the growth of thousands of layers. This is most probably caused by the rapid migration of Ga in a very low arsenic pressure.

The amplitude of RHEED oscillation in MEE is closely related to the amount of Ga supplied to the growing surface per unit cycle, i.e., the amplitude is maximum when the number of Ga atoms per cycle is equal to the number of surface sites. This finding is consistent with the 1 ML coverage criterion found in UHV ALE of CdTe (see Fig. 6.20a, [6.128]).

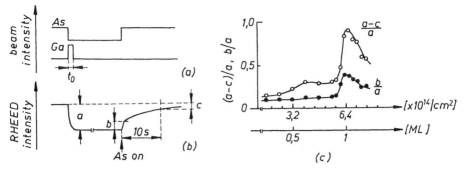

Fig. 6.23a–c. Recovery characteristic of RHEED specular beam intensity as a function of the number of surface Ga atoms. (a) Shutter operation sequence. (b) Corresponding RHEED intensity change. (c) Recovery as a function of the number of Ga atoms [6.181]

Figure 6.23 shows the RHEED intensity recovery at 580° C after reopening the As cell shutter as a function of the amount of Ga. The experimental shutter sequence is given in Fig. 6.23a. After the surface had been annealed in an As beam, the As cell shutter was closed and the Ga cell shutter opened simultaneously. The deposition of Ga atoms was continued for the time period t_0. After t_0 the Ga cell shutter was closed and the surface was annealed for about 20 s. Then the As cell shutter was reopened and the RHEED intensity recovery could be observed (Fig. 6.23b).

The results shown in Fig. 6.23 can be explained as follows. When the number of Ga atoms supplied to the surface is much less than the number of surface sites, these atoms are quite mobile on the surface and do not cohere with each other

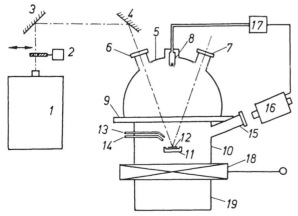

Fig. 6.24. Schematic diagram of the experimental setup used to study photostimulated molecular layer epitaxy. *(1)* Excimer laser; *(2)* light beam chopper; *(3, 4)* mirrors; *(5)* lamp housing; *(6, 7)* quartz windows; *(8)* high pressure Hg lamp; *(9)* quartz plate; *(10)* growth chamber; *(11)* quartz sample holder; *(12)* substrate; *(13, 14)* gas admittance nozzles; *(15)* sapphire window; *(16)* pyroscope; *(17)* temperature controller; *(18)* gate valve; *(19)* pumping unit [6.164]

to the surface site number, and if they are distributed in one atomic sheet on the surface, very quick RHEED intensity recovery can be expected following As cell shutter reopening. The characteristic shown in Fig. 6.23c is independent of the annealing time, i.e., even when the As_4 molecules are supplied immediately after closing the Ga cell shutter, no discernible change was observed in the RHEED intensity recovery time. This fact indicates a rapid migration of Ga atoms.

These experimental results suggest that the low-temperature growth of GaAs by MEE is possible if the number of Ga atoms supplied per unit cycle is chosen to be equal or close to the number of sites on the growing surface (compare this with the 1 ML coverage criterion found for CdTe UHV ALE [6.128]).

The quality of a GaAs layer grown at 200° C and of an AlAs-GaAs double quantum well structure grown at 300° C with MEE was demonstrated by photoluminescence. Only As_4 molecules were used for the arsenic source, such that arsenic was incorporated into the crystal through catalytic decomposition even at these low temperatures [6.181].

The MEE growth technique has been successfully applied by *Salokatve* et al. [6.182] for reduction of surface oval defects in GaAs grown in a two-step process. They have shown that if a thin GaAs buffer layer is deposited by supplying alternately Ga atoms and As_4 molecules to a GaAs substrate, prior to further growth by MBE, the density of the oval defects in the final layer is reduced reproducibly by a factor of 7, from about 490 to $70 \, cm^{-2}$, when compared with that obtained using MBE alone under closely similar conditions. The improved surface morphology produced by the pulsed beam method was thought to be related to initial film growth which proceeds in MEE most likely in a two-dimensional layer-by-layer fashion.

The same group has also demonstrated that MEE may cause beneficial effects in MBE growth of GaAs on Si substrates [6.183]. They grew GaAs films on Si(100) substrates using a two-step growth process of a 300° C GaAs buffer layer (by MEE) followed by a 600° C device layer (by conventional MBE). The films were examined by Rutherford backscattering and x-ray diffraction methods. A significant reduction in the defect density near the GaAs/Si interface and in the bulk of these films was observed when the buffer layer was deposited by supplying alternately Ga atoms and As_4 molecules to the substrate, rather than applying conventional MBE.

6.4.5 Molecular Layer Epitaxy

More complicated phenomena occur in the NSTL when the growth is realized by applying the other mode of ALE, namely MLE [6.163, 164, 184–187]. Molecular layer epitaxy is a crystal growth method using the chemical reaction of adsorbates on the semiconductor surface, where gas molecules containing one element of the compound semiconductor are introduced alternately into the growth chamber. GaAs molecular layer epitaxy using AsH_3 as an As source and TMGa as a Ga source was first demonstrated by *Nishizawa* et al. [6.163]. Later, instead

at this temperature. When the As molecules are supplied again, these Ga atoms react with arsenic to form Ga–As molecules. These molecules migrate on the surface to form islands. However, this process proceeds rather slowly, resulting in a slow recovery in RHEED intensity. If the number of Ga atoms is almost equal of TMGa, TEGa was used as a Ga source (Table 2.5), resulting in higher quality of epitaxial layers at lower crystal growth temperature [6.164]. An excimer laser which is operated with ArF (193 nm), KrCl (222 nm), KrF (249 nm), XeCl (308 nm), or XeF (350 nm) or a high pressure Hg lamp was used as the light source for irradiation during MLE. A schematic diagram of the experimental setup used for MLE with photostimulation facilities is given in Fig. 6.24. The experimental procedure was as follows [6.164]:

–   Semi-insulating GaAs substrates oriented at $(100) \pm 0.5°$ or $(111) \pm 0.5°$ were chosen which were suitably prepared before being mounted in the growth chamber. After evacuating the chamber to $p \leq 5 \times 10^{-8}$ Pa, the used substrate was heated to $600°$ C in order to remove the surface contaminants before the epitaxial growth started.

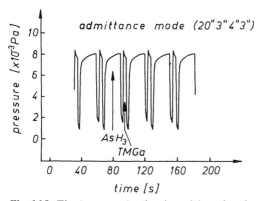

Fig. 6.25. The pressure in the deposition chamber during the gas admittance cycles of a MLE growth of GaAs from TMGa and AsH₃ sources. This periodic gas admittance mode is repeated up to the desired film thickness [6.163]

–   According to the applied deposition timing scheme (Fig. 6.25), AsH₃ was first introduced for 20 s into the chamber and then evacuated for 3 s. In the second step of the MLE deposition cycle MO gas (TMGa or TEGa) was introduced for 4 s and then evacuated for 3 s. The admittance pressures were chosen to be 6.7×10^{-3} and 6.7×10^{-3} Pa for TMGa and AsH₃, and 4×10^{-4} and 3×10^{-2} Pa for TEGa and AsH₃, at growth temperatures of $500°$ C for TMGa/AsH₃ and $300°$ C for TEGa/AsH₃.
–   A light beam from an excimer laser or from a high pressure Hg lamp was used to illuminate uniformly a substrate area of 1 cm² in order to cause photolytic effects during the growth process. The beam was mechanically

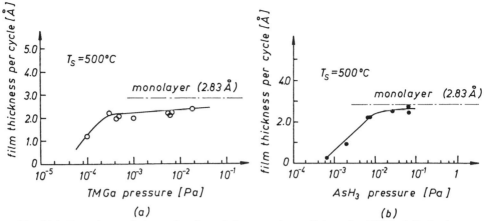

Fig. 6.26. Experimental determination of the growth conditions for MLE of GaAs from TMGa and AsH₃ by measuring the film thickness per deposition cycle as a function of (a) TMGa, and (b) AsH₃ pressures at a substrate temperature of 500° C with the admittance mode shown in Fig. 6.25 [6.163, 164]

chopped and so the substrate irradiation was synchronized with the gas admittance modes.

The given growth conditions (admittance pressures of the reactants and growth temperature) have been optimized experimentally by measuring the film thickness per deposition cycle [6.186] as a function of reactant gas pressure (Fig. 6.26) and substrate temperature, and determining the conditions which guarantee one monolayer coverage in each cycle [6.163, 164]. The lower growth temperature obtained when GaAs is grown from TEGa and AsH₃ results from the fact that the binding energy of C_2H_5-Ga in TEGa is lower than the binding energy of CH_3-Ga in TMGa [6.164].

The effect of substrate irradiation with UV light during film growth by MLE has been studied. The results may be summarized in general by stating that photoirradiation during the MLE growth greatly improves the surface morphology (Fig. 6.27) as well as the electrical properties of the films (Table 6.5). These effects have been ascribed by Nishizawa et al. to enhanced chemical decomposition [6.187] and enhanced surface migration [6.164] of adsorbates caused by UV light irradiation. It has been suggested that the essential migrating species is a Ga complex adsorbate like $Ga(CH_3)_x$ but not the As complex [6.164]. It should be emphasized that the irradiation effects do not result from surface reactions promoted by increased hole concentrations caused by electron–hole pair generation. If the electron–hole pair generation process were responsible for the surface reactions, then even near infrared irradiation ($\lambda = 890\,nm$) would be enough. However, no changes in the characteristics of the grown layers have been observed when infrared radiation has been applied.

With photoassisted MLE four types of irradiation modes have been tested. These were irradiation sychnronized with AsH₃ admittance, its admittance pause

Fig. 6.27. Surface morphology of epitaxial films prepared by MLE on the GaAs(100) surface without irradiation (**a**) and with KrF (249 nm) irradiation (**b**); and on the GaAs(111)B surface without irradiation (**c**) and with Hg lamp irradiation (**d**) [6.164]

(a)   (b)

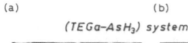
(TEGa–AsH₃) system

(c)   (d)   50 μm

Table 6.5. Electrical properties of GaAs films prepared by MLE and photo-MLE from TMGa and AsH$_3$ under the same gas admittance conditions [6.163]

T_s [°C]	Light source	μ_h [cm^2 V^{-1}s^{-1}]	Carrier density [cm^{-3}]	Type
600	no illumination	56	1.9×10^{19}	p
600	Hg lamp	103	1.3×10^{18}	p
700	no illumination	68	1.2×10^{19}	p
500	Hg lamp	103	3.1×10^{18}	p
500	Laser (257.3 nm)	84	4.6×10^{18}	p
500	Laser (257.3 + 514.5 nm)	110	2.4×10^{18}	p

(dead time in the deposition timing scheme shown in Fig. 6.25), MO-gas admittance and its admittance pause. If the substrate is synchronously irradiated with the admittance of the gaseous reactants, the adsorption steps and the following reaction steps may be specifically enhanced. With the irradiation during the admittance pauses, only the reaction or the migration processes of adsorbates are enhanced.

The mechanism of energy transfer from the photons to the crystal growth steps are difficult to identify. One of the possibilities is to measure the electrical properties of the grown films, namely the Hall mobilities and carrier concentrations.

The following experimental results have been obtained.

1. Irradiation (249 nm) synchronized with TMGa admittance increases the impurity level in the GaAs layer growth with MLE at the optimized growth

257

conditions, while synchronization of the irradiation with AsH_3 admittance causes a decrease in impurity level, which is, however, still higher than that obtained with no irradiation.

2.   Irradiation synchronized with admittance pauses causes an intermediate impurity level C lower than in the case of TMGa synchronization, but higher than in the case of AsH_3 synchronization.

These results demonstrate that some of the reaction or migration steps were indeed separately enhanced by irradiation, which is evidence that the light irradiation effect is not only that of annealing, even if there are some heating effects [6.164].

6.5 Doping During MBE Processes

Dopant incorporation during MBE belongs to the most important problems from the viewpoint of application of this growth technique for the fabrication of semiconductor devices. The incorporated impurities determine directly the electrical and optoelectronic properties of the prepared device structures. Therefore the ability to control the incorporation process of definite doping elements is an indispensable factor for the device maker using MBE. The problem should be divided into two parts, concerning respectively unintentional doping, which may be caused by the residual gas species present in the deposition chamber, and intentional doping, realized by a thermal beam co-deposition processes or by more sophisticated doping techniques applying ionized particle beams or laser irradiation.

6.5.1 Unintentional Doping

The combination of high pumping speeds and cryopaneling leads to residual gas pressures routinely achieved in present day MBE systems (Chap. 3) of about 5×10^{-11} Torr. At these low pressures the residual gas contains predominantly H_2 molecules, but H_2O, N_2 and CO_x molecules have also been found which exhibit very low partial pressures (less than 5×10^{-13} Torr) [6.188].

It is not exactly clear which effects cause the unintentionally introduced contaminants of the epilayer to form different doping levels. According to theoretical considerations [6.189, 190] only aluminum has an affinity for the listed residual species. It is therefore possible that the residual carbon level often found in grown epilayers are either introduced via the substrate or through thermal generation, for example, at hot filaments of measuring devices or at hot surfaces around the beam sources.

In order to reduce the incorporation of background-gas-related impurities, the density and the excitation state of background molecules such as carbon monoxide etc. should be reduced to a minimum. To help achieve this, all hot filaments not necessary during epitaxy, i.e., ion gauges, quadrupole mass spectrometers, and most importantly reflection electron diffraction and Auger electron spectrometer guns should be turned off [6.191]. Reduction of the temperatures of the ambient gas and quiescent cells which are not being used in the particular epitaxial run also helps. In this context it is also imperative that temporal or spatial variations in temperature of LN_2 cryopumping areas do not fluctuate during the epitaxial growth, as this would result in variations in liberation and adsorption of background gas species. This can be achieved by constantly checking the LN_2 flow and reducing the total power dissipation inside the deposition chamber of the MBE machine.

The unequivocal identification of the impurities incorporated unintentionally during the MBE process is not an easy task. Most frequently low temperature photoluminescence (Sect. 5.4.1) or electrical characterization techniques, like Hall effect measurements or deep level transient spectroscopy (DLTS) (Sect. 5.5.2), are used.

Among unintentionally introduced impurities, two groups may be distinguished. These are the shallow-level and the deep-level impurities. In this classification, the criterion according to which impurities are ascribed to one of these groups is the position of the energy level characteristic of the specific impurity in the band gap relative to the edges of the conduction or valence bands of this semiconductor (the ionization energy of the impurity).

The following rule is known from semiconductor physics [6.193]. An impurity is denoted a shallow-level impurity, or hydrogen-type impurity if its ionization energy E_j in the semiconductor is close to the Rydberg energy Ry = 13.6 eV multiplied by the impurity effective mass ratio m^*/m_0 and divided by the square of the dielectric constant of the semiconductor. This means that impurities with ionization energies $E_j \sim \mathrm{Ry}(m^*/m_0)\varepsilon^{-2}$ are counted as shallow-level impurities, while those with ionization energies $E_j \gg \mathrm{Ry}(m^*/m_0)\varepsilon^{-2}$ are counted as deep-level impurities. In the relations for ionization energy, m_0 stands for the free electron rest mass ($m_0 = 9.10956 \times 10^{-31}$ kg) while m^* denotes the effective mass of electrons (in the case of donors) or holes (in the case of acceptors) in the semiconductor. For example, in most III–V semiconductors the ionization energies of shallow-level impurities are in the range 2–10 meV for donors and 15–50 meV for acceptors.

Until now, the most complete set of data on unintentional doping during MBE has been collected for GaAs [6.190–192]. Three shallow acceptors have been identified in this compound as unintentional impurities present in high-purity samples (acceptor concentration equal to 4×10^{13} cm^{-3}. These are carbon, silicon and manganese [6.191].

The source of carbon remains unclear, however, reaction of carbon monoxide or dioxide with either surface arsenic or gallium, liberating oxygen, either as a volatile arsenic oxide or gallium suboxide (Ga_2O), and a free carbon atom is most probable.

There is no source of silicon in the MBE process, per se. Therefore, the only cause of the observed doping of GaAs layers with silicon is the thermal generation of this element in effusion cells charged with polycrystalline GaAs, as a source for As_2 molecules. This suggestion seems to be reasonable, because the introduction of cracking cells as sources of As (Sect. 2.3.2) has eliminated Si as an unintentional shallow-level acceptor present in GaAs grown by MBE.

Manganese in MBE films is believed to arise from surface accumulations which are formed during substrate heat treatment prior to epitaxial growth when semi-insulating chromium-doped GaAs substrates are used in the MBE process. This has been confirmed by SIMS and photoluminescence measurements. Application of semi-insulating GaAs substrates grown without chromium (and thus without associated manganese) has nearly completely eliminated manganese from MBE-grown GaAs epilayers [6.191].

Lead, sulphur, and amphoteric silicon have been identified as unintentional shallow-level donors in GaAs grown with MBE. However, no unequivocal conclusions may be drawn at present concerning the sources of these impurities. Hall measurements indicated only that the carrier concentrations related to these impurities are certainly below $10^{14} cm^{-3}$ [6.191].

The most consistent sources of deep-level impurities in MBE-grown GaAs are the substrates used. Chromium, copper and in certain cases iron are present in films grown on semi-insulating GaAs. These impurities are referred to as hole traps. In addition to these elements, four electron traps, designated the $M1$-$M4$ levels, have been found in GaAs by DLTS (Sect. 5.5.2). It should be noted that the density of hole traps generally exceeds that of electron traps by approximately two orders of magnitude [6.191]. Usually good correlation between substrate impurities and most of the mentioned deep-level carrier traps may be found in specific MBE growth arrangements.

Unintentional doping is an apparatus-related effect. Therefore, no general rules, valid for example for all MBE growth modifications and for all MBE-grown material systems, may be formulated, but there are some relatively new experimental findings concerning this problem. The first is the dependence of unintentional doping on the crystallographic orientation of the used substrate, which has been discussed in Sect. 6.2.2. The next is the beneficial effect on the general quality of the epilayer caused by growing a buffer layer by a low temperature pulsed beam deposition technique, e.g., by MEE, and subsequently growing the desired epitaxial structure by the conventional MBE technique at a higher temperature (Sect. 6.4.4).

6.5.2 Thermodynamics of Doping by Co-deposition

It has been demonstrated by *Heckingbottom* et al. [6.190, 194] that some understanding of the principles concerning intentional doping may be gained by applying thermodynamics to this problem. Thermodynamical considerations allow one to determine whether a desired incorporation reaction is possible or not under MBE conditions. This is important for doping reactions doomed to failure by thermodynamic arguments, because a waste of experimental time can thus be avoided. One can also determine which possible reactions can compete with the desired one. This knowledge may be helpful in the selection of dopants which are more easily used. In a more detailed study of a particular reaction, the comparison of the thermodynamic framework with experimental data allows the identification of any kinetic barriers and hence leads to a clearer understanding of the reaction mechanism.

The main chemical reactions for doping of a semiconducting compound $A^i B^j$ – where A, B stand for elements, and i, j for numbers denoting columns in the periodic table of elements – may be written as follows.

1. Replacement of the cation element atoms (A) by doping element atoms (C) of the group ($i - 1$) of the periodic table to give p-type doping:

$$C_{(g)}^{(i-1)} + V_{Ai} \leftrightarrow C_{Ai}^{(i-1)}[-] + h \quad , \tag{6.16}$$

where V_{Ai} is a vacant A site in the $A^i B^j$ compound, $C_{Ai}^{(i-1)}[-]$ is an ionized C atom on an A site in the compound (an ionized acceptor), h is a free hole in the valence band of $A^i B^j$, and the index (g) denotes the gas phase of the element.

2. Replacement of the anion element atoms (B) by doping element atoms (D) of the group ($j + 1$) of the periodic table, to give n-type doping:

$$\tfrac{1}{2} D_{2(g)}^{(j+1)} + V_{Bj} \leftrightarrow D_{Bj}^{(j-1)}[+] + e \quad , \tag{6.17}$$

where e is a free electron in the conduction band of $A^i B^j$, $D_{Bj}^{(j+1)}[+]$ is an ionized D atom on a B site in the compound (an ionized donor), and the other symbols have similar meanings to in (6.16).

3. Replacement of the cation element atoms by doping element atoms (D) of the group ($i + j$)/2 of the periodic table, to give n-type doping:

$$D_{(g)}^{(i+j)/2} + V_{Ai} \leftrightarrow D_{Ai}^{(i+j)/2}[+] + \left(\frac{j-i}{2}\right)e \quad . \tag{6.18}$$

4. Replacement of the anion element atoms by doping element atoms (C) of the group ($i + j$)/2, to give p-type doping:

$$C_{(g)}^{(i+j)/2} + V_{Bj} \leftrightarrow C_{Bj}^{(i+j)/2}[-] + \left(\frac{j-i}{2}\right)h \quad . \tag{6.19}$$

A direct quantitative analysis of the thermodynamics of the reactions (6.16–19) requires a knowledge of such quantities as the enthalpy and entropy of the considered system of defects in the crystal lattice (vacancies and ionized doping elements). Unfortunately, these quantities are not known with any precision for MBE-grown films [6.190]. As shown by *Heckingbottom* et al. [6.194], the lacking thermodynamical data concerning doping of semiconductor films during MBE growth may, however, be taken from other epitaxial growth techniques, like LPE or VPE. This procedure allows the formulation of theoretical predictions concerning doping of MBE-grown films which are in reasonable agreement with experimental data. The success of this approach, which is rather unexpected, relies on the following argumentation.

The first device-quality epitaxial layers of semiconductors were grown with near-equilibrium growth techniques, i.e., LPE and VPE. In these layers the doping level is a well-controlled parameter. If MBE-grown doped epilayers achieve a similar quality, then they must approach equilibrium conditions during growth to a comparable extent and should not contain large additional nonequilibrium concentrations of defects. It is commonly accepted, and evidenced by many examples of sophisticated device structures of highest quality (see, for example, [6.195, 196]) that MBE-grown layers are suitable for device purposes. This fact implies that approach to equilibrium during growth and dopant incorporation is comaparable in MBE, VPE and LPE [6.194].

The thermodynamic approach of Heckingbottom et al. has been applied successfully to predict the behavior of the common dopants Si, Ge, Sn, Pb, Mn, S, Se, Te, Zn, Cd, and Mg in GaAs under typical MBE growth conditions. Good agreement with experimentally observed behavior has been found for all cases except Pb. Thus, it became evident that the described method should be predictive for application of MBE to new material systems.

As an example of how the thermodynamic approach may be used, let us consider the sulfur doping of MBE-grown GaAs [6.190, 194]. The doping reaction is described by (6.17). This reaction is connected with a change in partial molar free energy of the system given by

$$\Delta G = -RT \ln K \tag{6.20}$$

with the equilibrium constant K [6.197] following from (6.17):

$$K = \frac{[S_{As}^+]n}{[V_{As}]\sqrt{p_{S_2}}} \quad , \tag{6.21}$$

where the brackets [ ] indicate concentrations of the point defects in the crystal lattice, n denotes the concentration of electrons in the conduction band of the semidconductor, and p_{S_2} is the beam pressure of the sulfur dopant in Torrs. One of the consequences of the reaction (6.17) is that $n = [S_{As}^+]$. Substituting this into (6.21) leads to the expression

$$n = [S_{As}^+] = \sqrt{K[V_{As}]\sqrt{p_{S_2}}} \quad . \tag{6.22}$$

Experiments on doping of GaAs with sulphur in VPE [6.198] lead to the conclusion that, although on cooling to room temperature $n = [S_{As}^+]$, the amount of S_{As}^+ is proportional to $\sqrt{p_{S_2}}$ present at the growth temperature. Hence at the growth temperature n must be constant and greater than $[S_{As}^+]$, so that according to (6.21)

$$[S_{As}^+] = \frac{K[V_{As}]\sqrt{p_{S_2}}}{n} \quad . \tag{6.23}$$

The observation that $n \neq [S_{As}^+]$ is typical for several dopants in both VPE and LPE of GaAs and at least two important models for explaining this behavior have been described [6.199, 200]. The alternative explanation that the discrepancy between experimental results and (6.22) is due to kinetic barriers has also been examined. While kinetic effects have been clearly identified, for example by comparing growth and doping on different faces of GaAs [6.201, 202], they typically lead to variations in growth rate or doping efficiency of factors of ≤ 5 [6.202]. It is also considered unlikely that kinetic limitations involving chlorides or hydrides (VPE) or Ga solutions (LPE) would all lead to the same basic modification, and an explanation in terms of an underlying aspect like a change in the relevant defect model is preferred. Equation (6.23) is therefore adopted in the following treatment. It remains to replace $[V_{As}]$ by an experimental parameter through the reaction

$$\tfrac{1}{4} As_4 + V_{As} \leftrightarrow 0 \tag{6.24}$$

so that the equilibrium constant for this reaction is given by

$$K_1 = \frac{1}{[V_{As}] \sqrt[4]{p_{As_4}}} \quad . \tag{6.25}$$

Rearranging this expression, and substituting in (6.23) for $[V_{As}]$ leads to

$$[S_{As}^+] = K_2 \frac{\sqrt{p_{S_2}}}{\sqrt[4]{p_{As_4}}} \quad , \tag{6.26}$$

where $K_2 = K/(n K_1)$.

Using (6.26) and the known doping achieved under VPE conditions one may now calculate the required conditions in MBE. In typical VPE growth [6.203] sulfur is introduced as H_2S in a large excess of H_2, allowing the equilibrium value of $\sqrt{p_{S_2}}$ to be calculated from

$$\tfrac{1}{2} S_2 + H_2 \leftrightarrow H_2S \quad , \tag{6.27}$$

giving

$$\sqrt{p_{S_2}} = \frac{p_{H_2S}}{K_3 p_{H_2}} \quad , \tag{6.28}$$

where K_3 can be calculated from standard thermodynamic data [6.190].

For $p_{H_2S} = 7.6 \times 10^{-5}$ Torr and $p_{H_2} = 760$ Torr a corresponding doping level of $\simeq 10^{17}$ cm^{-3} is obtained. Similarly the equivalent value of p_{As_4} in VPE is typically 7.6×10^{-1} Torr [6.204]. The resulting extrapolation to MBE conditions indicates that $[S_{As}(+)] = 2.2 \times 10^{17}$ cm^{-3} is in equilibrium with $p_{S_2} = 3.2 \times 10^{-21}$ Torr at a typical growth temperature of 560° C. It should be noted that the answer is approximate as (a) the VPE results apply at 750° C while the MBE is at 560° C, (b) the value of p_{As_4} used in MBE is not an equilibrium value, (c) it is doubtful whether the dependence of p_{As_4} is as great as assumed in (6.24) [6.200], and (d) only two-term equations were used in calculating equilibrium constants. However, if (c) applies it will serve to minimize the effects of (b). It is most unlikely that any of these approximations will affect the conclusion that sulfur doping should occur at useful levels in MBE. This result is used in the following way.

In MBE, straightforward control of doping is sought by control of the growth rate and the arrival rate of the doping species. Thus, in growth of GaAs(001), where all the Ga condenses and so governs the growth rate, the doping is given simply by $(F_S/F_{Ga}) \times$ (Ga atoms cm^{-3} in GaAs) where F_S and F_{Ga} represent the fluxes of sulfur and gallium atoms respectively. If the flux ratio is approximated by the beam pressure ratio, then noting that GaAs has 2.2×10^{22} atoms cm^{-3} of Ga, $p_{Ga} = 10^{-6}$ Torr and $p_{S_2} = 5 \times 10^{-12}$ Torr will lead to $[S^+_{As}] = 2.2 \times 10^{17}$ cm^{-3}. The thermodynamic calculations are used to establish that under such conditions $[S_{As}(+)]$ is stable and hence should form. It will be clear that the operating pressure is $\simeq 10^9$ above the stability boundary and an approximate assessment is therefore quite sufficient. The experimental results [6.205] confirm that this simple description is essentially correct at 560° C, with no evidence of any kinetic barriers to dopant incorporation to complicate the situation.

In a later study [6.190], it was found that at higher temperatures ($\simeq 600°$ C) sulfur incorporation is much reduced as the temperature is increased. The most likely reaction causing this observed loss of sulfur is

$$2Ga_{(l)} + \tfrac{1}{2}S_{2(g)} \leftrightarrow Ga_2S_{(g)} \quad , \tag{6.29}$$

where $Ga_{(l)}$ is a good approximation to the chemical potential of Ga in GaAs grown under LPE or MBE conditions [subscript (l) indicates the liquid phase]. The equilibrium constant for (6.29) is given by

$$K_4 = \frac{p_{Ga_2S}}{\sqrt{p_{S_2}}} \tag{6.30}$$

and from available thermodynamic data [6.206] the calculated values of K_4 are 2.54×10^{12} $\sqrt{\text{Torr}}$ at 560° C and 6.9×10^{11} $\sqrt{\text{Torr}}$ at 650° C. Substituting the value of $\sqrt{Sp_{S_2}}$ in equilibrium with $[S^+_{As}] = 2.2 \times 10^{18}$ cm^{-3} (3.2×10^{-19} Torr) in (6.30) gives $p_{Ga_2S} = 1.44 \times 10^3$ Torr at the lower temperature. It is clear that the value will not change substantially throughout the range of growth temperatures up to 650° C. Hence essentially all the sulfur should be lost as Ga_2S, according to the thermodynamic framework, under all typical MBE conditions. Any observed doping must therefore be due to a kinetic barrier to the formation of Ga_2S on the GaAs(001) surface. To emphasize the completeness of this situation one may recast the figures in line with the assumption that virtually all the sulfur is converted to Ga_2S. Then for a beam pressure of $p_{S_2} = 5 \times 10^{-11}$ Torr, converted to $p_{Ga_2S} = 1 \times 10^{-10}$ Torr, (6.30) and (6.26) lead to a value of $[S^+_{As}] = 1.5 \times 10^5$ cm^{-3}, i.e., the approximation that all the sulfur is converted to Ga_2S is accurate to $\simeq 1 \times 10^{-12}$.

In summary, therefore, there is only local equilibrium at 560° C. The reaction leading to the lowest free energy (Ga$_2$S formation) is essentially completely hindered kinetically, while the route to the metastable position [S$_{As}^+$] is kinetically facile. As the temperature is raised it is also found that the formation of Ga$_2$S is hindered by an increase in the p_{As_4}/p_{Ga} ratio. In fact, the temperature dependence of sulfur loss is similar to that for arsenic loss as measured by *Panish* [6.207], using the border between the arsenic-stabilized GaAs(2 × 4) and GaAs(3 × 1) surface structures as a criterion. These results suggest that the situation is dominated by the surface arsenic population. As the arsenic coverage is reduced, an increasing surface gallium population becomes free to bind to sulfur to form Ga$_2$S via a much lower energy transition state [6.190].

So far, doping with sulfur has been discussed only in relation to GaAs. The incorporation behavior of this element as related to doping of the ternary compound Al$_x$G$_{1-x}$As is more complicated, and is described in detail in [6.208].

6.5.3 Delta-Function-Like Doping Profiles

The term delta-function-like doping, or simply δ-doping, stands for the confinement of dopant atoms during MBE growth to a two-dimensional (2D) plane one atomic layer thick [6.209]. This doping concept, also known as atomic-plane doping, sheet doping, or spike doping during MBE growth, was originally proposed to improve the doping profile of Ge-doped n-type GaAs layers [6.191, 210]. Later, the δ-function-like doping profiles using Si donors and Be acceptors [6.211] were employed to generate symmetric V-shaped potential wells in GaAs with a quasi-2D electron (hole) gas [6.212] and to create a sawtooth doping superlattice [6.213].

The basic concept of δ-doping in GaAs is illustrated in Fig. 6.28. The Si donors are located in an atomic monolayer of the (100)-oriented GaAs host material, and the ionized donors provide a continuous positive sheet of charge. The fractional coverage of the available Ga sites in the (100) plane can reach several percent. The impurity charge distribution N_D can mathematically be described by the Dirac delta function, i.e., $N_D = N_D^{2D} \delta(z)$, where N_D^{2D} is the 2D donor concentration. Due to electrostatic attraction, the electrons remain close to their parent ionized donors and form a quasi-2D electron gas (2DEG) in the V-shaped potential well produced by the positive sheet of charge. In the narrow potential well the electron energies for motion perpendicular to the (100) growth surface are quantized into 2D subbands. The subband levels have been calculated by a simple approximation [6.214], taking into account band bending of the conduction band due to localized ionized impurities and to free carriers, and also in a more elaborate way self-consistently, taking into account the nonparabolicity of the conduction band [6.212]. Direct evidence for the formation of 2D subbands in δ-doped GaAs came from Shubnikov–de Haas (SdH) oscillations observed during magnetotransport measurements with the magnetic field perpendicular to the sample [6.212].

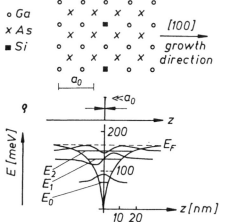

Fig. 6.28. Schematic illustration of δ-doping in GaAs. The Si donors are located in an atomic monolayer of the (100)GaAs host crystal *(top)*. A V-shaped potential well is formed *(bottom)*. The subband energies and wave functions are calculated self-consistently assuming a uniform sheet of positive charge [6.209]

The δ-doped GaAs samples were grown by MBE on (100)-oriented GaAs substrates using a growth rate of $1\,\mu m/h$ and a growth temperature $500 < T_S < 550°$ C. The δ-function-like Si and Be doping profiles were obtained by interrupting the growth of the GaAs host crystal by closing the Ga shutter and leaving the As shutter open. The As-stabilized surface reconstruction was thus maintained while the shutter of the respective dopant effusion cell was opened for a certain time interval (up to several minutes for high δ-doping levels). In this impurity growth mode the host crystal does not grow. To continue GaAs growth, the dopant shutter was closed and the Ga shutter opened again. For single-type-δ-doping layers in GaAs the measured doping density is in good agreement with the number of supplied dopant atoms up to very high doping densities (Fig. 6.29). The fractional coverage of the available Ga sites in the (100) plane reaches about 5 % if one assumes that the GaAs(100) surface contains 6.25×10^{14} Ga atoms cm^{-2}. At these high doping levels the surface reconstruction displayed in the RHEED pattern changes from the As-stabilized (2×4) to a (1×3) reconstruction. When low growth temperatures are employed, which is particularly important for Be δ-doping, the surface morphology of the epilayers remains atomically smooth also at high doping densities.

In Be δ-doped GaAs, hole densities as high as $2 \times 10^{13}\,cm^{-3}$ were easily achieved without any indication of saturation. A certain carrier freeze-out was observed in samples cooled to 77 K (no further freeze-out occurs down to 4 K). This phenomenon is not yet fully understood. In Si δ-doped GaAs, a deviation of measured electron density from the supplied dopant atoms beyond $1 \times 10^{13}\,cm^{-2}$ has been observed [6.209]. At these high doping densities the electrons in the V-shaped potential well populate high-index subbands so that the Fermi energy is increased by more than $300\,meV$ [6.214]. This energy shift is comparable with the energy separation between Γ and L conduction band minima in GaAs. Electrons may thus populate the L minimum at high Si doping levels.

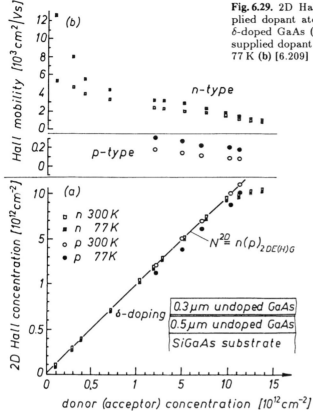

Fig. 6.29. 2D Hall concentration versus supplied dopant atoms in single n- and p-type δ-doped GaAs (a) and Hall mobility versus supplied dopant atoms measured at 300 and 77 K (b) [6.209]

No carrier freeze-out occurs for Si δ-doping, in agreement with homogeneous doping, also not in the saturation regime.

The confinement of the Si dopant atoms to an atomic layer in the GaAs host material has been confirmed by capacitance-voltage (C-V) measurements of the concentration profile of a δ-doped GaAs layer grown on n^+-GaAs substrate [6.209].

6.5.4 In-Growth Doping with Ionized Beams

In the described intentional doping techniques the dopant species have been generated thermally in effusion cells and delivered to the growing interface as neutral atomic or molecular beams with suitable selected impingement rates. In many cases of doping of MBE grown layers, more effective and better-controlled incorporation of the dopant atoms may be achieved by applying ionized beams of the doping species (see, e.g., [Ref. 6.113, Chap. 11]). In doing this, ion sources are used instead of, or together with, effusion cells.

Doping using ionized beams may be performed either as a post-growth process, realized usually with focused high energy (about 100 keV) ion beams in a separate processing chamber, or as an in growth process, realized with low-energy ion beams (0.1–3 keV [6.215], see also [Ref. 6.113, Chap. 11]) introduced to, or generated in, the deposition chamber. The first case of doping has been described in detail in Sect. 3.2.2, and in [6.216, 217]. Therefore, in this section we will concentrate only on in-growth doping.

Two modes of in-growth doping may be distinguished, i.e., the direct implantation mode and the secondary implantation mode. In the case of direct implantation, the low energy ion beam of dopant atoms is deposited on the substrate surface simultaneously with the beams of the constituent elements. Thus, the dopant atoms are continuously incorporated into the growing crystal lattice. Neither surface reevaporation nor surface segregation of dopant atoms is present during this growth-and-doping process [6.218]. This direct implantation technique is sometimes referred to as ion implanted molecular beam epitaxy, or I^2MBE [6.219]. The secondary implantation mode is based on a recoil process. The dopant-atom beam is generated thermally in an effusion cell and deposited on the substrate prior to the beginning of epitaxial growth of the layer to be doped. After a submonolayer of adsorbed dopant atoms has been formed on the substrate surface, the constituent-element beams, of which at least one consists of ionized atoms, are allowed to impinge on the substrate, which causes the start of MBE growth. The substrate is electrically biased during growth by application of a potential of 0.1–1 kV, negative in relation to the electrically grounded beam sources. Consequently, the ionized atoms of the constituent element beam are accelerated towards the growing interface. At this interface the ions collide with accumulated dopant adatoms, knocking a fraction of them a short distance into the layer. Buried below the surface of the epilayer the dopants can no longer segregate or reevaporate [6.218].

There are two reasons why ion implantation is employed to introduce impurities into the MBE-growing layers. The first is for accurate control of the doping levels of those dopant species which exhibit strong surface segregation, e.g., antimony in silicon. The second reason is that, in order to obtain doping levels higher than 10^{18} cm^{-3}, the sticking probability of dopants needs to be increased. However, in the case of Zn in GaAs there is no other way of introducing the doping element atoms into the MBE growing epilayer than ion implantation, because the sticking coefficient of neutral atoms of the doping element is zero at MBE growth temperatures.

Takahashi and co-workers [6.220, 221] were the first to introduce the direct implantation doping technique in order to increase the sticking probability of Zn atoms on GaAs. They have shown that an effective sticking coefficient of 0.01–0.03 on GaAs can be achieved using Zn$^+$ ions with an energy of 0.2–1.5 keV. Subsequently *Bean* and *Dingle* [6.222] established that the effective sticking coefficient of Zn$^+$ was in fact close to unity. The use of direct implantation for in-growth doping has been extended to ternary compound films of $Al_xGa_{1-x}As$

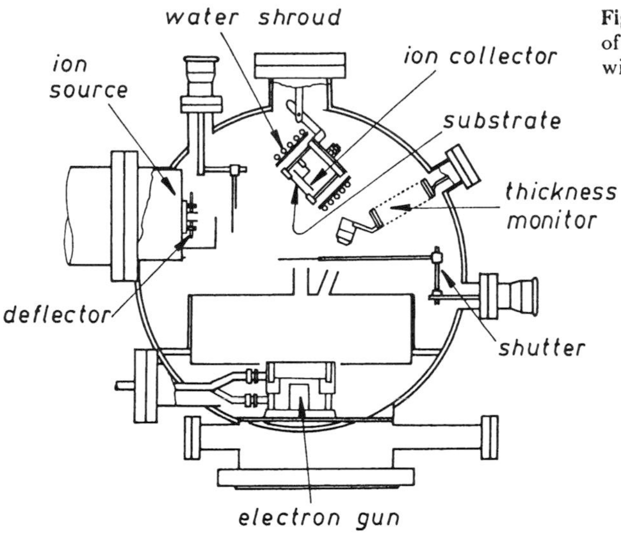

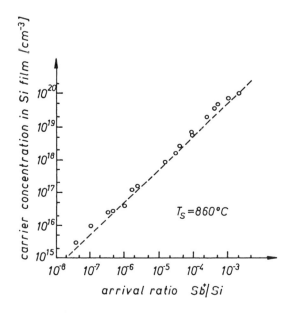

Fig. 6.30. Schematic diagram of a Si MBE system equipped with an Sb ion source [6.224]

water shroud

ion source

ion collector

substrate

thickness monitor

deflector

shutter

electron gun

doped with Zn^+ [6.221] and to GaP_yAs_{1-y} doped with N_2^+ [6.223]. It has become especially useful in MBE of Si and related compounds [6.218] doped with As^+ [6.219] or Sb^+ [6.224].

Sugiura has succeeded in achieving accurate control of the Sb doping level over the range $10^{16}-10^{20}\,\mathrm{cm}^{-3}$ by using the Si MBE system equipped with an Sb ion source shown in Fig. 6.30 [6.224]. The relationship between the carrier concentration in the films and the Sb^+/Si flux ratio at a substrate temperature of 860° C is shown in Fig. 6.31. It can be seen that the number of doped Sb atoms is proportional to the Sb^+/Si flux ratio over four decades. The dashed

carrier concentration in Si film [cm⁻³]

10^{20}
10^{19}
10^{18}
10^{17}
10^{16}
10^{15}

$T_S = 860°C$

10^{-8} 10^{-7} 10^{-6} 10^{-5} 10^{-4} 10^{-3}

arrival ratio $\overset{\cdot}{Sb}/Si$

Fig. 6.31. Carrier concentration in Sb-ion-doped films as a function of Sb^+/Si flux ratio. Dashed line indicates 100 % Sb doping efficiency [6.224]

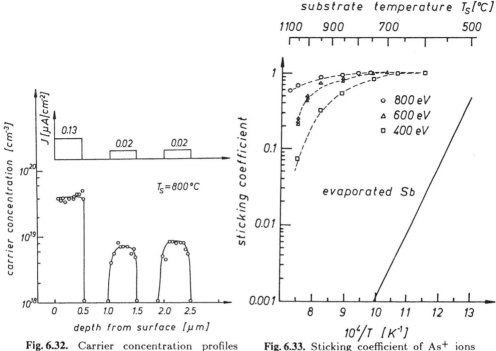

Fig. 6.32. Carrier concentration profiles in a film with three doped regions. The profile was determined by four-point probe and anodic sectioning techniques [6.224]. Upper trace gives Sb$^+$ ion current density

Fig. 6.33. Sticking coefficient of As$^+$ ions versus the Si substrate temperature for different ion energies [6.219]. The sticking coefficient of neutral Sb atoms on a Si substrate is also shown for comparison

line in this figure represents 100 % doping efficiency, and, as can be seen, all the data agree well with this. This efficiency is two orders of magnitude greater than that for evaporation doping and this improvement is believed to be due to the relative high kinetic energy of the dopant ions (130–1000 eV [6.224]). Extremely sharp doping profiles can be obtained, as demonstrated in Fig. 6.32, by changing the Sb ion current density. This figure shows the carrier concentration depth profile for a film grown according to the schedule of Sb ion current density shown in the upper trace. The carrier concentration is seen to follow the ion current changes faithfully. Moreover, no Sb atom surface segregation is detected in the ionization doping even at high doping levels. Similar results with accurate doping profile control were reported by *Ota* [6.219] who used As ions and an ion implantation machine coupled as a separate stage to the Si MBE deposition chamber. Although the sticking probability is significantly improved by the ionization doping technique, it should be noticed that the doping efficiency is still dependent on the growth temperature and the ion energy [6.219]. Figure 6.33 shows the sticking coefficient of As$^+$ ions with three different energies, 400, 600, and 800 eV, as a function of substrate temperature. The sticking coefficient of neutral Sb is also shown for comparison. It is obvious from the data that the sticking probability slightly decreases with increasing growth temperature and that this decline becomes more apparent as the ion energy increases.

The quality of the ion-doped layers has been evaluated by several methods, although significant differences in the crystallographic defect density between doped and undoped films were not looked for [Ref. 6.113, Chap. 11]). The mobility of ion-doped films is in good agreement with the bulk single-crystal values [6.224]. As is well known, ion implantation generally causes radiation damage even if the acceleration energy is low. Therefore, there exists a critical temperature above which radiation-damage-free epitaxial films can be obtained [6.219]. This temperature again depends on ion energy and it increases with increasing energy. For 800 eV As$^+$ ions, the critical temperature at which bulk-like mobilities are obtained was found to be around 850° C. Moreover, below 650° C, the radiation damage causes crystal growth to deteriorate and polycrystalline silicon layers are formed. However, if proper conditions are employed, it is shown that high-quality doped epitaxial layers can be grown and the doping level can be precisely controlled over the range 10^{14}–10^{20} cm^{-3}.

doping by secondary implantation

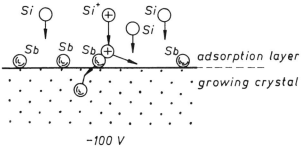

Fig. 6.34. Doping by secondary implantation. The adsorbed Sb atoms are implanted by recoil momentum from Si ions impinging on the growing crystal. Incorporation depth is only some atomic distances [6.47]

The secondary implantation technique [6.47, 225–227] has been introduced for doping MBE-grown Si with Sb atoms, which exhibit a strong surface segregation tendency and so are difficult to incorporate into the growing epilayer with sufficient control over the doping profile. The idea of this technique is illustrated in Fig. 6.34. The scheme of the Si MBE growth system in which secondary implantation has been applied for the growth of modulation-doped Si/Si$_{1-x}$Ge$_x$ strained layer superlattices with enhanced electron mobilities [6.47] is shown in Fig. 6.35. A carrier concentration profile evaluated by spreading resistance is shown in Fig. 6.36. After growing a nonintentionally doped buffer layer of Si, Sb atoms were deposited during MBE growth by opening the Sb cell shutter (see the upper part of Fig. 6.36). This establishes an Sb adsorption layer [6.228, 229], leading to doping in the 10^{16} cm^{-3} range by spontaneous incorporation. After

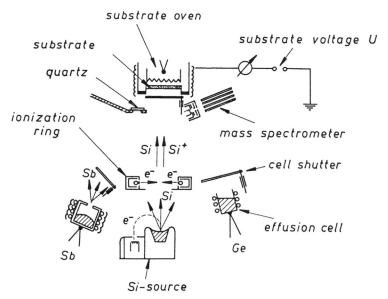

Fig. 6.35. Schematic illustration of the Si MBE system used for growth of modulation-doped $Si/Si_{1-x}Ge_x$ strained layer superlattices using the secondary implantation doping technique [6.47]

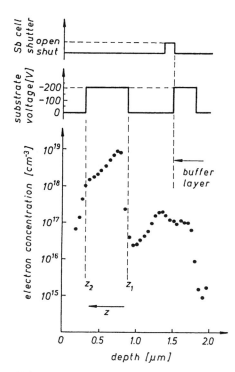

Fig. 6.36. Carrier concentration versus layer depth evaluated by spreading resistance. The growth temperature was 650° C. The program of the Sb cell shutter and the applied substrate voltage is given in the upper part [6.225]

switching on the substrate voltage Si^+ ions are accelerated towards the substrate resulting in a considerable increase of dopant concentration by three orders of magnitude from about $10^{16}\,cm^{-3}$ up to $10^{19}\,cm^{-3}$. The subsequent exponential decrease to $10^{18}\,cm^{-3}$ is caused by the loss of dopant adatom density by incorporation itself.

In order to investigate recoil yields in this doping technique, thin Si layers were grown with only partial incorporation of the Sb adlayer [6.225]. Prior to incorporation by secondary implantation Sb was deposited up to different adatom densities n_s (Fig. 6.37). Subsequently different Si^+ ion doses were employed at an energy of 0.5 keV. Figure 6.37 shows the total number N_B of incorporated Sb atoms versus Si^+ ion dose at several preadjusted dopant adlayer densities n_s. N_B increases linearly with increasing adlayer density as well as with increasing ion dose. At an ion energy of 0.5 keV the constant of proportionality is estimated to be $(5 \pm 2) \times 10^{-16}\,cm^2$ [6.225].

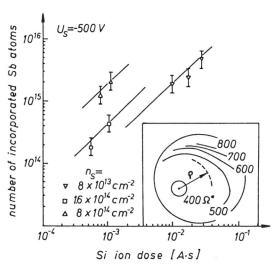

Fig. 6.37. Number of incorporated Sb adatoms versus Si^+ ion dose at different preadjusted Sb adatom densities. The inset shows the sheet resistance distribution on the wafer, where ϱ is the lateral separation from the maximum doping point on the substrate. The excentric position of the maximum is due to a small deviation from the coaxial arrangement of Si source, ionization ring, and wafer [6.225]

A slightly modified version of the secondary implantation technique (without any pregrowth deposition of dopant atoms) has been introduced under the name potential-enhanced doping (PED) to the Si MBE technology by *Kubiak* et al. [6.230–232].

Enhanced incorporation of dopants was induced by application of a voltage (U_s) via a Ta brush in contact with the rotating assembly onto which the substrate is mounted for growth. A schematic representation of the experimental configuration used is given in Fig. 6.38 [6.230].

The efficacy of PED on dopant incorporation during growth was initially investigated by applying staircase voltage programs of the type shown in Fig. 6.39a, since the resulting Sb doping profile, as shown in Fig. 6.39b, is optimized for profile evaluation using the electrochemical-CV method [6.233]. Data obtained for both negative and positive voltages from five epilayers all grown at the same

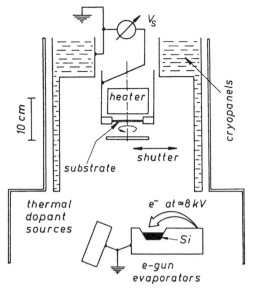

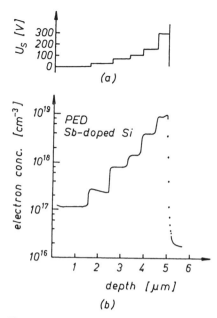

Fig. 6.38. Schematic representation of a Si MBE deposition chamber used to investigate PED. The chamber is fitted with two electron beam evaporators and up to three thermal evaporation sources spaced ~ 28 cm from the substrate [6.230]

Fig. 6.39. A typical voltage program (a) and resulting Sb doping profile (b) used to establish the efficacy of PED [6.231]

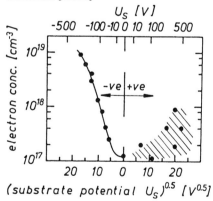

Fig. 6.40. Variation of Sb doping with applied potential (U_s) obtained from a series of epilayes all grown at the same temperature and Sb flux as used in Fig. 6.39. For convenience of presentation, the data are presented versus the square root of U_s [6.231]

substrate temperature and incident Sb-flux as used in Fig. 6.39 are presented in Fig. 6.40. The main features to be noted are (i) negative potentials are significantly more effective at sustaining enhancement than positive potentials, and (ii) only enhancement obtained under negative potentials was reproducible. As a result, only negative voltage PED will be discussed further.

Figure 6.41 demonstrates PED enhancement obtained over a wide range of growth temperatures (730°–850° C) and Sb cell temperatures (315°–390° C), conditions which yielded a zero-potential doping range $n(0) = 1.3 \times 10^{16}$ to

$1.1 \times 10^{17} \, \text{cm}^{-3}$. The ordinate has been normalized as an enhancement ratio of doping level at potential U compared with that at $0 \, \text{V}$: $n(U)/n(0)$. It is apparent that PED : Sb doping is highly reproducible. Within the range of growth conditions investigated, a calibration obtained under one set of growth conditions (pertaining to a certain Sb adlayer coverage) can be translated to any other condition. The use of InSb as an Sb_2 source yielded data [6.230] that coincided with those presented in Fig. 6.41, indicating that enhanced dopant incorporation was not dependent on the nature of the evaporated Sb species.

Antimony forms a two-dimensional surface adlayer with the surface coverage depending on growth temperature and incident Sb flux [6.228]. Based on reported data, the Sb surface coverages obtained under the range of growth conditions employed in the study of *Kubiak* et al. [6.231] are considerably less than a monolayer, but still sufficiently high to be detected by Auger electron spectroscopy (AES).

The efficiency of PED appears to be independent of adlayer coverage, and hence of the Sb flux, over the range of growth conditions employed in [6.231]. Although current models for incorporation from dopant adlayers do not specify the surface species present, the similarity of enhancement obtained from Sb_2 (InSb source) and Sb_4 (elemental source) suggests a common surface species from which incorporation occurs.

The efficacy of PED : As doping was reported by *Kubiak* et al. [6.230] to be less pronounced than in the case of Sb doping. Figure 6.42 shows the enhance-

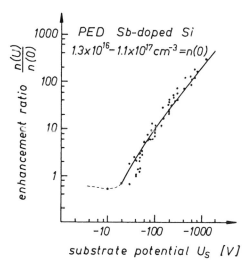

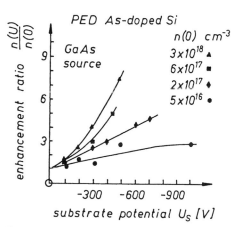

Fig. 6.41. PED enhancement over a range of growth temperatures and Sb fluxes (generated from elemental Sb and InSb) corresponding to a zero-potential doping range of 1.3×10^{16} to $1.1 \times 10^{17} \, \text{cm}^{-3}$. The ordinate is normalized as an enhancement ratio of doping level at potential U to that at $0 \, \text{V}$ [6.231]

Fig. 6.42. PED enhancement ratio over a range of As fluxes generated using a GaAs source at constant growth temperature. $n(U)/n(0)$ increases at a given potential with increasing As flux, unlike the case of Sb presented in Fig. 6.41 [6.231]

275

ment ratio obtained over a range of GaAs source temperatures of 550°–712° C [corresponding to the indicated $n(0)$ doping level] at a fixed growth temperature of $\sim 810°$ C. Unlike with PED : Sb doping, the presented data indicate that the enhancement ratio increases with As coverage. This result may indicate either that As adlayer formation is less simple than that of Sb, or, more probably, that the simultaneous presence of a Ga adlayer from a GaAs source affects As incorporation. The latter suggestion is supported by the following observations:

1. SIMS profiles have indicated that Ga is also incorporated during doping from the GaAs source, although the relative levels have not been quantified.
2. The increasing enhancement ratio with GaAs source temperature is consistent with the reported thermal decomposition behavior of GaAs, i.e., an increasing As/Ga flux ratio with increasing temperature [6.234].
3. Preliminary investigations using InAs indicate a higher enhancement ratio (factor of ~ 2) than that obtained from GaAs at a given $n(0)$ value, suggesting that In interferes less with As incorporation than does Ga.

Although further work is necessary to clarify the effect of the accompanying group III fluxes obtained from III–V sources, the conflicting responses of the enhancement ratio to arsenide and antimonide sources may imply a shorter surface residence time or a higher spontaneous incorporation efficiency for As than for Sb. Notwithstanding the difficulty of interpreting the effects of the accompanying group III fluxes from III–V sources, the negligible response of PED : Ga doping (from an elemental source) confirms a species dependence for PED doping [6.230].

An interesting feature of Sb and Ga doping is that the doping level obtained is independent of growth rate. For example, all the data presented in Figs. 6.39–41 had no correction applied for growth rate over an experimental range of greater than ten orders, and growth of epilayers in which changes in rate were intentionally introduced (with and without PED) confirmed this observation.

Kubiak et al. [6.231] have proposed the following mechanism of doping with the PED technique. The use of electron-beam evaporators for generation of the Si flux also leads to generation of charged particles in the MBE system; secondary (low energy) and backscattered (high energy) electrons are produced, as is a small fraction of low energy Si^+ ions caused by ionizing interactions between the Si flux and the primary electron beam. The application of a potential to the substrate would induce an electrostatic field within the MBE system, which would modify the trajectories of these particles and change the energy with which they impinge on the substrate. The high efficacy of negative potential PED implicates Si^+ ions as the species responsible for the phenomenon, and this is consistent with experiments where dopant incorporation has been enhanced by additional ionization of the Si growth flux by using an electron impact ionizer (Fig. 6.37) [6.225].

Enhancement of doping by Si^+ ions would be induced either by attraction of more Si^+ ions to the substrate by the electric field, thus increasing the number of ion-dopant interaction events, or by the increasing energy of the ions producing an increased cross section for interaction. To distinguish between these possible mechanisms, an experiment was performed to restrict the flux of Si^+ ions within the MBE chamber to those with direct linear trajectories to the substrate, by using an apertured shroud [6.231]. Since the enhancement ratio of Sb using this procedure was identical to that presented in Fig. 6.41 one may surmise that acceleration of Si^+ ions generated by the evaporators was responsible for the enhancement of incorporation associated with PED.

This statements means that both modifications of doping techniques described, namely PED and secondary implantation doping, are based on identical processes.

7. Material-Related Growth Characteristics in MBE

Different materials exhibit usually different properties from the point of view of growth peculiarities. The characteristic features which distinguish the materials are usually connected with a specific chemical element, e.g., Si, As, P, or Hg. The presence of this element in the material to be grown demands special technological precautions because of, for example, high evaporation temperature, high volatility, or extraordinary chemical reactivity. The special properties of the constituent elements also frequently bring about quite different growth mechanisms of the compounds crystallized with MBE.

In this chapter we will briefly present the currently known growth characteristics of the most important groups of materials grown by MBE. The subsequent sections of this chapter will be devoted to MBE of

- Si and IV–IV heterostructures
- GaAs and As-containing compounds
- narrow-gap II–VI compounds containing Hg

When discussing MBE processes related to different material groups we will preserve the same sequence of problems considered. First the substrate preparation procedure and the homoepitaxial growth peculiarities related to the material will be discussed, then the growth of structures with heterointerfaces. Each section will conclude with a presentation of some selected devices made of the considered material system.

7.1 Si and IV–IV Heterostructures

Integrated circuits based on silicon as a semiconductor material are currently manufactured nearly perfectly. Nevertheless, the performance of microelectronic devices and circuits is improving steadily with time. The forces behind this development are shrinkage of lateral and vertical dimensions, sophisticated fabrication methods, and advanced material systems. The introduction of semiconductor superlattices, i.e., new manmade semiconductors, as well as quantum effect devices, has already exerted a strong impact on electronics and become a challenge to manufacturing methods. Especially interesting are the silicon-based heterostructures, because they offer the potential of monolithic integration of conventional integrated circuits with superlattice structures. The superlattice regions in these devices yield the high performance core and high-speed links between different conventional parts of the integrated circuit [7.1].

At present, Si MBE is capable of growing epitaxial layers on large area substrates (125 mm in diameter) with deposition uniformity better than $\pm 1.5\%$ [7.2], high crystal quality [7.3], and arbitrary, well-controlled doping profiles [7.4]. Also the quality of Ge_xSi_{1-x}/Si heterostructures has been considerably improved by using a RHEED oscillation monitoring and controlling technique during MBE growth [7.5, 6].

Some interest arose also in group IV–IV heterostructures with Sn as constituent [7.7, 8]. The motivation for research into these heterostructures is the desire to integrate directly optoelectronic elements (lasers, photodetectors, etc.) with controlling logic devices onto a single chip. At present the preferred system for this integration is GaAs on Si. However, other promising candidates are alloys of Sn with Si and Ge [7.9] since there are theoretical predictions of a direct band gap in these systems [7.7]. Executing the integration with group IV elements only, i.e., Si, Ge, or Sn, would have certain advantages from the technological point of view. These would be the use of a proven technology (the Si technology) and avoidance of cross-doping problems.

Before discussing MBE growth of IV–IV heterostructures and the MBE-grown Si-based devices, we will discuss the problem concerning the MBE growth of Si epilayers on Si substrates.

7.1.1 Si Substrate Preparation Procedures

Thin Si layers of high crystalline quality are usually crystallized with MBE at temperatures below 950° C, lying far below the usual Si epitaxy range (1150°–1200° C) [7.10]. One of the fundamental problems in Si MBE is the appropriate substrate preparation. In contrast to the conventional crystal growth techniques, MBE cannot employ melt-back or chemical etch-back to generate a clean, well-ordered substrate surface.

In the mid 1970s two substrate preparation techniques were used to produce the first MBE-grown epilayers of Si exhibiting high crystallographic and electrical quality [7.11]. With one technique, flash evaporation, the substrate is simply heated to 1250° C for about 30 s, evaporating both volatile contaminants (such as oxides) and more stable species (such as carbon). The second technique, inert gas sputtering [7.12], abrades away about 10 surface atomic layers along with contaminants. The surface is then recrystallized and inert gas expelled by a moderate, 800° C anneal. Although the above techniques have been demonstrated to be quite effective, substrate preparation is still one of the most active areas of investigation in Si MBE [7.3]. The challenge is to develop a simple, inexpensive technique of substrate preparation that will yield consistent, uniform, large area silicon overgrowth with defect densities of less than $100\,\mathrm{cm}^{-2}$.

Recent work has concentrated on two areas: A reexamination of the established procedures of ex situ chemical cleaning and in situ sputter etching;

developments of new procedures such as UV ozone processing and in situ reduction of surface oxides by silicon deposition.

It has been known for some time that a perfectly clean silicon surface has a strong affinity for carbon, and that once formed these bonds are extremely hard to break [7.13]. Chemical cleaning procedures are, therefore, aimed at generating a passivating surface oxide layer rather than a clean silicon surface. These procedures generally employ hot basic or acidic peroxide solution and go by the names RCA clean, Piranha etch, or Henderson etch [7.3]. All yield a surface SiO_2 layer on the order of 10–30 Å thick. In early MBE work it was shown that these oxides can be removed in a MBE chamber at 800°–900° C [7.14, 15]. Residual carbon was then diffused away from the surface by a brief "flash" to $\sim 1150°$ C. In 1982, Ishizaka et al. proposed a modified procedure that has been shown to yield a thin nonstoichiometric oxide [7.16]. This suboxide has an enhanced volatility and can be removed by vacuum anneals at 700° C. It was also claimed that when the anneals were preceded by multiple nitric acid baths carbon could be effectively eliminated. The stated 700° C requirement was critical in that at temperatures below $\sim 800°$ C silicon dopant diffusion and wafer deformation are absent.

The efficacy of the Ishizaka procedure has been closely examined by *Xie* et al. [7.17]. In contrast to earlier work, highly sensitive deep level transient spectroscopy (DLTS) (see Sect. 5.5.2 for a description of the method) and secondary ion mass spectroscopy (SIMS) were used to examine epilayer substrate boundaries. Results from DLTS indicate the existence of traps 0.58 and 0.59 eV below the conduction band edge. These traps persist in significant concentrations in substrates annealed at 700° C and decline to the limit of detection only after 950° C processing. Data from SIMS suggest a correlation with residual interfacial carbon that is presumably dispersed by the higher temperature anneal. If the need for 950° C processing is confirmed, this would reintroduce diffusion and deformation problems, seriously undercutting the apparent advantage of the Ishizaka proposal.

The early Ar^+ sputter cleaning and annealing procedure has also been reexamined [7.12]. The sputter step is generally timed to remove ~ 100 Å of material within the MBE chamber. This is followed by an anneal at $\sim 800°$ C to remove ion beam damage. The procedure has yielded the highest published MBE silicon minority carrier lifetimes of $\sim 100 \mu$s. Nevertheless, doubts persist about the possibility of residual ion beam damage.

Hull and co-workers [7.18] have conducted a detailed examination of sputter cleaned epilayer substrate interfaces using transmission electron microscopy (see Sect. 5.6.1 for a description of the method). For certain samples, sputter times were shortened to produce removal of as little as 10 Å of material. It was thought that this would minimize ion beam damage and surface roughening, and might thus produce the highest quality interface. This was not the case. Longer sputter times produce interfaces and overgrowth free of observable defects. Shorter sputter times produce interfaces with small oxygen bubbles. It is apparent that ion bombardment knocks a small fraction of the surface oxygen deeper into the silicon. A number of atomic layers must thus be removed before this gradually decaying tail is eliminated. While this suggests the use of the longer sputtering times, these interfacial bubbles might be turned to advantage. Surprisingly, these bubbles did not nucleate threading dislocations in the overlying epilayer. As such, they might be employed as buried metal gettering centers.

New or significantly modified substrate preparation procedures have been proposed by *Kugimiya* et al. [7.19] and *Tatsumi* et al. [7.20]. They added in situ Si deposition steps to the chemical oxidation procedure. The aim was to enhance the removal of the protective oxide by using Si to reduce it to volatile silicon monoxide. This was done either by first coating the cool oxidized wafer with 5–20 Å of amorphous Si and heating or by heating the oxidized wafer under a low intensity silicon flux. Both procedures appeared effective and, in particular, minimum annealing temperature could be reduced to $\sim 800°$ C and yet yield very low defect density (100) growth.

On the (111) orientation, Tatsumi et al. could not completely eliminate stacking faults. Building on the report of *Tabe* [7.21], an ozone chemical cleaning step was added. Tabe had shown a dramatic reduction in defect densities of both (100) and (111) wafers exposed to ozone (produced by UV irradiation of air) and annealed in vacuum at $T > 900°$ C. He speculated that the decrease was due to removal of residual carbon by ozone. Tatsumi et al. employed a modified procedure wherein ozone was formed by electric discharge of high purity oxygen gas and then bubbled into the wet chemical oxidizing solution. When this was combined with the above in situ deposition of amorphous Si, (100) growth with undetectable line and stacking fault densities, and (111) growth with line densities of 500–600 cm^{-2} and stacking faults of 800–900 cm^{-2} resulted.

As mentioned in Sect. 1.2.2, one of the milestones in the development of MBE in the 1980s was the observation of the intensity oscillations of the RHEED pattern of growth control during MBE. This effect was applied to Si MBE for the first time in 1985 by *Sakamoto* et al. [7.22, 23]. The most important conclusions from the RHEED oscillation studies related to Si substrate preparation may be formulated as follows [7.24, 25]:

— The substrate must be annealed prior to MBE growth at a temperature much higher than the growth temperature in order to obtain stable RHEED intensity oscillations.

— With an increase of annealing temperature and time, the amplitude of the oscillations increases and their period doubles. That means a change of the reconstruction of the annealed surface from a two-domain structure (2 × 1 and 1 × 2) to a (2 × 1) single domain structure.

7.1.2 Homoepitaxy of Si Films

The arrangement used for Si MBE growth is shown in Fig. 3.38. As the Si beam source a high-purity Si single crystal is preferred, which is evaporated with a

270° C deflected electron beam generated in an electron gun, similar to that shown in Fig. 2.22. The substrate wafer used for the MBE growth is prepared according to one of the procedures described in Sect. 7.1.1. Doping during growth may be realized either by ion beam implantation (Sect. 6.5.4) or by co-deposition of thermally generated dopant beams.

The two-dimensional growth mode, preferred in MBE, occurs in Si-homoepitaxy under definite growth conditions. These have been studied experimentally by *Kasper* [7.10] and *Sakamoto* et al. [7.25, 26], and predicted theoretically by *Fuenzalida* and *Eisele* [7.27]. As an example, the investigations based on RHEED pattern intensity oscillations will be presented here, following Sakamoto et al.

Experiments were made in an ion-pumped MBE system. The base pressure and the pressure during growth in the growth chamber were 1×10^{-7} and 1×10^{-6} Pa, respectively. The acceleration voltage of the incident electron beam was 40 keV. The incident angle of the beam on the substrate surface was usually 0.9° (off-Bragg condition). High purity single-crystalline Si was evaporated using a 2 kW electron gun. Silicon substrates of well-oriented (001), 4 cm, $47 \times 10 \times 0.5$ mm^3 in size were degreased and finally boiled in HCl : H$_2$O$_2$: H$_2$O (1 : 1 : 4) solution to form a protective thin oxide film [7.28]. Then they were loaded into the MBE system. To evaporate the thin oxide film, a Si beam of 3×10^{13} atoms/cm^{2}s irradiated the substrate which was heated by passing a current directly through the sample to a temperature of 800° C [7.19]. After the buffer layer growth, the substrates were annealed under various conditions. Figure 7.1 shows the RHEED intensity oscillations after the anneal-

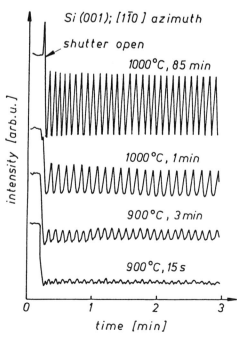

Fig. 7.1. RHEED intensity oscillations of the specular beam during Si MBE growth with various annealing conditions on the Si(001) surface after buffer layer growth; [1$\bar{1}$0] azimuth, $T_s = 400°$ C [7.25]

ing which were observed in the [1$\bar{1}$0] azimuth at a substrate temperature of 400° C. With a weak annealing condition of 900° C for 15 s, a small asymmetric monatomic-layer mode oscillation was observed. With an increase of the annealing temperature and time, the amplitude of the oscillations increased rapidly and their period doubled. This drastic change in the RHEED intensity oscillations after the high temperature annealing indicates some constructive change in the annealed surface. In order to explain the observed surface structure reformation, RHEED patterns have been checked after heat treatment under various conditions. After annealing at 900° C for 1 min two kinds of half-order diffraction spots originating from the (2 × 1) and the (1 × 2) surface reconstructions were both observed on the half-order Laue zone [7.24]. However, annealing at 1000° C for 20 min made one of the series of the half-order spots disappear. One may therefore conclude that on annealing at 1000° C for 20 min, the surface reconstruction changed from the two-domain structure to the single-domain (001)-(2 × 1) structure, i.e., to the surface with only the (2 × 1) reconstruction, or at least with a preponderance of the (2 × 1) reconstruction. This means that small domains separated by monatomic-layer-height steps in the as-grown surface were developed to larger domains separated by biatomic-layer-height steps by the high temperature annealing.

Kaplan [7.29] pointed out in his studies using LEED that the steps of a biatomic-layer height could be observed in a vicinal Si(001) surface since they are energetically more stable than the monatomic ones. More recently, *Aizaki* and *Tatsumi* [7.30] observed the single-domain Si(001)-(2 × 1) structure only during the growth on a 0.5° off-(001) surface, not after the growth and not when growing on a well-oriented (001) surface. However, a stable single-domain Si(001)-(2 × 1) structure on a well-oriented (001) surface has been observed by *Sakamoto* and *Hashiguchi* [7.24].

The peculiar biatomic-layer mode oscillations can be explained by considering the growth on the single-domain structure. The growth model considered for the homoepitaxial growth on the (1 × 2) single-domain structure and the RHEED intensity oscillations observed from different azimuths during the growth are illustrated in Fig. 7.2. Although this model is analogous to that of GaAs (Fig. 1.9), there is a difference in that elongated islands of monatomic-layer height ($a_0/4 = 1.36$ Å) with their longest dimension parallel to alternately the [110] and [1$\bar{1}$0] azimuths appear for Si, while islands of biatomic-layer height (Ga+As layer; $a_0/2 = 2.83$ Å) with the longest dimension parallel to [110] appear for GaAs. The biatomic-layer mode oscillation can be related to the reflectivity difference between the (1 × 2) and (2 × 1) reconstructed surfaces [7.31].

To investigate the alternating surface reconstructions, the RHEED intensity oscillation were observed during the homoepitaxial growth of Si using three sets of optical fiber system simultaneously. Figure 7.3 shows the oscillations taken from three different diffraction spots in the RHEED pattern of the [010] azimuth during the growth at a substrate temperature of 500° C. The insets represent schematics of the reciprocal lattice and the corresponding diffraction pattern. One

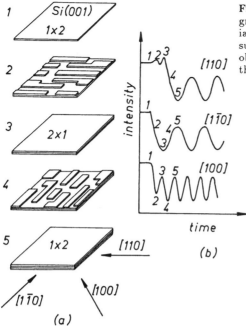

Fig. 7.2. (a) Schematic illustration of the growth model considered for homoepitaxial growth on the (1 × 2) single-domain surface. (b) RHEED intensity oscillations observed from different azimuths during the growth [7.25, 26]

period of the oscillations in trace a corresponds to the growth of one monatomic-layer. In traces b and c, one period of the oscillations corresponds to the growth of the biatomic-layer height. The phase difference between b and c is 180° (antiphase). These experimental results indicate that although the (2 × 1) and (1 × 2) surface reconstructions appear alternately during the growth (Fig. 7.2a), the specular beam intensity in the [010] azimuth is not affected by them. That is, the oscillations in this azimuth were caused by the reflectivity change due to the smooth-rough transition of the surfaces [7.31].

It is plausible to assume that elongated islands of monatomic height are formed with the longest dimension parallel to the [110] or [$\bar{1}$10] direction when growth starts on a flat surface. Since in the [010] azimuth the angles between the growth directions of the surface steps and the incident electron beam direction are +45° or −45° during the growth, the change of surface reconstruction during growth should not cause any intrinsic differences in the specular beam.

On the other hand, in the [$\bar{1}$10] azimuth, the period of the RHEED intensity oscillation of the specular beam depends on the glancing angle [7.31]. This may be understood as follows: It is expected from a multiple-scattering theory [7.32] that the intensity of the specular beam from stepped surfaces strongly depends on the incident azimuth with respect to the longer step edges. When the incident beam azimuth is parallel to the longer step edges, the intensity is several times lower than the intensity when the azimuth is perpendicular to the step edges. This means that the amplitude of RHEED intensity oscillations is larger in the former case than in the latter case. If there are alternating surface reconstructions of (2 × 1) and (1 × 2), the longest dimension of the step edges or islands is different for the two surfaces. We expect that the amplitude of the RHEED intensity oscillation of the specular beam is different for the two surfaces. Moreover, the specular beam intensity from a flat surface is different for the two surfaces because of the geometrical difference of the surfaces. Then one period of the RHEED intensity oscillation becomes twice as large as the period of monolayer formation: a biatomic-layer-mode oscillation can be expected. Details of theoretical calculations are given in [7.33, 34].

If we assume that the surface after growth is exactly the same as that before growth, the diffraction spots originating from the surface reconstruction should have the same intensity. However, in Fig. 7.3 neither b nor c reproduces the intensity observed before growth. This result coincides with observations that the half-order reconstruction spots which should be missing are still weakly observed [7.31]. These results suggest that the second layer starts to grow before completion of the first layer.

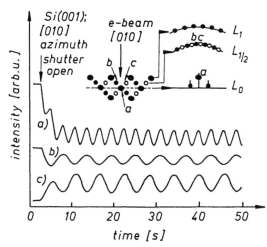

Fig. 7.3. RHEED intensity oscillations taken from three different diffraction spots in the RHEED patterns of the [010] azimuth. The insets schematically represent a reciprocal lattice and the corresponding diffraction pattern. Large filled circles (a), small filled circles (b), and small circles (c) represent the specular (and bulk) diffraction spots, and the (2 × 1) and (1 × 2) reconstruction-related spots, respectively [7.31]

Further investigation was carried out using growth interruption experiments [7.25]. Figure 7.4 shows the RHEED intensity variation of specular beam in the [1$\bar{1}$0] azimuth during growth at a substrate temperature of 400° C. The growth was interrupted for 30 s at various phases of the intensity oscillations. When the growth was interrupted at the maximum of the oscillations, as indicated by arrow (a) in Fig. 7.4, the intensity increased after interruption. On the other hand, when growth was interrupted at the minimum of the oscillations, as indicated by arrow (b), the intensity decreased. According to the growth model shown in Fig. 7.2, each maximum and minimum of the oscillations roughly corresponds to the point when the surface is mostly covered by the (1 × 2) and (2 × 1) reconstruction, respectively. Because the surface should be more or less smooth, whatever the phase of the growth interruption may be, the surface mostly covered by the (1 × 2) and (2 × 1) reconstructions will become the single-domain surface, which will cause the intensity to increase and to decrease, respectively.

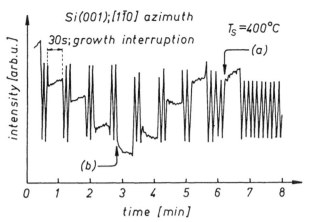

Fig. 7.4. RHEED intensity oscillations of the specular beam taken from the [1$\bar{1}$0] azimuth of the Si(100) substrate at a substrate temperature of 400° C. The growth was interrupted at various phases of the oscillations for 30 s [7.25]

The MBE growth of Si homoepitaxial films has also been investigated using a high energy ion scattering/channeling technique (see Sect. 5.6.2 for a description of the technique) [7.35, 36]. The reason for these investigations was to study quantitatively the initial stages of interface formation (overlayer thickness up to 10 Å) during Si MBE. Atoms in the surface monolayer(s) of a clean material are usually positioned not in their bulk-like sites but in the sites of a reconstructed surface. The MBE process, on the other hand, results in an interface with atoms at bulk-like sites. This means that a reordering process occurs during the growth. This reordering phenomenon is the origin of the considerable influence of reconstruction on epitaxial growth.

The high energy ion scattering/channeling experiments have shown that the room temperature deposition of Si on Si substrates causes a reordering process on the Si(100)-(2 × 1) surface but not on the Si(111)-(7 × 7) surface. On both surfaces, however, deposition at 300 K results in a highly imperfect overlayer. To obtain higher-quality Si layers by MBE a temperature of about 790 K is needed for Si(111)-(7 × 7) substrates and of about 570 K for Si(100)-(2 × 1) substrates. The fact that growth temperatures for high-quality MBE of Si grown on Si are lower for Si(100) substrates than for Si(111) substrates has been known from the very beginning of Si MBE. It results from the evident influence of the substrate reconstruction on MBE growth [7.37].

7.1.3 Heteroepitaxy of Ge and Sn on Si Substrates

The lattice constant differences between the group IV elements Ge, Sn, and Si are in the range of several percent [a_0(Si) = 0.543 nm, a_0(Ge) = 0.566 nm, a_0(Sn$_{gray}$) = 0.649 nm] [7.38], which means that the heteroepitaxial layers of Ge or Sn grown on Si substrates are highly strained. This fact influences considerably the growth mechanism in MBE. It has been shown theoretically [7.39] that in a system with any nonzero misfit, a uniform film with a thickness greater than several monolayers is not in the equilibrium state. This system can, however, lower the chemical potential by formation of clusters (this is also called the islanding effect). Clusters will form on either the bare substrate (Volmer-Weber growth mode) or on a few layers of uniform film (Stranski-Krastanov growth mode), depending on the misfit between the two crystal lattices and on the value of the film–substrate interaction energy (Sect. 6.2.1).

Let us first consider Ge/Si MBE growth. The initial stages of interface formation during MBE growth of Ge films on Si(100)-(2 × 1) and Si(111)-(7 × 7) substrates have been investigated by high energy ion backscattering/channeling, Auger electron spectroscopy, and low energy electron diffraction by *Gossmann* et al. [7.37, 40].

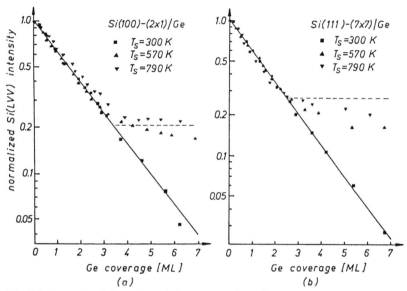

Fig. 7.5. Normalized intensity of the 92 eV Si(LVV) Auger line as a function of coverage for Ge deposition on **(a)** Si(100)-(2 × 1) and **(b)** Si(111)-(7 × 7) at (■) 300 K, (▲) 570 K, and (▼) 790 K. The solid lines are the result of an exponential least squares fit to the 300 K data. The dashed lines are described in the text [7.40]

As can be seen from the results of AES experiments shown in Fig. 7.5, growth at elevated temperatures (triangles) initially still proceeds in a simple way, the data points following the exponential decay curve (solid lines) very closely. At about 3 ML, however, a break occurs and the Si AES intensity stays practically constant thereafter. In principle this deviation from exponential decay, also observed by *Chen* et al. [7.41], can be due to islanding (Volmer-Weber mode) or indiffusion (compound formation). However, considering the AES together with the ion backscattering data leads to the conclusion that only simple growth up to a particular coverage followed by indiffusion is compatible with the data.

This interpretation is different from the suggestion of *Narusawa* and *Gibson* [7.42] who, on the basis of normal incidence channeling data, proposed a Stranski-Krastanov growth mode (layer-by-layer growth up to 3ML, followed by islanding) for the deposition of Ge on Si(111)-(7 × 7) at 620 K.

The practically constant value of the Si AES intensity after its breakaway from the exponential decay curve can be explained within an island growth model only if the islands do not grow laterally in the coverage range investigated, i.e., if the islands do not coalesce. This implies that the island fraction f_i, i.e., the fraction of the sample surface covered with islands, is constant. To keep the discussion as general as possible one should take into account the possibility that islanding occurs on top of Θ_{SK} monolayers of layer-by-layer grown Ge (Stranski-Krastanov growth). Considering the finite acceptance angle of the electron energy analyzer (Sect. 5.2.1) and using the inelastic mean free path obtained from the 300 K data then allows the normalized Si(LVV) intensity to be calculated as a function of Ge coverage. The two parameters Θ_{SK} and f_i can be determined from a least squares fit of the calculated intensity to the actual data. For all cases investigated in [7.40] [Si(100) and Si(111) substrates, deposition temperatures of 570 and 790 K] always $f_i > 0.35$ and $\Theta_{SK} > 1.9ML$, i.e., the thickest Ge regions were always less than ~ 20 Å thick. Even if the island heights are not all of the same value, as was assumed in the fitting procedure, but are statistically distributed, this is still less than the depth resolution in the ion backscattering experiments (~ 50 Å) and would thus lead to only a small broadening of the Ge backscattering peak. Contrary to this expectation, however, one can observe very extensive tailing of the Ge peak for deposition of Ge at elevated temperatures, corresponding to scattering from Ge atoms up to 200 Å deep in the sample. Therefore islanding is not compatible with the ion backscattering data.

Island formation has been reported in previous works on Ge on Si epitaxy [7.43–45]. However, only *Cullis* and *Booker* [7.44] performed their experiment under ultra-high-vacuum conditions, while *Krause* [7.43] had previously shown that small amounts of carbon can completely change the growth morphology. *Aleksandrov* et al. [7.45] used a vicinal surface adjoining the (111). All work in [7.43–45] was carried out with deposition rates at least one order of magnitude larger than in the work [7.40] and the thinnest overlayer investigated had a thickness of > 10 Å, already outside the range of coverages in the work [7.40].

On the basis of the described experiments one may conclude that Ge deposition at 570 and 790 K under carefully controlled experimental conditions, basically proceeds with a simple growth mode, comparable to deposition at 300 K, up to a critical coverage where indiffusion effectively removes additional Ge from the probing depth of AES. Since no significant diffusion is expected on the basis of bulk diffusion coefficients [7.46], this process should be ascribed to the strain at the interface. The critical coverage should then be identified with the coverage at which the substrate surface has completely reordered (the surface reconstruction positions of the Si atoms have been replaced by their bulk-like positions).

The strain at the Ge/Si(100) interface associated with ultrathin (2–6 monolayers) epitaxial films has been determined experimentally with an ion backscattering and channeling technique to be at the level of 3 %–4 % [7.47], exceeding that accessible in bulk materials. These values, in both single-layer films and Ge-Si superlattices, are in approximate agreement with bulk elastic properties, suggesting that Poisson's ratio is applicable at the level of epilayers which are a few monolayers thin.

The surface structure and film defects in the Ge/Si MBE-grown heterostructures have been studied with low energy electron diffraction [7.48, 49], RHEED [7.50], electron microscopy, and x-ray diffraction [7.51]. It has been discovered that a correlation exists between lateral compressive strain in Ge films MBE-grown on Si(111) substrates and the reconstructed state of the surface of these films. The in-plane lattice constant varies continuously with film thickness while surface symmetry changes from $c(2 \times 8)$ to (7×7). These results indicate the influence of lateral compressive stress on the (7×7) reconstruction [7.48]. The experimental data obtained with RHEED investigations can be systematized and presented in the form of a so-called kinetic T_s-h diagram [7.50]. As may be seen from Fig. 7.6, many different superstructures appear on the growing surface during Ge heteroepitaxy on Si(111) substrates, depending on the growth temperature (T_s) and the layer thickness (h).

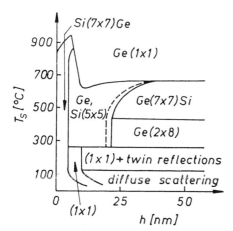

Fig. 7.6. Kinetic diagram of structural transformation on the film surface during Ge heteroepitaxy on Si(111) substrates [7.50]

Growth, nucleation and electrical properties of MBE-grown Ge heteroepitaxial layers doped with As to electron concentrations as high as $2.5 \times 10^{20} \, \mathrm{cm}^{-3}$ have been studied by *Sheldon* et al. [7.52]. It has been found that at the Si (100) substrate temperatures examined (300°–600° C), heteroepitaxy of doped Ge on Si favors three-dimensional growth, which is enhanced at higher growth temperatures.

Investigations on heteroepitaxy of Sn grown by MBE on Si substrates [7.8, 53] or on Ge substrates [7.7, 54] are still in their infancy. However, some experimental results concerning the growth of Sn on Si(111) and on Si(100) surfaces seems worth quoting [7.53].

The morphology of the Sn/Si system changes strongly at a critical tin coverage θ_c [$\theta_c = (1.25 \pm 0.1) \times 10^{15}$ cm^{-2} on Si(111) and $(1.2 \pm 0.2) \times 10^{15}$ cm^{-2} on Si(100)]. Below this critical coverage layered growth occurs. Measurements of the Si-surface peak of the Auger Si(LVV) line as a function of tin coverage ($\theta < \theta_c$) indicate an adsorbate-induced reordering of the Si reconstruction in good agreement with previous measurements on Si/Si and Ge/Si [7.35, 37, 40]. Above θ_c clear indications of islanding have been found. The islanding process exhibits the following features:

1. The volume of the islands grows at a constant rate for both Si(111) and Si(100) surfaces, but the rate is different for the two surfaces at the same temperature.
2. Smaller islands decompose in order to fulfill the conservation of tin atoms on the surface in this postdeposition process.
3. During the whole growth process the distribution of the islands is sharply peaked about an average value, i.e., the smaller an island the faster it decomposes.
4. The growth rate is independent of the starting tin coverage on both surfaces.
5. The growth rates vary for different surfaces and surface preparations.
6. The evaporation rate of Sn in an island is a factor ~ 10 greater than in the Sn layer bonded to the Si surface (compare this result with the NSTL model described in Sect. 6.3.2).

These results are in accord with the Stranski-Krastanov (SK) growth mode, where θ_c represents the uniform film thickness under the clusters. It was found that the value of θ_c is temperature independent and this initial layer does not undergo islanding to temperatures of 1150 K.

The main result from these data is that the growth of the islands is not limited by the surface diffusive transport of tin from one island to another over the SK layer. Therefore, by use of the Arrhenius equation, an average activation energy can be derived for the limiting factor in island growth. The values are 0.32 ± 0.04 eV for sputter/annealed Si(111) and 1.00 ± 0.20 eV for sputter/annealed Si(100). Assuming that the energy state for tin within the islands is independent of the size and substrate, one can interpret these differences as due to the different energy states of a single tin adatom on the uniform SK layer, i.e., the bonding and structure dependence of this first layer. These results may be compared to the molecular dynamics simulations of *Grabow* and *Gilmer* [7.39]. These simulations yield predictions of the type of growth mode, SK (clusters on a few layers of uniform film) or Volmer-Weber (clusters on a bare substrate) as a function of substrate-adsorbate potential and lattice mismatch. Using a value of 1.35 [7.53] for the ratio of the Sn-Si interaction to the Sn-Sn interaction and a 19.5 % mismatch, the predicted growth mode is barely within the SK region, in agreement with the observations of *Zinke-Allmang* et al. [7.8, 53].

7.1.4 Ge_xSi_{1-x}/Si Heterostructures and Superlattices

The heteroepitaxy of Ge_xSi_{1-x}/Si structures has attracted much attention since the process used in Si MBE made it possible to achieve a pseudomorphic growth of these heterostructures up to 1 μm in thickness on Si substrates [7.55]. Electric properties of epitaxial films have been improved and modulation-doped Ge_xSi_{1-x}/Si strained-layer heterostructures show two-dimensional carrier gas properties and enhanced mobilities for electrons [7.56] and holes [7.57]. Recently, n-channel [7.58] and p-channel [7.59] modulation-doped n-doped field effect transistors were successfully fabricated. In this way, progress in the heteroepitaxy of Ge_xSi_{1-x}/Si has made it possible to begin band-gap engineering of Si-based materials.

Because the electric and optical properties of heterostructures are very sensitive to crystal microstructure such as the layer thickness, alloy composition, impurity doping and heterointerface roughness, it is necessary to monitor and control crystal growth with atomic-order precision. This can be done using RHEED pattern intensity oscillations. The first observation of such oscillations during strained-layer heteroepitaxy of Ge_xSi_{1-x} on a Si substrate was reported by *Sakamoto* et al. in 1987 [7.5]. The same authors also described the growth of Ge_xSi_{1-x}/Si superlattices using the phase-locked epitaxy technique.

Several types of information concerning crystal growth may be inferred from RHEED intensity oscillations. First the precise layer thickness may be determined by counting the number of oscillations. One period of oscillation during the Ge_xSi_{1-x} growth observed from the [110] azimuth corresponds to a biatomic-layer height ($a_0/2$ = 0.27 nm) growth. This bilayer-mode oscillation was first found during Si growth, where it was caused by the alternating growth of (2 × 1) and (1 × 2) reconstructed surfaces [7.22, 31]. Another in situ characteristic is the alloy composition. Since the sticking coefficients of Si and Ge under these growth conditions are almost unity, the mole fraction x of Ge in the Ge_xSi_{1-x} layer is easily obtained from the oscillation frequencies as [7.5]

$$x = \frac{f(\text{GeSi}) - f(\text{Si})}{f(\text{GeSi})} \quad , \qquad (7.1)$$

where $f(\text{Si})$ and $f(\text{GeSi})$ denote the RHEED intensity oscillation frequencies during Si layer and Ge_xSi_{1-x} layer growth (keeping the Si evaporation rate constant), respectively.

The Ge_xSi_{1-x}/Si heterostructures and superlattices may be grown in typical Si MBE systems, such as those shown schematically in Figs. 3.38 and 6.35. The details of MBE growth of these structures are given in [7.1, 5, 60]. In presenting the most important data on this subject we will follow the report of *Sakamoto* et al. [7.5].

Films of Ge_xSi_{1-x} with various Ge mole fractions were grown by controlling the evaporation rate of a Si source and that of a Ge source. Typical RHEED intensity oscillations of a specular beam observed from the [1$\bar{1}$0] az-

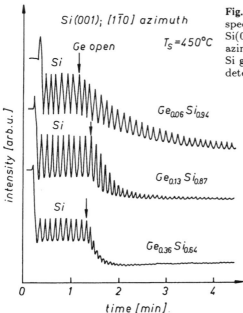

Si(001); [1̄10] azimuth

Ge open

$T_S = 450°C$

Si

Si

$Ge_{0.06}Si_{0.94}$

$Ge_{0.13}Si_{0.87}$

Si

$Ge_{0.36}Si_{0.64}$

intensity [arb.u.]

time [min]

Fig. 7.7. RHEED intensity oscillations of the specular beam during Ge_xSi_{1-x} growth on Si(001) at 450°C taken from the [1̄10] azimuth for various Ge mole fractions. The Si growth preceded the Ge_xSi_{1-x} growth to determine alloy compositions [7.5]

imuth at 450°C are shown in Fig. 7.7. The growth of Ge_xSi_{1-x} was preceded by Si growth for several oscillations in order to determine the alloy composition.

While the oscillation amplitude and the average intensity were almost unchanged during Si growth, they began to decrease after Ge_xSi_{1-x} growth started. The decrease was more rapid with an increase in the Ge mole fraction in the grown film. In the growth of Ge on Si at 450°C, i.e., $x = 1.0$, only two periods of oscillation were observed.

The damping behavior of the oscillations could be explained by clustering during Ge_xSi_{1-x} growth as follows. When the Ge mole fraction was small, the RHEED pattern during Ge_xSi_{1-x} growth was almost the same as that during Si growth. However, with an increase in the Ge mole fraction, the RHEED pattern rapidly changed from the initial streaky type to the final spotty type. It was concluded from these observations of RHEED patterns that the decrease of the oscillation amplitude during Ge_xSi_{1-x} growth was a result of a change in the growth mode from a two-dimensional layer-by-layer type to a three-dimensional type which is caused by a clustering of Ge_xSi_{1-x}. The reason why three-dimensional growth was promoted by an increase in the Ge mole fraction is not clear. Considering the fact, however, that more than ten periods of oscillations were observed during the homoepitaxial growth of Ge at 350°C (in contrast with the poor oscillatory property of the Ge/Si strained-layer growth), it cannot be said that Ge itself has a strong tendency to cluster. These results suggest that the motive force for clustering has some relation to a lattice mismatch between Si substrates and Ge_xSi_{1-x} epilayers [7.39].

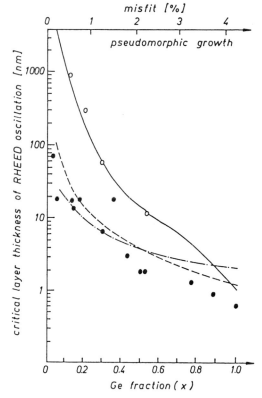

misfit [%]

pseudomorphic growth

o experiment (Bean et al.)
——·—— theory (Van der Merwe)
————— (Matthews et al.)
———— (People et al.)

• RHEED oscillation

Si(001); [1̄10] azimuth
$T_s = 450°C$

Fig. 7.8. Filled circles summarize experimental data concerning the critical thickness t_c(osc) of a $Ge_x Si_{1-x}$ layer grown at 450° C at which RHEED intensity oscillations almost fade away [7.5]. Data on the critical layer thickness for the pseudomorphic growth t_c(psd) are also shown: (o) *Bean* et al. (experiment) [7.55]; (——) *People* et al. (theory [7.63]; (---) *Matthews* and *Blakeslee* (theory) [7.62]; (-·-) *Van der Merwe* (theory) [7.61]

It has been concluded from the RHEED investigations that a lower substrate temperature, which causes a lowering of surface migration, is preferable for the layer-by-layer growth of Ge, since Ge atoms on a Si surface have a relatively large cohesive energy. This large value of cohesive energy is the reason for the occurrence of three-dimensional growth.

Figure 7.8 summarizes experimental data concerning the critical of layer thickness t_c(osc) of $Ge_x Si_{1-x}$ at which the RHEED intensity oscillations almost faded away for various Ge mole fractions. Studies concerning the critical layer thickness of the pseudomorphic growth t_c(psd) obtained from both calculations [7.61–63] and experiments [7.55] are also shown in the figure. It should be noted that the magnitude of t_c(osc) was roughly equal to that of t_c(psd) derived from earlier calculations [7.61,62], however, it was significantly smaller than that of the experimental t_c(psd) of *Bean* et al. [7.55]. The relationship between t_c(osc) and t_c(psd), which are the critical layer thicknesses of physically different phenomena, has not yet been made clear. However, the coincidence of the experimental t_c(osc) and the calculated t_c(psd) suggests that the damping behavior of the RHEED intensity oscillations is closely related to a misfit stress. Considering that the three-dimensional growth took over from the two-dimensional layer-by-layer growth at t_c(osc), it can be said that the three-dimensional growth, which would cause a microscopically inhomogeneous strain, might promote the formation of dislocations in the epilayer.

Another interesting feature of Ge_xSi_{1-x}/Si heterostructures, reported in [7.5], was the smoothening of the surface during Si overgrowth. The Ge growth was stopped and the Si source shutter was opened after the RHEED pattern changed to a spotty type indicating a rough Ge surface. Although no oscillations could be observed at the beginning of the Si overgrowth, a weak oscillation appeared and gradually gained in amplitude with an increase in the Si-overgrown layer thickness. One may consider from this result that the atomically rough surface from the three-dimensional growth of Ge was smoothed during the initial stage of Si overgrowth by Si atoms; then, the RHEED intensity oscillation was observed on the smoothed Si-overgrown surface.

With the help of the oscillation recovery, *Sakamoto* et al. [7.5] were able to apply a "phase-locking technique" to the growth of a Ge_xSi_{1-x}/Si strained-layer superlattice. They obtained 26 periods of a $(Ge_{0.25}Si_{0.75})_{10}/Si_{10}$ strained-layer superlattice, i.e., 520 atomic layers (70 nm), while monitoring the RHEED intensity oscillation (Fig. 7.9). The total thickness of the $Ge_{0.25}Si_{0.75}$ layers (35 nm) was much larger than the critical thickness of the RHEED intensity oscillation for the Ge mole fraction (~ 8 nm).

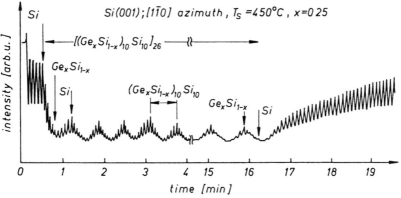

Fig. 7.9. RHEED intensity oscillations during the growth of a $(Ge_{0.25}Si_{0.75})_{10}/Si_{10}$ strained-layer-superlattice structure on Si(001) at 450° C. Using a "phase-locked epitaxy technique", 26 periods of strained-layer superlattice were grown without any growth interruption [7.5]

The Ge_xSi_{1-x}/Si interfaces in the strained-layer superlattice grown on the Si(001) substrate were examined by high resolution transmission electron microscopy (HRTEM). The RHEED intensity oscillation and the HRTEM image showed a difference in roughness between a Ge_xSi_{1-x} on Si interface and a Si on Ge_xSi_{1-x} interface. This suggests that it is necessary to use a Ge_xSi_{1-x} on Si interface in order to improve the electrical properties of a modulation-doped structure. Similar results were obtained on the strained-layer heterostructure of $Ge_xSi_{1-x}/Si(111)$.

As described in Sect. 6.4.2, growth interruption at the growth of GaAs/AlAs superlattices made the heterointerfaces smooth and the RHEED intensity oscillation recover. However, the RHEED intensity did not recover during the growth of a Ge_xSi_{1-x}/Si strained-layer heterostructure with a growth interruption at 450° C. This indicated that the surface was not made smooth. The search for a growth procedure which will make the Ge_xSi_{1-x} surface smooth and at the same time prevent the degradation of the crystallinity and the diffusion of Ge at the Ge_xSi_{1-x}/Si interface is left for future studies. Despite this, the experimental results reported in [7.5] are evidence that the door to atomic-order control of Ge_xSi_{1-x}/Si heteroepitaxy has been opened.

7.1.5 Devices Grown by Si MBE

The technological results presented above have been used to prepare a number of Si-based device structures. These devices have demonstrated not only the potential of the considered material system but also a surprising robustness of the MBE-grown materials under conventional device processing [7.64]. Modulation-doped transistors have been fabricated with both p-type [7.59] and n-type [7.58] channels.

The radical reduction of band gap has also been exploited in a series of Ge_xSi_{1-x}/Si photodetectors for fiber optic applications [7.65–67]. As shown in Fig. 7.10, a strained-layer superlattice can serve as a light-guiding absorbing

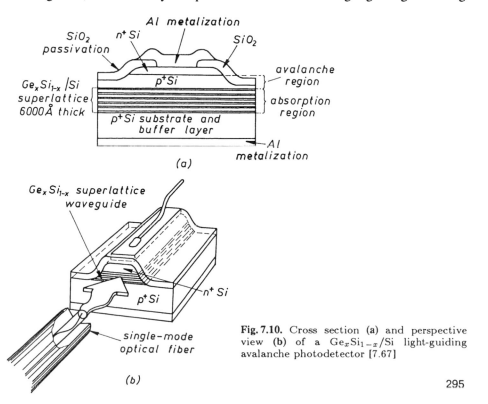

Fig. 7.10. Cross section (a) and perspective view (b) of a Ge_xSi_{1-x}/Si light-guiding avalanche photodetector [7.67]

layer within a silicon device. The light-guiding geometry not only accommodates the restrictions of absorption length and strained-layer critical thickness, it also minimizes the device dimensions to provide for high-speed operation. Indeed, avalanche versions of this device have demonstrated a 3 dB bandwidth of over 8 GHz with a high-frequency gain of 6. Over a 45 km optical fiber link, 1.3 μm laser radiation has been detected at 800 Mbyte/s with error rates of less than one part in 10^9 [7.64].

At present, it seems evident that the Si MBE technology has achieved its greatest successes in fabrication of high speed devices [7.68–72], although some interesting results have also been reported in the field of monolithic integration [7.73].

We will describe here the present state of the art of the application of MBE to fabrication of $Ge_x Si_{1-x}$/Si high speed devices. In doing this we will follow the paper of *Daembkes* [7.72].

Investigations on MBE-grown modulation-doped $Ge_x Si_{1-x}$/Si heterostructures demonstrated the existence of a two-dimensional hole gas (2DHG) and of a two-dimensional electron gas (2DEG) at the $Ge_x Si_{1-x}$/Si heterointerface, each of them exhibiting an enhanced mobility due to the reduction of impurity scattering [7.74, 75] (for a description of modulation doping and physical effects related to two-dimensional carrier enhanced mobility see [7.76] or Sect. 5.5.1). Further experiments confirmed that, depending on the strain, two different types of energy band alignment are observed. When grown on a Si substrate or buffer layer, the $Ge_x Si_{1-x}$/Si heterostructure is compressed and exhibits a type I band alignment as shown in Fig. 7.11a. However, the same heterostructure, equally strained but grown on a $Ge_x Si_{1-x}$ buffer layer, which means that the Si layer is dilated to match the binary compound buffer and the $Ge_x Si_{1-x}$ layer is compressed to match the strained Si layer, exhibits a type II band alignment, as shown in Fig. 7.11b [7.77]. It has also been confirmed experimentally that the 2DEG is formed inside the Si layer for a band alignment of type II (Fig. 7.11b), and that the 2DHG is formed inside the $Ge_x Si_{1-x}$ layer for a type I configuration (Fig. 7.11a).

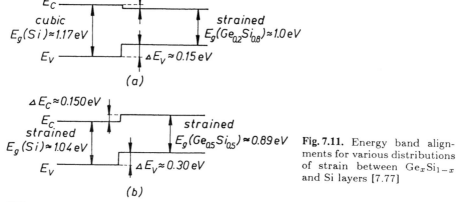

Fig. 7.11. Energy band alignments for various distributions of strain between $Ge_x Si_{1-x}$ and Si layers [7.77]

Bell researchers were the first to demonstrate p-channel Ge_xSi_{1-x}/Si modulation-doped field effect transistors (p-MODFETs) [7.59]. According to their results concerning the confinement of holes in Ge_xSi_{1-x} [7.74] they used the following layer structure: On a p^--Si substrate an undoped buffer layer was grown, followed by an undoped 25 nm $Ge_{0.2}Si_{0.8}$ confinement layer. The doping was incorporated in the following 50 nm thick p^--Si layer as a modulation-doped layer of ~ 10 nm thickness which is typically set back from the heterointerface by about 10 nm. The thin doping layer was produced during MBE growth by a simultaneous implantation of boron into the undoped Si layer [7.78]. Typical implantation energies were around 1 keV and the dose would give a bulk concentration of around 2×10^{18} cm^{-3}. The typical MBE structure and the proposed band structure are shown in Fig. 7.12. As can be seen, two potential wells are formed at the heterointerfaces, but it is assumed that the majority of the carriers are confined at the side of the well near the boron doping. Only near the pinch-off condition will a distinct amount of the current flow through the lower channel. This fact might lead to a slightly degraded pinch-off behavior.

The FETs are fabricated using standard Si VLSI processing technologies. Ohmic contact regions were prepared using BF_2 implantation. The annealing temperature was kept at modest levels (700°–750° C) in order to limit the defect formation due to strain relaxation and to minimize possible Ge diffusion. As shown in Fig. 7.12a, Al is used for ohmic contacts and Ti to form the Schottky gate. Layers were thinned using the reactive-ion etching technique [7.59].

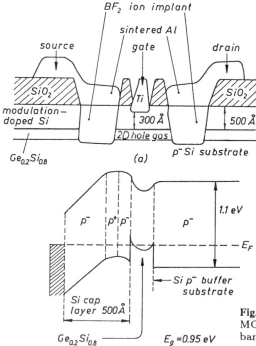

Fig. 7.12. Cross section of a p-channel MODFET (a), and the schematic energy band alignment of this device (b) [7.72]

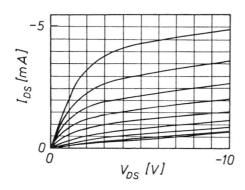

Fig. 7.13. Direct current characteristics of an enhancement-mode p-channel Ge_xSi_{1-x}/Si MODFET [7.59]

By adjusting the layer thickness under the gate contact both enhancement mode and depletion mode p-channel MODFETs were fabricated. Typical characteristics of a depletion mode transistor are shown in Fig. 7.13. Maximum values of the transconductance are reported to be nearly 10 mS/mm. The achievable values are reduced by the fairly high parasitic access resistances of the FET structure used. The reasons for this are the low mobility of the holes, the low sheet carrier concentration of about 2.5×10^{11} cm^{-2}, and the long source-to-drain spacing.

The hole transport properties were extracted from the direct current characteristics of the transistors. At zero gate bias the drift velocity of the holes in the channel was determined to be $V_D = 1.1 \times 10^7$ cm/s which is in good agreement with expected values [7.59].

The enhanced mobility of electrons constituting the 2DEG inside the Si layer of modulation doped structures is only achievable if the concept of strain symmetrization [7.1, 79] is applied. Following the results of the AEG group on enhanced electron mobility in multi-quantum-well structures [7.75] a simple structure for an n-channel Ge_xSi_{1-x}/Si MODFET has been developed by *Daembkes* and co-workers [7.58]. This structure is shown in Fig. 7.14a. On a (100) high resistivity Si substrate a 200 nm $Ge_{0.3}Si_{0.7}$ buffer layer is grown. Its thickness is well above the critical value for stress accommodation by misfit dislocations. It is intended that the strain at this heterointerface be released by the creation of interfacial misfit dislocations. By TEM investigations it was confirmed that a dense misfit dislocation network exists at the plane between the Si substrate and the buffer layer [7.80]. The following layers, however, are virtually free of these defects. The Ge content of the buffer layer was chosen to symmetrize the strain in the following layers.

On top of the buffer layer a 20 nm undoped Si layer follows in which the quantum well will be formed. Then a 10 nm $Si_{0.5}Ge_{0.5}$ layer is grown. This layer contains the antimony doping spike of about 2 nm thickness (see Sect. 6.5.3 for the definition of spike doping). To avoid the formation of a second quantum well a graded SiGe layer of 10 nm thickness forms the transition to the undoped Si top layer. The Si top layer is grown to allow the formation of high quality Schottky contacts for the gate electrode and avoid disadvantageous surface currents.

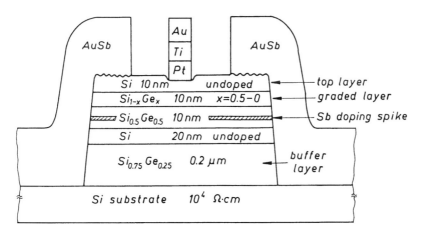

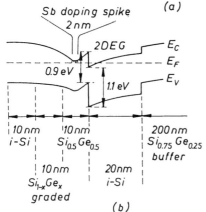

Fig. 7.14. Cross section of an n-channel MODFET (a), and the schematic energy band alignment of this device (b) [7.72]

The layers for the first series of devices were grown in standard SiGe MBE equipment at a substrate temperature of 600° C and a growth rate of about 0.14 μm/h on high resistivity Si substrates (Fig. 6.35). All the layers except for the δ-doped SiGe layer are undoped. The Sb doping spike was produced using the method of doping by secondary implantation (Sect. 6.5.4) during the growth of the $Ge_{0.5}Si_{0.5}$ layer. Since then a series of layers has been grown at extremely low temperatures down to 240° C. In these experiments the $Ge_{0.5}Si_{0.5}$ layer was doped by spontaneous incorporation of the Sb atoms at temperatures below 350° C.

The band structure of this n-channel structure is depicted in Fig. 7.14b. It is assumed that only one conductive channel is formed at the Ge_xSi_{1-x}/Si heterointerface due to the band bending of the buffer layer.

Further technological steps were made in analogy to well-proven MODFET technology. Special care was taken to avoid any high temperature treatment, which might degrade the abruptness of the heterointerface or promote diffusion. Conventional contact lithography was used to define the device levels. The individual devices were isolated by a mesa technique. A dry etching process using $CF_4 + 3.9\%$ O_2 was applied. Ohmic contacts were formed by thermal evaporation of $0.3\,\mu$m AuSb. Due to the undoped Si top layer and the omission of any n-implantation or diffusion the ohmic contacts have to be alloyed. Alloying at $330°$C under protective gas for 30 s leads to low resistivity ohmic contacts with smooth edges and nearly perfect morphology. The gate was formed by e-gun evaporation of a Pt/Ti/Au sandwich of 100 nm/10 nm/150 nm thickness. Platinum is chosen as a Schottky metal as it establishes one of the highest Schottky barriers to Si. Immediately before the evaporation an oxygen plasma dip was performed to remove organic residuals, followed by a HF dip to remove oxides.

A number of different wafers with fairly low carrier concentrations in the range of $(1-6) \times 10^{11}\,\mathrm{cm}^{-2}$ were grown and processed. All of the investigated samples resulted in good FETs. Typical dc characteristics of a SiGe/Si n-channel MODFET are shown in Fig. 7.15.

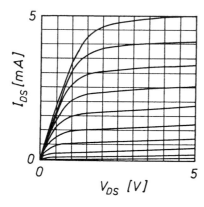

Fig. 7.15. Typical characteristics of an n-channel Ge_xSi_{1-x}/Si MODFET [7.72]. $L = 2\,\mu$m, $W = 90\,\mu$m. Upper trace corresponds to $V_{GS} = 0$ V, $\Delta V_{GS} = 0.2$ V

All devices showed good characteristics with a complete pinch-off behavior, and distinct ohmic and saturated regions. No looping was detected. In some cases a certain bias-dependent shift of the characteristics towards higher currents was observed, which might indicate the presence of some traps.

The following statements may serve as a conclusion to the considerations concerning devices grown with Si MBE [7.72]:

- The Ge_xSi_{1-x}/Si heterosystem has been demonstrated to be a most interesting material, because it clearly extends the present potential of the homoepitaxial silicon-based devices. The system's immanent strain and the band gap discontinuities are two important additional parameters which add much more flexibility to device design and allow the development of completely new structures for device applications.
- The enhanced transport properties in this heterosystem are already exploited in modulation-doped FETs. First results of millimeter-wave diodes have also been published. One of the most important applications of this material system, however, will be the fabrication of heterobipolar transistors. In this area a strong industrial interest is expected.

7.2 GaAs- and As-Containing Compounds

Compound semiconductors appear to be among the most prominent materials for the generation and detection of light for photonic communication devices. One of the fabrication processes used for such materials is MBE. As the most studied compound semiconductor grown with MBE, GaAs provides a prototype for understanding the detailed nature of that growth process. Currently, the (100) surface is the favored growth substrate, partly because of the ease with which it may be cleaned and partly because of the difficulty of achieving similarly good MBE growth on other surfaces such as ($\bar{1}\bar{1}\bar{1}$) [7.81].

In recent years Si has drawn increasing attention as a substrate material for GaAs MBE [7.82]. One of the reasons for this is the desire to use a substrate which has a high thermal conductivity, large diameter and low cost. Another reason is the possibility of monolithically combining Si integrated circuits with high speed and/or optical GaAs devices. There are, however, a number of difficulties in MBE of the GaAs-on-Si system caused by the considerable lattice mismatch (4 %) between the two materials and by the growth of a polar compound semiconductor on a nonpolar elemental one. These differences should introduce large densities of dislocations and antiphase disorder in the epitaxial GaAs layer.

Several studies have suggested that these difficulties can be overcome reasonably well by either (a) using buffer layers such as Ge [7.83], $GaAs_x P_{1-x}$ [7.84], and InGaAs/GaAs strained-layer superlattices [7.85], or (b) substrates should be tilted away from the (100) orientation [7.86–88], or (c) using a low-temperature growth procedure at the initial stage of the GaAs growth (two-step growth method) [7.89, 90], or a combination. However, these methods have some drawbacks. The growth of a buffer layer makes the process complicated and expensive. Silicon substrates tilted away from the (100) orientation are not suitable for the integration of Si and GaAs devices because Si(100) wafers are most widely used, especially for Si MOSFET LSIs. With respect to the two-step growth method, it is difficult in experiments to find the proper growth temperature and layer thickness at the initial growth step, with good reproducibility.

In the work reported by *Noge* et al. [7.91], a good morphological GaAs layer free from antiphase domains (APDs) has been obtained on an accurately (100)-oriented Si substrate by MBE. The growth process of a GaAs layer was the same as that for the usual growth on a GaAs substrate and the two-step growth was not used.

In the MBE growth of compound semiconductors, not only must the purity of starting material be maintained and the substrate temperature be carefully controlled, but also the relative flux of two or more elements has to be carefully governed. On GaAs(100) and related III–V(100) surfaces, an optimal-growth region has been established that is characterized by a (2 × 4) diffraction pattern observed with RHEED. The growth conditions under which this region occurs involve substrate temperatures between 500° and 700° C and As_4 pressures in the 10^{-6}–10^{-5} Torr range. Higher As fluxes produce a $c(4 \times 4)$ structure and lower fluxes a (4 × 2) structure, neither of which produce as high quality material as that grown on a (2 × 4) substrate [7.81]. These diffraction patterns have also been observed under static, UHV conditions, and on the basis of these UHV studies a full monolayer of As atoms in the terminating layer has been assigned to the (2 × 4) structure [7.92, 93].

RHEED has proven to be an important technique for studying MBE growth on the GaAs(100) and related surfaces. Detailed monitoring of the intensity of RHEED features shows a cyclical variation which correlates exactly with the epitaxial layer-by-layer growth. *Briones* et al. [7.94] have used these RHEED oscillations to determine that the initial surface stoichiometry, presumably corresponding to the most As-rich form of the (2 × 4) structure, varies by one-quarter of a monolayer of Ga atoms before converting into either the (4 × 2) or $c(4 × 4)$ structures. In recent work by *Larsen* and *Chadi* [7.95] structural models characterized by missing rows of surface As dimers have been proposed for the (2 × 4) and $c(2 × 8)$ As-stabilized surfaces of GaAs. These researchers have shown by experiments with LEED and angle resolved UPS, as well as by theoretical considerations, that a model with monolayer As coverage and symmetric or asymmetric dimers is inadequate. Their total-energy calculations have led to a conclusion that the optimal surface unit cell for the GaAs(001) surface is the (2 × 4) cell for the surface coverages $\theta_{As} = 0.5$ and $\theta_{As} = 0.75$.

7.2.1 Preparation of the GaAs(100) Substrate Surface

Most of the MBE growth of GaAs, $Al_xGa_{1-x}As$, and other III–V compounds has been performed on (100)-oriented ($\pm 0.1°$) substrate slices 200–500 μm thick. The general procedure for preparation of (100) GaAs substrates involves [7.96] polishing with diamond paste to remove the initial saw-cut damage followed by etch-polishing on an abrasive-free lens paper soaked with sodium hypochloride or bromine/methanol solution leaving a mirror-like finish. The slice is then successively rinsed in trichloroethylene, methanol, and double-distilled water. After boiling in hydrochloric acid twice, the substrate is free-etched in a stagnant solution of $H_2SO_4 : H_2O_2 : H_2O = 3 : 1 : 1$ at 48° C for 1 min (*Cho* [7.97] reported slightly different etching conditions: $H_2SO_4 : H_2O_2 : H_2O = 4 : 1 : 1$ at 60° C for 10 min). Finally, the wafer is flooded with double-distilled water to stop the etchant, rinsed in deionized water for passivation [7.97] and blown dry with filtered nitrogen gas just before mounting it with (or without [7.98]) liquid In to a preheated (160° C) Mo plate under dust-free conditions. Surface analysis using AES [7.96] shows that at this stage the hydrophobic GaAs surface is generally contaminated with both oxygen and a small amount of carbon, their relative surface concentrations depending on the precise etching procedure used. The oxygen can easily be removed by heating in vacuum to 540° C [7.97], a temperature well within the congruent evaporation region of GaAs. *Massies* and *Contour* [7.99] have investigated, by x-ray photoelectron spectroscopy, the effects of the chemical etching of the GaAs(001) surface by the $H_2SO_4/H_2O_2/H_2O$ solution. They have demonstrated that, in contrast to what was generally believed, rinsing in running deionized water after etching does not produce any passivating oxide film on the surface. The surface-oxidized phases are only due to the sample manipulation in air after etching. This oxidation process is enhanced by the sample heating for indium soldering on the sample holder. It has been shown that the surface-oxidized phases can be avoided by handling the sample in an inert atmosphere.

Fronius et al. [7.100] have developed a new time-saving method of preparing (100) GaAs substrates prior to MBE growth. The substrate wafer is first polished with diamond paste to remove saw-cut damage, and then etch-polished on an abrasive-free lens paper soaked with NaOCl solution until a mirror-like finish has been achieved. After this the GaAs wafer is simply placed twice in concentrated H_2SO_4 kept at $300 K$ and stirred ultrasonically. The slice is then repeatedly rinsed in double-distilled water to remove all SO_4^{2-} ions and finally blown dry with filtered N_2 gas. Careful ultrasonic stirring in water may accelerate SO_4^{2-} removal from the surface.

The important final step for surface preparation is heating of the wafer to $250°-300° C$ in air under dust-free conditions for 3–5 min. During this heating a stable pore-free surface oxide on the GaAs (100) substrate is reproducibly generated, which protects the GaAs surface from carbon and other contamination. The passivated wafer is then either soldered with liquid In to a conventional Mo substrate plate or it is fixed to a special Mo substrate holder designed for direct radiation substrate heating [7.98].

GaAs substrate wafers prepared in the manner described can be stored in air under dust-free conditions for more than four weeks without any noticeable degradation of the passivated surface. After being transferred from the loading to the preparation chamber (base pressure $< 10^{-10}$ Torr), the surface oxide of freshly prepared and air-stored wafers is thermally desorbed by heating to $540° C$ in a flux of arsenic.

According to *Massies* et al. [7.99], during the soldering process (it lasts several minutes), a thin oxide layer is generated on the GaAs(100) substrate, which protects the surface from carbon contamination. After transfer to the MBE growth chamber, the oxide layer is thermally desorbed by heating the substrate wafers to $530° C$ in a flux of As_4 molecules. When the oxide desorption process is finished, a clear (2×4) surface reconstruction is observed in the RHEED pattern.

The efficiency of the preparation procedure of Fronius et al. was demonstrated by the high quality of modulation-doped n-$Al_xGa_{1-x}As$/GaAs heterostructures with a high-mobility two-dimensional electron gas. These structures were grown with MBE on GaAs(100) substrates prepared according to the described procedure [7.101].

Fujiwara et al. [7.102] studied the consequences of different substrate preparation procedures for the generation of macroscopic surface defects in GaAs epilayers grown by MBE. These defects [7.103, 104] have been an ubiquitous problem for practical applications. Several studies have indicated that the most common type of surface morphological defect is the ovally shaped defect with the density range from 10^2 to $10^5\,\mathrm{cm}^{-2}$. Furthermore, several kinds of defects of the oval type have been reported. Their origins were investigated by a number of researchers. Experimentally the defect formation was related to the substrate preparation [7.105–107], the Ga source cell (Ga splitting [7.103] and Ga oxides [7.108]), and the As source cell [7.106]. However, the factors which cause the major oval defects are still controversial, and it is likely that there is not just a single cause but several, depending on the particular growth conditions employed. There are various kinds of oval defects, and the densities appear to vary greatly in a complex way with the growth conditions. Because of the density variations due to many reasons, it seems, in general, to be difficult to correlate the defects and their origins. Therefore, it is important to classify the oval defects before we indicate the trends when discussing the causes of the oval defects. Figure 7.16 shows the two representative kinds of oval defects which are of primary importance. These are the smaller ($< 10\,\mu\mathrm{m}$) oval defect without microscopic core particulates (called type α hereafter) and larger ($\geq 10\,\mu\mathrm{m}$) oval-shaped hillocks with a nucleus at their center (type β). The density of α-type defects was typically $\geq 10^4\,\mathrm{cm}^{-2}$ when observed, while that of the β-type was distributed between 10^2 and $10^4\,\mathrm{cm}^{-2}$.

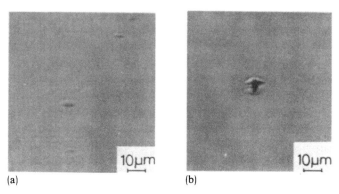

(a)   (b)

Fig. 7.16a,b. Nomarski phase contrast micrographs of GaAs oval defects. (a) α-type without macroscopic core particulates. (b) β-type with macroscopic core particulates [7.102]

It has been demonstrated experimentally [7.102] that surface microscopic contaminations (probably carbon) due to the contaminants in the cleaning solutions and/or in the vacuum can cause the α-type oval defect. The β-type oval defect is attributed to macroscopic contaminations on the substrate surface. By elimination of both microscopic and macroscopic contaminations through careful substrate preparations, Fujiwara et al. have shown that the total defect density may be reduced to about $300\,\mathrm{cm}^{-2}$ for $1\,\mu\mathrm{m}$ thick GaAs layers without significantly modifying the epitaxial source parameters.

In addition to oval defects, another kind of defect, which may be called "pair" defects have been observed on MBE-grown GaAs epilayers [7.109]. The name for these defects comes from the fact that they look like pairs of oval defects positioned side by side. In some cases the pair defects outnumber the oval defects, however, the pair defects can be completely eliminated by a suitable substrate preparation procedure. *Chai* et al. [7.109] have demonstrated that the pair defects originate from sulfur on the substrate surface caused by the $4:1:1$-$H_2SO_4 : H_2O_2 : H_2O$ etch. They have also demonstrated that the right preparation step to avoid the pair defects is the dipping of the GaAs(100) substrate into 30 % HCl solution for 5 min following the sulfuric acid etch.

The idea of treating the GaAs(100) substrates with HCl has been adapted by *Massies* and *Contour* [7.110] to in situ deoxidation of these substrates. After GaAs substrates had been etched in $5:1:1$-$H_2SO_4 : H_2O_2 : H_2O$ mixture, they were cleaned by gaseous HCl at room temperature in a partial pressure range of HCl extending from 0.05 to 50 Torr. The effect of HCl gas on GaAs(001) substrates has been studied by x-ray photoelectron spectroscopy. Oxide-free surfaces are obtained for HCl pressures $p \sim 0.1$ Torr after 15 min reaction time. The residual adsorbed chlorine could be completely removed by annealing at 450° C under vacuum. A reconstruction characteristic of a clean GaAs surface has then been observed with LEED. The 1 μm thick GaAs layers grown on HCl-treated substrates indicated a concentration of oval defects of size above 1 μm equal to 1.8×10^3 cm^{-2} (this is an average value over three growth runs) [7.110].

The as yet lowest level of oval defect surface density, less than 100 cm^{-2}, has been reported by *Fronius* et al. [7.111]. This success is based on the NaOCl polishing procedure described above [7.100].

7.2.2 Growth of GaAs on GaAs(100) Substrates

The most up-to-date review at the time of writing on the physical nature of the MBE growth of GaAs on GaAs is that by *Madhukar* and *Ghaisas* [7.112]. This review provides a critical account of the present status of the atomistic models of MBE growth, computer simulations of the growth process, and the resulting structural and chemical characteristics of the surfaces and interfaces grown with GaAs MBE. Emphasis is placed on conceptual understanding of the issues and consequences rather than on quantitative comparisons with data, such as for the RHEED intensity dynamics. However, qualitative comparisons are possible and are performed by the authors of the review. We will not dwell longer on this review here, because many ideas presented in it have been considered in detail in Chap. 6 and in Sect. 1.1.2.

A coherent growth mechanism for MBE on III–V(100) surfaces in general and the GaAs(100) surface in particular has been proposed by *Farrell* et al. [7.81]. This model for the growth mechanism is based on experimental results concerning RHEED intensity oscillations [7.94] high resolution electron energy loss spectroscopy studies [7.113], angle-resolved photoemission and LEED investigations [7.95], and tight-binding-based total-energy calculations [7.114] of the atomic structure of the GaAs(100)-(2 × 4) reconstructed surface. We will describe this mechanism following the considerations of *Farrell* et al. [7.81].

Two assumptions should be noted before beginning the discussion of various stages in the MBE process. The first assumption deals with the nature of the unreacted species. While both As_2 and As_4 molecules may be used in the incident flux, it is generally observed that As_2 is the predominant species desorbed from the surface. Therefore, for simplicity, it is assumed that As_2 is the acutal diffusing species present on the surface; Ga is assumed to be present in a monatomic form prior to actual chemisorption.

A second major assumption is that surface species or complexes with the longest residence time will tend to be those that are electronically stable, i.e., not in an excited electronic state. While many excited species are expected to form during growth, it is assumed that they will be labile unless stabilized within a relatively short period of time (e.g., some number of vibrational cycles). Therefore, the surface complexes that characterize a particular stage of growth will be those that closely resemble structures seen under static-UHV conditions.

We will first describe the surface structure in the presence of an As flux but before the Ga flux is turned on to begin the growth process. Growth will then be followed through one full cycle representing the deposition of a complete bilayer.

Barring multiple-scattering effects, when the RHEED oscillations show their maximum intensity the surface is most perfect. This is seen when the surface is allowed to reach equilibrium in the presence of an As flux but in the absence of a Ga flux, prior to the beginning of epitaxial growth. When the As flux is high enough to produce a RHEED intensity maximum, but not so high as to change the surface to a $c(4 × 4)$ structure, then the GaAs surface exhibits a structure shown in Fig. 7.17. This structure is characterized by having the dangling and bonding As orbitals completely filled and created by three As dimers and one dimer vacancy per unit cell [7.113]. The dimer vacancy exposes Ga atoms in the

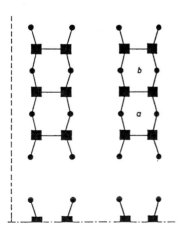

Fig. 7.17. Planar view of the As-rich GaAs(100)-(2 × 4) surface unit cell. The As atoms (■) form three dimers per unit cell. The dimer vacancy exposes four Ga atoms (●) in the second layer. A Ga dimer can be inserted at either a or b. The size of the symbol indicates the proximity of the layer to the surface [7.81]

second layer. The bonding of the surface As atoms approaches $s(p)$ rather than sp^3 hybridization as is found in the bulk. (The notation $s(p)$ indicates that the As $4s$ state splits off from the three As $4p$ states. Here, the $4p$ orbitals are the bonding orbitals, being used for the dimer and backbonds, and the $4s$ orbital is the "nonbonding" orbital, but lies well below the Fermi level.) As a consequence, the surface is dilated or slightly relaxed outward, and the dimer bond contracts the dimerized As atoms together to an interatomic distance that is considerably shorter (e.g., 2.6 Å) [7.115] than the projected bulk positions (4.0 Å).

When the Ga flux is turned on, the Ga atoms presumably arrive at the surface species. A single Ga atom might attempt to bond to one As atom via an As dangling orbital, but this would be a rather short-lived species having only one covalent bond and three electrons in the Ga nonbonding orbitals. (Here, a total of five electrons, two from the As dangling orbital and three from the Ga itself, are shared among one bonding orbital and two dangling orbitals). A somewhat more stable species is produced, however, when a Ga atom is inserted into an As-dimer bond. Here, two covalent bonds are formed and only one electron is left in a Ga nonbonding orbital. Both species are expected to be more or less labile, and a given Ga atom may enter into a number of these unstable surface complexes before it reaches its final bonding site.

The insertion of two Ga atoms into two adjacent As-dimer bonds, however, does allow the formation of a stable Ga species on the surface. This is a Ga dimer with five covalent bonds and no electrons in the Ga dangling orbitals. (The two single electrons left in the nonbonding Ga orbitals of each of the two inserted Ga atoms combine to form a covalent bond between the two inserted Ga atoms.) Note that there are two sites per unit cell where such a Ga dimer could form (Fig. 7.17a or b). Unlike the As dimer, the Ga dimer has approximately sp^2 bonding, with the consequence that it is relaxed inward toward the bulk. This will relax the underlying As atoms that it is bonded to back towards, and perhaps even beyond, the projected bulk positions. This configuration is shown in Fig. 7.18 and the relaxation of the As atoms serves an important function in

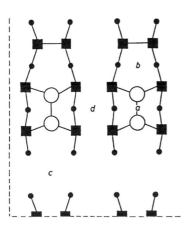

Fig. 7.18. After the chemisorption of the 1/4 monolayer of Ga (o), one Ga dimer per unit cell initiates growth of the second layer. In the right-hand cell the dimer may be aligned with that in the left-hand adjacent cell, as shown, inserted at a, or staggered if inserted at b. Insertion of a second Ga dimer at point d necessitates the filling of the As vacancy at point c [7.81]

setting up the following stages. (Note that the insertion of either one or a pair of Ga atoms into As-dimer bonds is more stable than chemisorption across As dangling orbitals at this stage, not only on the basis of electronic considerations but also on the basis of steric considerations, as the dimer sites are initially separated by about 2.6 Å while the other sites are about 5.4 Å apart.)

When approximately one-quarter of a monolayer of Ga has been deposited, there will, on the average, be one Ga dimer per unit cell. As there are two sites per unit cells, these Ga dimers may be either next to those in adjacent cell, as in Fig. 7.18, or offset by one atomic spacing as would be the case if the dimer at point a had been inserted at point b instead. Because of the As-stabilized conditions used in MBE growth, while these Ga dimers are being formed there is still an As flux to the surface that greatly exceeds the Ga flux. However, at this stage of growth there are essentially no stable binding sites for As, and short-lived species presumably form and decay with the subsequent diffusion and desorption of As_2.

For the growth process to proceed further, four Ga atoms must be chemisorbed contiguously as two adjacent dimers to provide a bonding site for an As_2 dimer. In addition, the dimer vacancy in the first layer must be filled to provide a basis for the chemisorption of further Ga atoms. As will be seen, these two steps (the chemisorption of an As dimer to fill the vacancy and the chemisorption of a second pair of Ga atoms) must occur essentially simultaneously within a given unit cell to achieve an electrically stable situation. This step, and a similar one near the end of the cycle, appear to be the limiting steps for epitaxial growth to occur. In addition, the necessity of having an As_2 dimer (or its monatomic constituents) available for concurrent chemisorption may explain, in part, the requirement of a much larger As than Ga flux to the surface for successful growth to occur.

For the configuration shown in Fig. 7.18, spatially the only reasonable bonding site for a Ga atom is between four of the As atoms already bonded to Ga dimers, point d. Here is where the lateral relaxation of these As atoms becomes important, as it will allow the Ga atom to bridge this position forming two covalent bonds simultaneously (rather than only one as would have been true in the very first stage of Ga chemisorption). However, while this site is geometrically feasible, and even desirable, when only one Ga atom is adsorbed it produces electrically unstable species with three electrons in the Ga dangling orbitals. The chemisorption of a second Ga atom in the adjacent site, with the concomitant formation of a dimer bond between the two Ga's, eases the situation only slightly as it reduces the number of electrons in Ga dangling orbitals to only two per atom. Even so, this species is still expected to be extremely labile as it contributes a total of four electrons to the unit cell that are nominally above the Fermi level and therefore may be considered to be in an excited state. These four extra electrons would seem to present an enormous barrier to the growth process were it not for another chemisorption process that actually requires such an excess.

As noted above, the second difficult step at this stage of growth is the filling of the dimer vacancy in the original surface layer. This vacancy exposes four Ga atoms to which the new As dimer must bond. In the structures shown in Figs. 7.17 and 7.18, these Ga atoms have empty dangling orbitals. However, for a physisorbed or gaseous As_2 species to be bonded stably to this site, four extra electrons must be provided by the substrate. The reasons for this are as follows. An unbonded As_2 molecule (either in the gaseous state or thermally diffusing on the surface) has ten valence electrons, five from each As. A bonded As dimer requires 14 electrons, two for the dimer bond, eight for the four backbonds, and two each in the two dangling s-like orbitals. Therefore, the substrate must provide four additional electrons for the dimer vacancy to be filled in other than an ephemeral fashion.

On this basis, it is eminently plausible that these two chemisorption processes, one producing an excess of four electrons and one requiring four additional electrons, are closely coupled in both time and space. The resulting structure is shown in Fig. 7.19 for several adjacent unit cells. However, it may now be seen that there are additional binding sites for As_2 molecules, shown as positions e and f in Fig. 7.19. In fact, given the excess As flux, it is probable that almost as soon as two adjacent Ga dimers are formed, a new As surface dimer will be inserted into the Ga-dimer bonds. This is shown schematically in Fig. 7.20. This process is expected to be relatively rapid (e.g., < 0.1 s) on the time scale of a bilayer growth cycle, but much slower than the residence time of labile Ga species, (e.g., $< 10^{-9}$ s).

The chemisorption of these As dimers initiates the growth of the top of the bilayer. Note that there is now a top or surface layer with one-quarter of a monolayer coverage, and a second layer with one-half of a monolayer coverage.

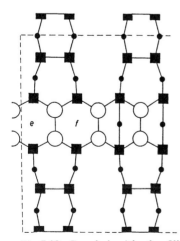

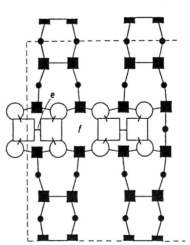

Fig. 7.19. Coupled with the filling of the As dimer vacancy in the lower As layers, the chemisorption of a second Ga dimer per unit cell opens up a new site for the chemisorption of an As dimer centered at either e or f. Note that island formation is extended along the $[1\bar{1}0]$ direction [7.81]

Fig. 7.20. The chemisorption of an As dimer ($\square$-$\square$) initiates growth of the top of the surface bilayer and completes the second stage of the growth cycle [7.81]

This represents the roughest atomic configuration, and, not surprisingly, occurs at the middle of the bilayer growth cycle where the experimentally observed RHEED intensities reach their minimum value. This suggests that one possible method for checking this model is to investigate the curvature of the RHEED intensities in the vicinity of the minima as a function of the incident As flux.

Figure 7.20 brings us to the middle of the bilayer growth cycle. It is worthwhile to pause at this point and explore some of the tacit assumptions underlaying the above model. First, let us consider the question whether or not the first and second steps may be reversed. In the first step, (Fig. 7.18) a Ga dimer is inserted across two As-dimer bonds. Here, the bonding sites are about 2.6 Å apart [7.115]. In the second step (Fig. 7.19), the Ga dimer is inserted between four As dangling orbitals. If the step order is inverted, then these dangling orbital sites are unphysically far apart (e.g., 5.4 Å). Therefore, at least for the initial or nucleation step within a given domain, it seems improbable that step 2 will precede step 1. However, the possibility of spontaneous dimer dissociation at growth temperatures may allow an invertion of step order and the resulting As dimers would be greatly weakened, allowing for easy insertion of Ga atoms. It is probable that both processes occur in parallel on the surface.

A second question that needs to be addressed is what happens when the Ga dimers in Fig. 7.18 are staggered rather than perfectly aligned to enhance the Ga dimer insertion illustrated in Fig. 7.19. The easy answer, that growth only proceeds within a domain of properly aligned Ga dimers, is confounded by two facts. The first is that as there are two apparently equivalent sites per unit cell, these domains would be quite small. The second is that growth is apparently characterized by the development of long islands or plateaus along the [1$\bar{1}$0] direction (as shown schematically in Figs. 7.19, 20 and the following). This indicates that the growth process in contiguous unit cells is correlated and points to the possibility that either atomic rearrangement occurs at domain boundaries during the development of the Ga half-monolayer (presumably a strongly exothermic process that could power either surface diffusion or rearrangement reactions), or that the first and second stages, as simplistically outlined above, may occur in a coherent fashion to extend growth along the [1$\bar{1}$0] direction after the initial, nucleation stage. (In the former case, thermally activated exchange of the Ga dimer between points a and b may be important.) The latter case could be envisioned as, for example, the insertion of two Ga dimers across As-dimer bonds, followed by the second stage, where the initial dimer vacancy is filled and a Ga dimer is inserted across the "dangling bonds positions", and culminating in a "domino theory" extension where laterally contiguous unit cells experience rearrangement of their initial Ga chemisorption under the influence of previously reacted sites. This picture, while conceptually more complicated actually describes what occurs on the surface (Fig. 7.19).

In any event, the surface structure at the midpoint of the growth cycle is expected to resemble that shown in Fig. 7.20 after the subsequent chemisorption of the first quarter-monolayer of As. Note that there are no As bonding sites

within this simple picture. Here again, we are faced with a large incident flux of As atoms with no place to go. There is, however, a site for chemisorbing a Ga dimer. As in the first stage, this presumably does not occur with the instantaneous chemisorption of a molecular Ga_2 species, but as a stepwise process involving two Ga atoms that may have to make a number of attempts via unstable adsorption sites before achieving stable pairing as is shown in Fig. 7.21.

We are now a full three-quarters of the way through the growth cycle but are again faced with a double dilemma; on the average, there are no available sites for As chemisorption in the top layer, and one dimer vacancy per unit cell must be filled in the next-to-last layer. As before, these two separate processes necessitate in one case (the insertion of an As dimer across four unoccupied Ga dangling orbitals) the provision by the substrate of four extra electrons per unit cell and in the second case (the filling of the Ga dimer vacancy) the development of four unwanted electrons per unit cell. This is the second rate-limiting step in the growth process. As before, it is proposed that these two steps occur essentially simultaneously with a variety of intermediates being formed and then dissociating until, statistically, a stable electronic situation results. Referring to Fig. 7.21, it can be seen that there is one site per unit cell g that corresponds to a Ga dimer vacancy. There are, however, four alternative sites per unit cell for the essentially simultaneous chemisorption of an As dimer (positions h, i, k, l in Fig. 7.21). The occupation of one such site, however, rules out the occupation of other sites within a given unit cell (e.g., h rules out i, or k rules out l, Fig. 7.22). Furthermore, the completion of the bottom, Ga half of the bilayer opens up one additional As

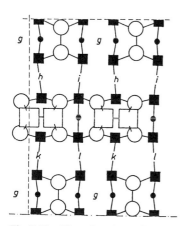

Fig. 7.21. The chemisorption of a third Ga dimer per unit cell completes the third stage of the growth cycle. Note that this structure is stabilized relative to a (4 × 2) unit cell by the presence of 1/4 of a monolayer of As in the outermost layer. Further growth necessitates the simultaneous filling of the Ga dimer vacancy site at point g and the chemisorption of one As dimer per unit cell at either points h or i or at points k or l [7.81]

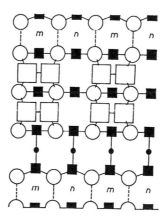

Fig. 7.22. The correlated chemisorption of a Ga dimer at point g and an As dimer at point h. This opens up one additional site for the adsorption of an As dimer. Site m is favored over site n on the basis of backbonding [7.81]

dimer binding site per unit cell. As in the second stage, this fourth or last stage of bilayer growth is proposed to involve the incorporation of a second As$_2$ unit still keeping all As dangling bonds filled and all Ga ones empty. While intermediate structure may occur, backbonding requirements indicate that the new structure within a given domain will be that shown in Fig. 7.22. There again, we find three As dimers and one dimer vacancy per unit cell. For variety, a $c(2 \times 8)$ rather than a (2×4) domain is shown (Fig. 7.23). Note that, otherwise, the atomic configuration is essentially that shown in Fig. 7.17, and we have completed the bilayer growth cycle.

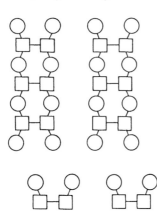

Fig. 7.23. The growth cycle is complete and the As-rich structure shown in Fig. 7.17 is recovered. This occurs when an As dimer and a Ga dimer are simultaneously chemisorbed on the structure shown in Fig. 7.21 followed by the chemisorption of the final As dimer. Here, for variety, a $c(2 \times 8)$ unit cell is shown [7.81]

It is appropriate to note a few words here on the basic $(2 \times 4)/c(2 \times 8)$ unit cell. The first point is one that may not be obvious from a casual perusal of Figs. 7.17 and 22. That is, that there is a little difference in free energy between these two structures. Given the presence of the dimer vacancy, and the covalent nature of the GaAs bonds, adjacent As dimer triplets are connected only via the third layer. In a sense, this may be regarded as a fifth-nearest-neighbor interaction, which is rather weak even in a directionally bonded system. Both unit cells involve three aligned As dimers per unit cell. Backbonding strain is minimized by having the dimers aligned along the [110] direction for any given triplet. The dimers are contracted along the [1$\bar{1}$0] direction, and as the As bonding approaching $s(p)$ hybridization, the underlaying Ga atoms are also displaced primarily along the [1$\bar{1}$0] direction. This picture is in contrast to that proposed by *Larsen* et al. [7.92, 93] where the second-layer Ga atoms are at their bulk positions with the As dimers asymmetrically bonded (with partially filled orbitals). The structure shown in Figs. 7.17 and 22 produces significantly less backbonding strain on the underlaying Ga atoms by not pulling them in two directions at once and allows the preferred $s(p)$ bonding of the As surface dimers.

Another interesting point to note is that the (2×4) structure is apparently preserved, albeit in a somewhat diffuse form, throughout the growth cycle. This is particularly interesting near the third stage where approximately 3/4 of a monolayer of Ga atoms has been chemisorbed. Earlier work of Farrell et al.

under static UHV, or nongrowth, conditions indicates that the stable structure in this region is the (4 × 2) or $c(8 × 2)$ unit cell that is characterized by three parallel Ga dimers and one dimer vacancy, per unit cell, in a fashion somewhat analogous to the As-rich (2 × 4) or $c(2 × 8)$ unit cells shown in Figs. 7.17 and 22 (though orthogonally oriented and with rather different surface relaxation). Under MBE growth conditions, however, the (2 × 4) structure is locked-in at the second stage by the chemisorption of an As dimer on top of the Ga atoms (Fig. 7.20). This chemisorption step prevents reconstruction at the 3/4 monolayer level (Fig. 7.21) and preserves the general nature of the (2 × 4)/$c(2 × 8)$ unit cell throughout the growth cycle. Hence one does not observe an oscillation back and forth between (2 × 4) and (4 × 2) as one successively adds monolayers of As then Ga then As and so forth; as one might at first expect. This is of course in agreement with what one finds experimentally.

On the basis of the presented growth mechanism, some specific modifications of the so-far practiced experimental growth procedures may be suggested [7.81]. One obvious suggestion is to vary the incident flux of one or more species during the growth cycle. For example, within this simple model, no As uptake occurs at the first and third steps, but an excess of As is crucial at the second and fourth stages. Therefore, it might be reasonable to pulse the As (or Ga) source synchronously with the monolayer growth cycle. Careful experimental investigation is needed, however, to refine such an approach. A simple variation on this idea has already been attempted via the sequential deposition of a full monolayer of As followed by a full monolayer of Ga. This is the migration enhanced epitaxy growth technique [7.116] which led to good quality epitaxial growth even at temperatures below 200° C. Obviously, this line of investigation provides very attractive possibilities and needs further study.

A second suggestion for facilitating the growth process involves the use of monochromatic light to selectively break certain bonds at specific stages during the growth process. For example, the initial stage involves the breaking of As-dimer bonds at the surface and the insertion of Ga atoms. If these dimer bonds could be broken photonically rather than thermally, lower reaction temperatures might be usable. Considerable information exists in the literature about the relative position of the dimer-associated surface states and the valence-band maximum (VBM) [7.92, 93]. Some of these states lie within 1 eV of the VBM and excitation from these states to above the gap, particularly to the appropriate, quasi-localized antibonding states, might be feasible. The use of polarized light to select between Ga and As dimers also opens up interesting possibilities. While many questions need to be investigated along this line, it is important to note that recent experiments with II−VI compounds show that the use of light allows growth of materials with previously unobtainable doping characteristics [7.117]. Therefore, we suggest that this approach also opens up a potentially valuable area for further investigation and practical applications.

7.2.3 Growth of $Al_xGa_{1-x}As$/GaAs Heterostructures

The MBE growth of the GaAs-$Al_xGa_{1-x}As$ system has been studied from the very beginning of the MBE technique (Chap. 1). Consequently, all of the currently known modifications of the MBE technique have been used to grow these materials. Thus, the GaAs-$Al_xGa_{1-x}As$ system has become a model for MBE growth. The growth techniques listed in the classification scheme shown in Fig. 1.14 may be compared by using as a criterion the quality of the grown $Al_xGa_{1-x}As$/GaAs heterostructures and the growth conditions applied. However, the parameters of devices which were grown with different modifications of the MBE technique may serve as more important comparison criteria.

One of the most important problems connected with the MBE growth of $Al_xGa_{1-x}As$/GaAs heterostructures is the structure of the heterointerface on the atomic scale [7.118–128]. This determines the quality of the heterointerfaces, which subsequently plays the most crucial role for the exploitation in optoelectronic devices (heterojunction lasers and superlattice avalanche photodetectors), as well as in high-speed 2DEG field effect transistors [7.123]. The structure of the heterointerfaces on an atomic scale has been most extensively investigated using photoluminescence (PL) (Sect. 5.4.3) and transmission electron microscopy (TEM) (Sect. 5.6.1).

Atomistic models of the interface structure have been presented by *Tanaka* and *Sakaki* [7.118] for $Al_xGa_{1-x}As$/GaAs/$Al_xGa_{1-x}As$ (x = 0.2–1) quantum wells (QWs). Using PL measurements they investigated the influence of growth interruption (Sect. 6.4.1) on the top (AlGaAs-on-GaAs) and bottom (GaAs-on-AlGaAs) interfaces of those heterostructures, evidencing the considerable differences in the MBE growth. Models of the interfaces are shown in Fig. 7.24. They

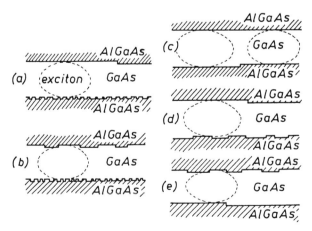

Fig. 7.24a–e. Atomic-scale models of interface structure of various GaAs-$Al_xGa_{1-x}As$ QWs (x = 0.2–1) prepared with and without growth interruption. (**a**) Model of type-*A* and type-*B* QWs with high Al content in the barriers (x > 0.5). (**b**) Type-*C* and type-*D* QWs. (**c**) Type-*A* QWs with low Al content (x < 0.3). (**d**) and (**e**) Type-*B* and type-*C* QWs, respectively, with low Al content [7.118]

are related to four different types of QWs. Type A is characterized by having both interfaces prepared with sufficiently long (90 s) growth interruption. Type-B QWs were prepared with an interruption of 90 s only at the top interface. Type-C QWs were prepared with the interruption of 90 s only at the bottom interface. Finally, type-D QWs have both interfaces formed without growth interruption.

Before beginning the discussion on the proposed models it is worthwhile recalling (Sect. 5.4.3) that the broadening of the PL peak due to interface roughness is mainly determined by the relative magnitude of the atomic step intervals (L_s in Fig. 5.28) and the lateral diameter D_{ex} of excitons, which are in fact the optical probe. When $L_s \gg D_{ex}$, interfaces "seen" by individual excitons are smooth, leading to sharp PL spectra. When L_s becomes comparable to D_{ex}, excitons will experience the fluctuation of the well width, leading to PL broadening. On the other hand, when $L_s \ll D_{ex}$, the interface roughness is hardly sensed by excitons. This "pseudo-smooth" nature of such interface can lead to the sharpening of PL.

The PL spectra measured by *Tanaka* and *Sakaki* [7.118] are shown in Figs. 7.25 and 26. Let us first discuss the QWs with high barriers (layers of $Al_xGa_{1-x}As$ with $x > 0.5$).

The fact that PL spectra and their linewidths (Fig. 7.25) are mainly determined by the interruption at the top interface indicates that the growth interruption smoothens out the roughness of the top interfaces but has very little influence on the structure of the bottom interfaces. Since type-C and type-D QWs, having

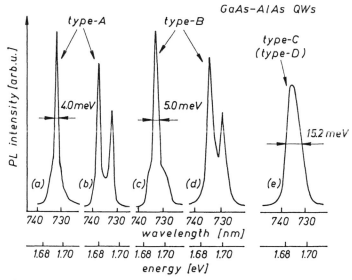

Fig. 7.25a–e. Photoluminescence spectra of GaAs-AlAs QWs ($x = 1$) with well width $L_z = 48.1$ Å and barrier width $L_b = 50.9$ Å measured at 77 K: **(a, b)** Spectra of type-A QWs. **(c, d)** Spectra of type-B QWs. The line shape and linewidth are quite similar for these two types of QWs. **(e)** Spectrum of type-C QWs. The spectra of type-D QWs are nearly identical to those of type-C [7.118]

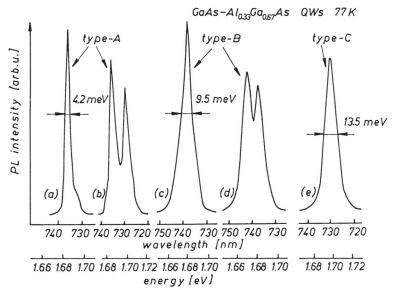

Fig. 7.26a–e. Photoluminescence spectra of GaAs-Al$_{0.33}$Ga$_{0.67}$As QWs ($x = 0.33$) with $L_z = 34.0$ Å and $L_b = 50.9$ Å measured at 77 K. **(a, b)** Spectra of type-A QWs. **(c, d)** Spectra of type-B QWs. **(e)** Spectrum of type-C QWs [7.118]

continuously grown top interface, show a broad PL, the top interface prepared by conventional MBE must consist of atomic steps of one monolayer height and the interval of such steps must be comparable to the exciton size, as shown in Fig. 7.24b. This leads to PL broadening. When the top interface is prepared with growth interruption, these atomic steps are smoothened, or equivalently, the step intervals can become much larger than the exciton size as shown in Fig. 7.24a. This leads to a sharp PL of type-A and type-B QWs.

The fact that type-A and type-B QWs, having a truly smooth top interface, show sharp PL indicates that the bottom interface also acts as a smooth (truly smooth or pseudo-smooth) interface when evaluated optically by excitons. Diffusion of the Al atoms along the growth front is expected to be much smaller than that of the Ga atoms along the GaAs surface because of the adherence of the Al atoms. Hence, the step intervals on the AlAs or Al$_x$Ga$_{1-x}$As ($x > 0.5$) surfaces are likely to be much smaller than the exciton size. This pseudo-smooth feature of the bottom interfaces remains virtually the same, irrespective of the growth interruption. In such a case, the excitons are no longer subject to inhomogeneous broadening. This pseudo-smooth nature of the bottom interface and truly smooth nature of the top interface can account for the sharp PL peaks of type-A and type-B QWs.

Note that this pseudo-smooth bottom interface has an important difference from the truly-smooth interface, in the sense that the effective well width is not always equal to an integer multiple of one atomic layer, but has some statistical variation. Actually, the wavelength of PL peaks from type-A and type-B QWs has a small but appreciable variation $\Delta\lambda$ over the wafer, being ± 5 and ± 20 Å, respectively.

The behavior of the QW interfaces during growth interruption changes in comparison to the presented data when the height of the barriers becomes lower (layers of $Al_xGa_{1-x}As$ with $x < 0.3$). The fact that the PL spectrum (Fig. 7.26) of type-B QWs becomes broad as x is reduced to 0.2 indicates that the surface of AlGaAs with $x < 0.3$ becomes quite similar to that of GaAs. Hence, type-A QWs, prepared with growth interruption at both interfaces, should have two truly smooth interfaces, as shown in Fig. 7.24c. In contrast, the bottom interface of type-B QWs and the top interface of type C, which are both formed without interruption, have atomic steps whose lateral size is comparable to the exciton size, as shown in Figs. 7.24d and e. For this reason the linewidth of QWs is sharp only when the growth interruptions are done at both interfaces.

In order to gain some quantitative understanding of the atomic-scale structures of heterointerfaces, one may use the theory of *Singh* et al. [7.124, 129, 130] to interpret the presented PL data in terms of interface roughness. We will follow here the consideration of *Tanaka* and *Sakaki* [7.118].

The interface of a QW is made up of islands and valleys whose step height is δ_1 and lateral sizes are δ_{2a} and δ_{2b}, respectively. When the average concentration of islands and valleys is C_a^0 and C_b^0, the probability of finding concentration fluctuations C_a and C_b over the exciton size D_{ex} is given by

$$P(C_a, C_b, D_{ex}) = \exp\left[-\left(\frac{D_{ex}^2}{\delta_{2a}^2}C_a \ln\frac{C_a}{C_a^0} + \frac{D_{ex}^2}{\delta_{2b}^2}C_b \ln\frac{C_b}{C_b^0}\right)\right] \quad . \tag{7.2}$$

and the width of the quantum well is given by

$$L_z = L_{z0} + \delta_1(C_a - C_a^0) \quad , \tag{7.3}$$

where L_{z0} is the average well size. In the following, we assume that the typical exciton size D_{ex} is 250 Å, the step height of islands and valleys $\delta_1 = 2.83$ Å (one atomic layer), $\delta_{2a} = \delta_{2b} = L_s$, and $C_a^0 = C_b^0 = 0.5$. Since the linewidths of low temperature PL from the wide QWs ($L_z > 100$ Å) are typically 1 meV, we estimate that the broadening associated with impurities and factors other than interface roughness is estimated to be 1 meV in the QWs.

Taking this into account, the PL linewidth of GaAs-AlAs QWs has been calculated as a function of well width L_{z0} for several different values of L_s. The results are shown in Fig. 7.27.

The previous discussion has shown that the PL linewidth of type-D (and type-C) QWs is mainly determined by the top interface, while the bottom interfaces are pseudo-smooth and have little effect on PL linewidth. We can estimate the lateral size of the atomic steps L_s on continuously grown top interfaces from the data of type-D or type-C QWs. Hence, L_s is evaluated to be about 200 Å. On the other hand, the PL linewidth of type-A (and type-B) is likely to be mainly determined by the bottom interface because the growth-interrupted top interface is truly smooth and does not contribute to the PL broadening. Hence, the lateral

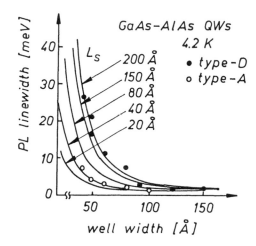

Fig. 7.27. Calculated PL linewidth due to interface roughness for various intervals of atomic steps L_s for GaAs-AlAs QWs with $L_b = 50.9$ Å as a function of well width. Open circles and solid circles are experimental data measured at 4.2 K for GaAs-AlAs QWs of type A and type D, respectively [7.118]

size of atomic steps L_s on the pseudo-smooth bottom interface is estimated to be about 40 Å. Now it is clarified that the effective interface roughness of 0.2 atomic layer in growth interrupted QWs, which has been previously reported by the study of PL and absorption linewidth [7.122, 131, 132], is mainly attributed to the pseudo-smoothness of the bottom interface. To achieve ever-narrower PL, one should improve the bottom interface either by making it truly smooth, or by making the step interval far smaller than 40 Å.

As already pointed out in Sect. 6.4.2, $Al_xGa_{1-x}As$/GaAs quantum well structures and superlattices with very smooth interfaces can be grown when using RHEED intensity oscillations for growth control, i.e., when growing them with phase-locked epitaxy [7.133]. During such growth procedures, RHEED intensity oscillations of the postgrowth process of the GaAs have been observed at high temperatures ($T_s > 690°$ C) by *Kojima* et al. [7.134], and *Van Hove* and *Cohen* [7.135], independently. Figure 7.28 shows growth and postgrowth RHEED intensity oscillations at (a) a low substrate temperature ($T_s = 580°$ C), (b) high substrate temperature ($T_s = 700°$ C). The As flux was applied throughout the experiments. The RHEED intensity oscillations for the growth started when a Ga flux was applied. During the growth, As-stabilized RHEED patterns gave the surface reconstruction of (2×4) at $T_s = 580°$ C and (3×1) at $T_s = 700°$ C. As mentioned previously, one period of the oscillations exactly corresponds to the growth of one monolayer of GaAs. After crystal growth was suspended, the surface reconstruction was essentially unchanged both at $T_s = 580°$ C and $T_s = 700°$ C. At the low substrate temperature the RHEED intensity recovers monotonically. However, at the higher substrate temperature, a new mode of the oscillations starts after the growth is suspended, i.e, the Ga flux is turned off while the As flux is maintained. As the sublimation of GaAs is reported at higher substrate temperatures, the postgrowth RHEED intensity oscillation can be attributed to the sublimation of GaAs. In order to confirm this idea, *Kojima* et al. [7.134] employed the new "deposition–sublimation" method utilizing the AlAs layer as

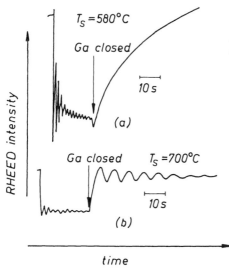

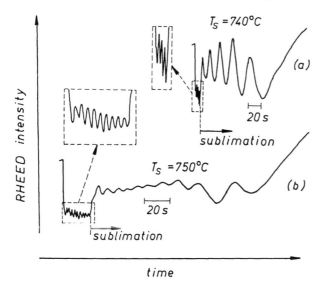

Fig. 7.28a,b.
Growth and postgrowth RHEED intensity oscillations for GaAs on a GaAs(001) surface. **(a)** At a low temperature ($T_s = 580°$ C). **(b)** At a high temperature ($T_s = 700°$ C). Arsenic flux is applied throughout [7.136]

Fig. 7.29a,b.
Growth and postgrowth RHEED intensity oscillations for GaAs on an AlAs sublimation stopper layer. **(a)** GaAs (5 monolayers) on AlAs at $T_s = 740°$ C; **(b)** GaAs (10 monolayers) on AlAs at $T_s = 750°$ C [7.136]

a sublimation stopper. In this method, as shown in Figs. 7.29a, b, 5–10 monolayers of GaAs are deposited by the PLE method on an AlAs sublimation-resistant buffer layer. When the PLE is finished, vacuum sublimation of GaAs is observed until the AlAs buffer is exposed and the postgrowth RHEED intensity oscillation terminates. The number of oscillation periods for the sublimation process always equals the number of layers deposited on the AlAs within experimental error. This result indicates that the postgrowth oscillation period corresponds to one monolayer sublimation of GaAs. It is also concluded that the vacuum sublimation should likewise occur in a layer-by-layer fashion as long as the postgrowth oscillations continue, as in the case of the growth process. The mechanism of the postgrowth oscillations is considered to be the reverse process of MBE growth. The difference between them lies in the controlling parameters of the phenomena. In the case of deposition, the oscillations are determined by the intensity of the Ga flux, while the sublimation is controlled by the substrate temperature and the impinging As flux. The postgrowth RHEED intensity oscillation method is a new microscopic probe for the vacuum sublimation of crystalline materials and can be used for etching with precise depth resolution [7.136].

An interesting growth procedure used for the preparation of specially shaped quantum wells in the GaAs-$Al_x Ga_{1-x} As$ system has been demonstrated by *Miller* et al. [7.137−140], using the pulsed beam mode of MBE [7.141]. This procedure is well illustrated by the growth of half-parabola wells [7.140]. The technique used was a computer-controlled long-period (24 Å) pulsed MBE in which the duty-cycle of Al deposition within each pulse period was made proportional to the desired Al content at that point of the quantum well profile. The growth was carried out under As_2-rich conditions at GaAs growth rates of ~ 200 Å/min. Chromium-doped GaAs substrates with (100) crystal orientation were used at growth temperatures of 650° C. The aluminum flux time profile during the shuttering sequences was calibrated in separate runs by an ion gauge at the growth position of the sample in the absence of other molecular beams. The integrated depositions of Al and Ga during the actual growth runs were determined independently by measurement of GaAs and AlAs film thicknesses adjacent to separate shadows of gallium and aluminum evaporant beams cast by a wire mask above the substrate. The individual thicknesses within each period were calculated on the basis of the individual deposition times. Another determination of layer thicknesses was made for several pulsed beam samples by transmission electron microscope imaging of cross sections, which allowed measurement of the deposited periods and quantum well thicknesses.

The deposition sequence for one of the half-parabolic well samples is shown in Fig. 7.30. It contained five wells with $L = 522 \pm 31$ Å separated by 229 Å $Al_{0.26}Ga_{0.74}As$ barriers, with each well comprising 43 alternate depositions, with depositions $i = 1, 3, 5, \ldots, 43$ being GaAs of thickness

$$\left(\frac{L}{22}\right)\left[1 - \left(\frac{44 - i}{44}\right)^2\right]$$

and depositions $i = 2, 4, 6, \ldots, 42$ being $Al_{0.26}Ga_{0.74}As$ of thickness

$$\sim \left(\frac{L}{22}\right)\left(\frac{43 - i}{44}\right)^2 .$$

Depositions less than ~ 0.1 Å (layers 40 and 42) were omitted because of shutter timing considerations. Model calculations suggest that this produces no observable change in energy levels. The film thickness measurements adjacent to the mask shadows permitted an Al-content determination accuracy of $\pm 5\%$. The ion-gauge flux monitoring of the Al beam showed an Al composition constancy

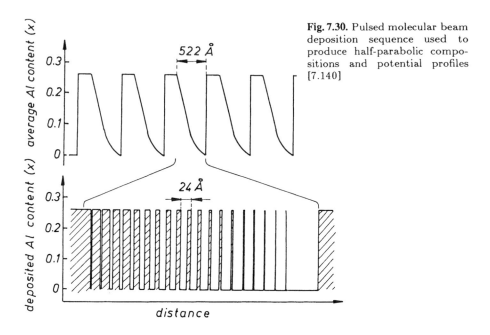

Fig. 7.30. Pulsed molecular beam deposition sequence used to produce half-parabolic compositions and potential profiles [7.140]

of $\pm 1\%$ in the barriers. The transmission electron microscope cross section confirmed the layer periodicity and its constancy to within 10 % [7.140].

7.2.4 Growth of GaAs on Si Substrates

Epitaxial growth of GaAs on Si substrates has attracted a great deal of attention because of the promising possibility of combining the optoelectronic performance of III–V materials with Si integrated circuit technology [7.142]. The quality of the GaAs epitaxial layer has been found to be strongly dependent on the crystallographic orientation of the Si substrate and the initial growth conditions [7.143–145].

A consistent presentation of the problems related to GaAs-on-Si epitaxy has been given by *Kroemer* [7.146]. Concerning these problems we will follow this review to a large extent.

The diamond structure in which Si crystallizes consists of two interpenetrating face-centered cubic sublattices. These sublattices differ from each other only in the spatial orientation of four tetrahedral bonds that connect each atom to its four nearest neighbors belonging to the other sublattice. For example, in Fig. 7.31a the atoms with the bond orientations indicated as A and B belong to different sublattices. There is no distinction between the two sublattices otherwise; both are occupied by the same atomic species.

In the zinc-blende structure, in which GaAs crystallizes, the sublattices are occupied by different atoms, in the case of GaAs, one by Ga atoms the other by As atoms. In a crystal without antiphase disorder the sublattice allocation is

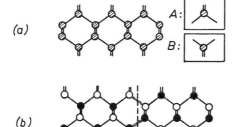

(a)

A:

B:

(b)

Fig. 7.31. (a) Two sublattices in a Si crystal, distinguished only by bond orientation in space. **(b)** Antiphase boundary (APB) formation in the zinc-blende structure, containing (in the case of GaAs) both Ga-Ga and As-As bonds. The configuration shown is the simplest possible one, a perfectly (110)-oriented APB, with alternating Ga-Ga and As-As bonds [7.146]

the same throughout the crystal. But if this allocation changes somewhere inside the crystal (Fig. 7.31b), the interface between domains with opposite sublattice allocation forms a two-dimensional structural defect called an antiphase boundary (APB). The domains themselves are called antiphase domains (APDs).

Such APBs can be expected to form when GaAs is grown on Si or Ge, especially on a (100)-oriented substrate, the most widely used crystallographic orientation for MBE growth. Inasmuch as As forms strong bonds with Si, whereas Ga does not, the first atomic layer bonding to the Si substrate should be expected to be an As layer. Now, any real (100) surface will always exhibit steps. At any step only one atomic layer high (or an odd number of layers high) the sublattice site allocation of Ga and As on opposite sides of the step is interchanged, and an APB results (Fig. 7.32).

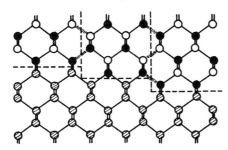

Fig. 7.32. Mechanism of APB formation during polar-on-nonpolar growth due to the presence of single-height steps on the substrate surface [7.146]

The APBs are structural defects which may also be found in GaAs containing Ga-Ga and As-As bonds. Such bonds represent electrically charged defects: A comparison of the number of bonding orbitals with the number of valence electrons available to fill them shows that Ga-Ga bonds act as acceptors, and As-As bonds as donors, with effective charges $\pm q/2$ per bond. In general, an APB will contain roughly equal numbers of both charges, thus acting as an extremely highly compensated doping sheet with very little net doping. The situation is least bad for an APB that follows exactly any (110) plane, as in Fig. 7.31b. In that case Ga-Ga and As-As bonds will alternate within each crystallographic unit cell, leading to perfect local charge compensation. But, for the deviations from this idealized arrangement, the lack of exact local charge balance will lead to potential fluctuations that will affect the electronic properties. Inasmuch as the initial Si surface steps are not likely to have the exact orientation within the surface plane that would lead to comparatively benign perfect-(110) APBs, the APBs actually resulting form the nucleation on a real surface must be expected to exhibit local charge fluctuations with large amplitude, and hence be harmful.

There are at least two approaches to the essentially complete avoidance of APBs: One involves a switch to a different crystallographic orientation on which APBs do not form in principle, even in the presence of steps, like the (211) orientation [7.147]. The other is somehow to enforce a perfect doubling of the height of all surface steps, an approach that has proven tenable since early 1985, contrary to all prior expectations [7.89, 148].

Most investigators working on GaAs-on-Si growth have preferred to continue to work with the conventional (100) orientation, or with wafers deliberately misoriented from the (100) orientation by a few degrees, relying on step doubling for the supression of APBs.

When a step on a Si(100) surface is an even number of atomic layers high, the two sublattices on the GaAs side are in registry again, and an APB will not occur at this step. Unfortunately, it is well established experimentally that for an "as polished" exactly (100)-oriented Si surface the most common step height is one atomic layer [7.29] and there is in fact ample evidence [7.147] that not only the growth of GaAs on Si, but also the growth of other III–V compounds, such as GaP, on exactly (100)-oriented Si or Ge substrates usually exhibits copious APBs. *Kaplan* [7.29] reported that on Si surfaces tilted by a few degrees from the (100) plane towards the (011) plane, most steps are two atoms high. However, in order to achieve APB-free growth, it is necessary that all steps be two atoms high, not just the majority of them. The most convincing direct evidence for perfect step doubling already on the pregrowth Si(100) surface is contained in the stunning work by *Sakamoto* and *Hashiguchi* [7.24], who showed that a nominally (100)-oriented Si surface would go from a single stepped surface to a doubly stepped surface during a prolonged high-temperature anneal (20 min at 1000° C), with all step terraces belonging to the same sublattice.

For the growth of GaAs on the Si(100) surface, *Kroemer* [7.146] proposed a nucleation model based on experimental data of *Fischer* et al. [7.86, 148, 149] and the following two postulates:

1.   Arriving Ga atoms bond to Si atoms only when at least one Ga-As bond can be formed along with every Ga-Si bond. The idea behind this postulate is that the 1/4 electron excess of the Ga-As bond is transferred to the Ga-Si bond, where it helps to form the latter bond.
2.   Even the formation of As-Si bonds is facilitated if at the same time Ga-Si bonds are formed, to take up the electron excess of the As-Si bond.

These postulates lead to the idea that the initial nucleation of GaAs on Si does not simply take the form of either As or – much less likely – Ga first bonding to the original Si surface, especially not in the presence of a sufficiently large nonbonded and hence mobile Ga concentration on the Si surface. Instead, one should therefore make the central postulate that:

3. The initially arriving As atoms will undergo an exchange reaction during which (Fig. 7.33) a Si atom from the last Si plane (plane 0) exchanges sites with an arriving As atom, with two Ga atoms simultaneously bonding to both the As atom and two adjacent Si atoms of the original surface.

[011] side view:

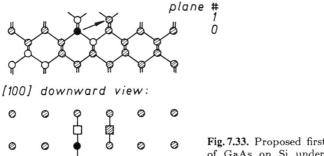

[100] downward view:

⊘ Si □ Ga *plane #1*

⊘ Si ● As *plane #0*

Fig. 7.33. Proposed first stage of the nucleation of GaAs on Si under Ga-rich conditions: an incoming As atom interchanges sites with a Si atom in plane 0 (the original top Si plane), simultaneously bonding two "waiting" Ga atoms. The ejected Si atom is placed on an adjacent site in plane 1. *Top:* [011] view; *bottom:* [100] (i.e., downward) view [7.146]

Note that in this initial nucleation step two As-Si bonds and two Ga-Si bonds are formed; hence the nucleus is an electrically neutral object. The number of Si-Si bonds does not change during the site exchange.

If we make the plausible assumption that the Si atom ejected from plane 0 is simply placed on one of the adjacent sites in plane 1, the site in plane 2 that has backbonds to both this Si atom and to the adjacent Ga atom will form a natural bonding site for another As atom, and an As-Ga pair will connect between the remaining bond of the Si atom in plane 1 and a Si atom in plane 0, as shown in Fig. 7.34. During this second nucleation step again two As-Si and two Ga-Si bonds are formed.

It is evident that this nucleation process automatically leads to a placement of all Ga atoms into plane 1 and of all As atoms into planes 0 and 2, the opposite choice from what would be present if the As-Si site exchange did not take place, but exactly the order observed by Fischer et al. under nucleation conditions of sufficiently high temperature and Ga flux. There can be little doubt that any subsequent lateral spreading of each nucleus thus formed will react in this sublattice order. This spreading can proceed either by adding As-Ga pairs, or by adding As atoms and performing an As-Si site exchange. Because the two

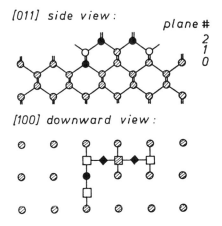

[011] side view:

plane #
2
1
0

[100] downward view:

Fig. 7.34. Proposed second stage of the nucleation of GaAs on Si under Ga-rich conditions: two As atoms bond to the Si atom in plane 1, and a Ge atom closes the bonding loop of the outer of the two As atoms to the Si surface [7.146]

◆ As   plane #2
▨ Si □ Ga   plane #1
⊘ Si ● As   plane #0

processes lead to opposite bond charges, a combination of both may be expected to take place, to aid in the establishment of electrical neutrality, but it is doubtful that perfect neutralization will take place.

We note, however, that, if perfect neutralization did take place, the resulting composition of the different planes adjacent to the interface would be exactly what is demanded by the two-layer reconstruction model of Harrison, Kraut, Waldrop and Grant (HKWG) [7.150].

The experiments of Fischer et al. [7.86, 148, 149] make it clear that under lower-temperature As-dominated conditions the As-Si site exchange postulated for Ga-dominated nucleation does not occur. The reason for this is probably one or both of the following: either the presence of an excess of Ga may be necessary to drive the exchange, or the temperature is too low for the exchange reaction to overcome an energetic reaction barrier that is very likely present, for example the As_2 dissociation barrier. Quite possibly both may play a role.

Nor is it clear whether or not the HKWG interface reconstruction process takes place at lower temperatures and, if not, exactly how electrical neutrality is subsequently established – if indeed it is. Quite possibly the HKWG mechanism occurs even then, but with the Si atoms ejected from plane 0 now replaced by Ga rather than As atoms, because of the prior formation of a tightly-bonded partial As coverage in plane 1. An alternative possibility would be that the top Si plane remains intact, but is neutralized by the formation of a very high concentration of Ga-on-As site antisite defects, as mentioned earlier. Or maybe the interface remains highly charged, being neutralized only by mobile electrons in the conduction band. This is evidently another fertile field for future research [7.146].

The high-temperature Ga-dominated nucleation model presented above is an idealization. In practice, one must expect numerous defects to occur, especially the following three:

1. Occasionally, As atoms may end up in plane 1, by bonding to Si plane 0 without undergoing the site exchange reaction. So long as these atoms

remain in the minority, they will simply appear as local antisite point defects (double donors) on what is otherwise a Ga sublattice, without causing actual finite-size antiphase domains.

2. As mentioned already, in those areas of the interface where the growth proceeds by lateral spreading from the initial nuclei, the neutralization of As-Si bonds by Ga-Si bonds formed may be incomplete, leading to a net doping of the interface, which may be either donor-like or acceptor-like, depending on the exact details of the process.

3. Finally, some of the Si atoms removed from plane 0 might not be incorporated into plane 1, but are taken up by the growing GaAs bulk instead. On purely thermodynamic grounds one should expect at least the GaAs layers closest to the interface to be Si-doped to the thermodynamic solubility limit.

Additional Si atoms may accumulate on the GaAs surface, from where they are gradually incorporated into the growing GaAs as bulk dopant.

Because of these various kinds of charge defects, the GaAs-on-Si(100) interface will almost certainly be one with a residual interface charge sufficiently large that it cannot be ignored for device purposes, combined with a heavy Si doping of at least the near-interface region of the crystal. The extent to which these two effects will take place will depend strongly on details of the exact growth procedure.

Up to a point, these defects are largely inconsequential, so long as they remain confined to within a few atomic monolayers of the original interface, and do not have a deleterious effect on the quality of subsequent layers. The defect structure near the interface is likely to be dominated by the very large density of misfit dislocations there, compared to which the other defects are a comparatively minor disturbance. Largely because of the misfit dislocations, the GaAs/Si interface itself is not likely to be usable as a part of the "intrinsic" device for most devices under current consideration, and its short-range properties are therefore not of primary concern. Apart from the dislocations, the defect making itself felt farthest from the interface is probably Si uptake by the growing GaAs, and its suppression probably deserves the highest priority after the suppression of the propagation of misfit dislocations [7.146].

The structural properties of GaAs epilayers in the early stages of the growth on Si(100) substrates have been investigated by an x-ray scattering technique [7.151], and high resolution transmission electron microscopy [7.151–154]. It has been demonstrated that in situ annealing of the GaAs epilayer during growth interruption, i.e., prior to final GaAs growth, imporves both the structural and the optical properties of the GaAs films [7.155]. Significant improvement in crystalline quality of MBE-grown GaAs on Si(100) substrates has also been observed after postgrowth rapid thermal annealing (RTA) [7.156, 157]. After RTA at 900° C for 10 s, the percentage of displaced atoms near the GaAs/Si interface, as estimated by Rutherford backscattering/channeling, decreased from 20 % to 7 %. Photoluminescence (PL) intensity after RTA increased significantly (as much as sixfold in some cases), with the actual increase varying from sample to sample and with annealing conditions. In some cases, the PL intensity after RTA was comparable to the PL intensity obtained from GaAs grown on a GaAs substrate under similar conditions. Rapid thermal annealing at 940° C resulted in a degradation of the PL intensity as compared to annealing at 900° C. Also, the PL peaks after RTA were found to shift to lower energies by 2–5 meV at low temperatures [7.156].

The electrical activity of defects in GaAs grown on Si by MBE has been examined by studying the diode characteristics and DLTS of Schottky barriers. The defects are not apparent from the forward-biased diode characteristics but they are indicated by large current and early breakdown under reverse-biased conditions. Postgrowth RTA has been found to improve the diode behavior significantly, making it almost comparable to GaAs on GaAs. The reverse current in the as-grown material shows a very weak temperature dependence, indicating that its origin is not thermionic emission or carrier generation. It is speculated that a large part of this current is due to defect-assisted tunneling, which is reduced by RTA. DLTS indicated only a modest increase in the concentrations of the well-known electron traps typical of GaAs MBE with no evidence for new levels in the upper half of the band gap.

In concluding the consideration of GaAs-on-Si (100) MBE the papers of *Kawabe* et al. [7.158–160] should be mentioned. They analyzed the initial stage of GaAs growth, and proposed a mechanism for self-annihilation of the APBs during the growth to explain the experimental results obtained with RHEED [7.160].

Growth of GaAs on other oriented Si substrates has been analyzed by *Uppal* and *Kroemer* [7.146, 147] [growth on the (211) surface], and by *Zinke-Allmang* et al. [7.161] [growth on the (111) surface].

7.2.5 Device Structures Grown by GaAs MBE

The capability of creating a wide variety of complex compositional and doping profiles in semiconductors, is an indispensable feature of the growth technique, from the point of view of its feasibility of being applied in device technology. The unique potential of MBE for device fabrication has been confirmed by many achievements in the field of optoelectronic and high-speed electronic devices. This is especially true when GaAs MBE is considered [7.162–166]. It is difficult to give a complete list of references containing data on exploitation parameters of all types of devices made of the GaAs-Al_xGa_{1-x}As system grown by MBE, because the number of references would be very large. Therefore, we will restrict ourselves to some selected examples.

1. *Hayakawa* et al. [7.167–169] have demonstrated interesting data concerning:

a)   room-temperature continuous wave operation of $(AlGaAs)_m (GaAs)_n$ superlattice quantum well lasers emitting in the visible (at 680 nm) with low threshold current densities [7.168], and

b)   near-ideal, low threshold behavior in (111)-oriented GaAs/AlGaAs quantum well lasers [7.167]. The threshold current density of the (111)-oriented laser with cavity length $L = 490 \, \mu m$ and quantum well width $L_z = 55 \, Å$ has been as low as 124 A/cm².

These authors have also investigated the effect of group V/III flux ratio γ on the reliability of GaAs/Al$_{0.3}$Ga$_{0.7}$As double-heterostructure lasers grown by MBE at 720° C. The threshold current does not change with γ, while the degradation rate strongly depends on γ, and has a minimum at $\gamma \sim 3$. It should be emphasized that in the case of $\gamma \sim 3$, the degradation rate of the MBE-grown lasers is lower than that of lasers grown by liquid phase epitaxy [7.169].

2. The GaAs/AlGaAs graded-refractive-index lasers grown on Si(100) substrates still exhibit much higher threshold currents than the similar lasers grown on GaAs(100) substrates. The progress is, however, remarkable. Broad area devices having a width of 110–120 μm and cavity lengths of $\sim$ 500–1200 μm exhibit threshold current densities as low as 600 A/cm² at room temperatures [7.170] and total efficiencies of as high as 0.75 W/A.

3. Growth and device characteristics of an index-guided GaAs/AlGaAs multi-quantum-well (MQW) laser, called a pair-groove-substrate (PGS) MQW laser, were described in detail in the paper of *Yuasa* et al. [7.171]. The laser structure is fabricated by using single-step MBE on a GaAs (001) substrate with a pair of etched grooves along the $\langle 1\bar{1}0 \rangle$ direction. A mesa between a pair of grooves, where the lasing action occurs, becomes narrow during growth, and the narrow mesa offers lateral wave-guiding that stabilizes a fundamental transverse mode. The superior crystalline quality of the mesa top, which can be examined by a microprobe photoluminescence technique, serves to lower the lasing threshold currents. The lasers with mesa widths below 2 μm show stable transverse mode operation with a low threshold current of 20 mA, as well as a high external differential quantum efficiency of 68 %. The low threshold and high characteristic temperature produce a high-temperature continuous-wave operation at 153° C for the lasers mounted on silicon heat sinks.

A schematic cross-sectional view of the PGS MQW laser is shown in Fig. 7.35. Lateral definition in the PGS waveguide structure is achieved by using the MBE growth over the grooves. *Yuasa* et al. have obtained the following results on the growth process of Si-doped n-type GaAs on the preferentially grooved (001) n-type GaAs by marking periodically with an n-type Al$_{0.2}$Ga$_{0.8}$As layer. Grooves aligned parallel to the $\langle 1\bar{1}0 \rangle$ and $\langle \bar{1}\bar{1}0 \rangle$ directions on the substrates were etched using a preferential chemical etching solution of $NH_4OH : H_2O_2 : H_2O$ (1 : 3 : 50). The slanting planes in the grooves intersect the (001) planes at about 55° for the two directions of the grooves, that is, the grooves exhibit (111)A and (111)B slanting planes corresponding to $\langle 1\bar{1}0 \rangle$- and $\langle \bar{1}\bar{1}0 \rangle$-oriented grooves, re-

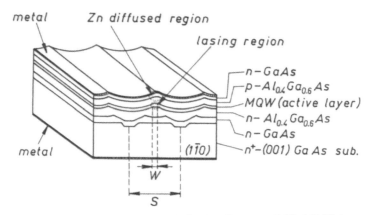

Fig. 7.35. Schematic diagram of the MBE-grown PGS MQW laser on a GaAs (001) substrate with a pair of grooves parallel to the ⟨1̄10⟩ direction. The mesa width W of the MQW active layer is a function of the grown layer thickness and the separation S (center to center of the grooves) [7.171]

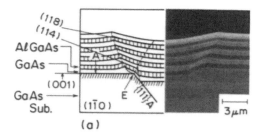

(a)

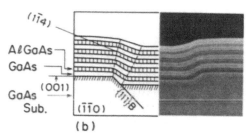

(b)

Fig. 7.36. SEM photographs and schematic diagrams of cross sections of alternating layers of GaAs and Al$_{0.4}$Ga$_{0.6}$As on GaAs (001) substrates with 1 μm deep grooves along the ⟨1̄10⟩ (a) and ⟨1̄1̄0⟩ (b) directions [7.171]

spectively. The growth was carried out in a Varian MBE/GEN II system at about 760° C with a growth rate of 0.9 μm/h for GaAs. The substrates were rotated at 5 rpm during growth to prevent asymmetrical growth.

Figures 7.36a and b show scanning electron micro-graphs of cleaved and strained planes of the grown layers on the substrates with the grooves along the ⟨1̄10⟩ and ⟨1̄1̄0⟩ directions, respectively. The groove depth was about 1 μm. The growth characteristics depend on the crystallographic orientation [7.172, 173]. The slanting planes in the ⟨1̄10⟩- and ⟨1̄1̄0⟩-oriented grooves propagate up through the grown layers, changing their indices from the initial (111)A and (111)B to (114) or (1̄1̄4), and (11̄4) or (1̄14), respectively. As the growth pro-

329

ceeds further, (118) or ($\bar{1}\bar{1}8$) planes are produced instead of (114) or ($\bar{1}\bar{1}4$) planes, as seen in Fig. 7.36a. The (001) area indicated by the arrow A in Fig. 7.36a, which is adjacent to the $\langle 1\bar{1}0\rangle$-oriented groove, becomes narrow with an increase in the grown layer thickness, while the (001) area adjacent to the $\langle \bar{1}\bar{1}0\rangle$-oriented groove spreads slightly, as shown in Fig. 7.36b. The growth processes can be explained by the orders of the growth rates as $[(111)A] < [(114)$ or $(\bar{1}\bar{1}4)] < [(001)] < [(118)$ or $(\bar{1}\bar{1}8)]$ and $[(111)B] < [(001)] < [(1\bar{1}4)$ or $(\bar{1}14)]$, which were determined by the propagation angle of the intersection of the adjacent planes in [7.173]. Here [ ] means the growth rate.

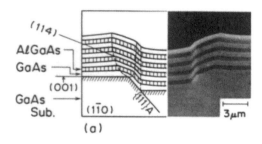

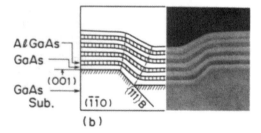

Fig. 7.37. SEM photographs and schematic diagrams of cross sections of GaAs/Al$_{0.4}$Ga$_{0.6}$As multilayers on GaAs (001) substrates with 2 μm deep grooves along the $\langle 1\bar{1}0\rangle$ (a) and $\langle \bar{1}\bar{1}0\rangle$ (b) directions [7.171]

The groove depth also affects the change in the indices of the slanting planes during growth. Figures 7.37a and b show SEM photographs of cleaved planes of the grown layers on the 2 μm-deep grooves along the $\langle 1\bar{1}0\rangle$ and $\langle \bar{1}\bar{1}0\rangle$ directions, respectively. The depth of the grooves is twice that in Fig. 7.36. The (111)A crystal plane is kept to a thickness above several micrometers, accompanying the (114) planes, as shown in Fig. 7.37a. In contrast, the (111)A plane in the shallow groove for Fig. 7.36a vanishes at an early stage of growth. The (111)B plane also changes during growth with a similar dependence of the index on the grown layer thickness, as shown in Figs. 7.36b and 7.37b. This difference in faceting originates from the fact that the change in the indices of the crystal planes begins at the edges, indicated by the arrow E in Fig. 7.36a. Since the new crystal planes are not readily produced in areas distant from the edges, the initial crystal planes on the longer ramps are kept for the thicker grown layers. The smoothness of the surfaces of the grown layers over the deep grooves is different from that over the shallow grooves. The PGS structure utilizes the growth characteristics of the epitaxial layers over the mesa between a pair of

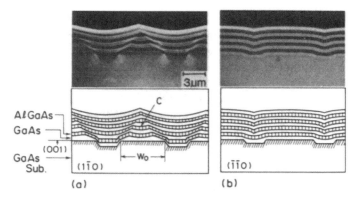

Fig. 7.38. SEM photographs and schematic drawings of the cross-sectional view of GaAs/Al$_{0.4}$Ga$_{0.6}$As multilayers on GaAs (001) substrates with a pair of grooves along the $\langle 1\bar{1}0 \rangle$ (a) and $\langle \bar{1}\bar{1}0 \rangle$ (b) directions [7.171]

grooves parallel to the $\langle 1\bar{1}0 \rangle$ direction. Figure 7.38 shows SEM photographs of the cleaved and strained multilayer structure grown on the GaAs (001) substrate with a pair of grooves. The grooves in Figs. 7.38a and b are aligned along the $\langle 1\bar{1}0 \rangle$ and $\langle \bar{1}\bar{1}0 \rangle$ directions, respectively. As the growth proceeds, the mesa width between the grooves in Fig. 7.38a decreases because of the faster growth of the (001) plane on the mesa top compared to the (114) or ($1\bar{1}4$) planes on the ramps [7.173]. The concaved shape of the mesa top, as indicated by the arrow C, is introduced by the change of the (001) plane into higher index crystal planes in the vicinity of the intersection with the (114) and (118) planes. On the other hand, the mesa in Fig. 7.38b spreads because of the slower growth of the (001) plane compared to the ($1\bar{1}4$) or ($\bar{1}14$) planes.

These growth processes by MBE on the grooved substrates are markedly different from those by LPE and OM VPE. In the case of LPE, since the crystal growth is much faster in the grooved region than on the flat surface, the surface of the grown layer over the grooves becomes planar-like with an increase in the thickness of the epitaxial layer [7.171]. In the case of OM VPE, since the growth rate on the (111)A ramps is slightly slower than on the (001) plane, and no change in the indices of the crystal planes occurs on the grooves, the mesa shape is almost unchanged during growth [7.171].

Narrowing of the mesa during MBE growth, as seen in Fig. 7.38a, is an advantage for the formation of the lateral waveguide. Actually, the width of the mesa-shaped structure, as narrow as 1 μm, is easily controlled by choosing the separation between the grooves and adjusting the grown layer thickness.

4. A semiconductor laser optical logic gate based on quenching and capable of performing the NOT, NOR, and NAND functions was described by *Grande* and *Tang* [7.174]. The device can be operated both pulsed and cw at room temperature. In addition, this logic gate can be monolithically integrated.

The device structure was grown by MBE on an n$^+$-GaAs substrate and consisted of five 120 Å GaAs quantum wells separated by 35 Å Al$_{0.2}$Ga$_{0.8}$As barriers. The wells and barriers were sandwiched by p- and n-$Al_{0.44}Ga_{0.56}As$ layers. A new one-step, two-level dry etching technique was used to fabricate the index-guided lasers and etched cavity facets that make up the logic gate.

The given examples of device structures made of the $GaAs$-$Al_xGa_{1-x}As$ systems have concerned heterojunction lasers. However, at present, many other optoelectronic devices, like solar cells [7.175, 176], doping superlattice photodetectors [7.177, 178], compositional superlattice avalanche photodiodes [7.179, 180], sawtooth superlattice light emitting diodes [7.181], are routinely made with MBE of this material system.

Among the nonoptoelectronic devices fabricated with GaAs MBE, the modulation-doped field effect transistors [7.164, 182], resonant tunneling devices, i.e., transistors and generators [7.180, 183–185], as well as tunneling hot electron transfer amplifiers [7.186, 187], and bipolar transistors [7.166] may be mentioned as examples.

7.3 Narrow-Gap II–VI Compounds Containing Hg

The narrow-gap II–VI compounds containing Hg form a fairly large group of semiconducting materials with interesting physical properties and a great potential for application in infrared detector technology and optical telecommunication systems [7.188–191]. Among these materials, $Hg_{1-x}Cd_xTe$ plays the dominant role.

$Hg_{1-x}Cd_xTe$ is a semiconducting alloy with a band gap that can be selected in the range from 0 to $1.5\,eV$ at room temperature by varying the mole fraction of CdTe. The variable band gap aspect and high electron mobilities make this material nearly ideal for infrared device applications. The need for large-area $Hg_{1-x}Cd_xTe$ material of good crystalline quality for detectors and advanced focal plane arrays has prompted research into thin-film structures of $Hg_{1-x}Cd_xTe$. The advantages of the backside-illuminated operation mode of the infrared photovoltaic detectors and the encouraging experimental results concerning superlattices as structures for infrared detectors have given rise to growing interest in heterostructures of this system [7.192–198].

While $Hg_{1-x}Cd_xTe$ continues to be an important material in this group, more and more interest is being directed toward all Hg-based semiconducting materials and structures, as well as to their associated substrate materials. This trend may be easily recognized in the proceedings of the latest conferences devoted to II–VI semiconductors [7.199] and to the physics and chemistry of mercury cadmium telluride [7.200–202].

The specific feature of these materials relevant to MBE is the high vapor pressure of Hg, which produces particular difficulties in the growth process. This is probably the reason why MBE of Hg compounds started only at the beginning of the 1980s, after Faurie and Million demonstrated the MBE growth feasibility of $Hg_{1-x}Cd_xTe$ [7.203, 204], and later the growth of CdTe-HgTe superlattices [7.205, 206].

These pioneering experiments were made on CdTe(111)B substrates (with Te atoms in the outermost monolayer). However, the lack of CdTe substrates sufficiently cheap in production yet exhibiting high structural perfection has stimulated interest in CdTe films grown on foreign substrates such as InSb(100) [7.207–209] or GaAs(100) [7.210–213]. Gallium arsenide has several advantages over the other possibilities. First, large area wafers are available with high crystal quality. Second, GaAs is transparent in the entire useful infrared spectral range and can therefore be used for backside-illuminated detectors. Third, because of the well-developed GaAs technology it can be an active part of the device in optical communication applications. Last, but certainly not least, are the economic considerations: GaAs is rather inexpensive, about 10 % of the CdTe price [7.188]. Thus, of all the foreign substrates, GaAs is certainly the most attractive.

The one disadvantage of GaAs is its large lattice mismatch (14.6 %) with CdTe. This large lattice mismatch creates a highly disordered interfacial zone which requires the use of a thick buffer layer. This concern has proven to be minimal since several groups have successfully grown CdTe on GaAs by both MBE and MO CVD [7.188]. Excellent results on growing CdTe epilayers on GaAs substrates have been recently reported by *Humenberger* [7.214], who used hot wall beam epitaxy (HWBE) with a specially designed effusion cell [7.215] (Sects. 2.2.3 and 3.2.1). The most important problems concerning substrates for MBE growth of Hg-containing compounds will be discussed in the following section.

7.3.1 Substrates for MBE of Hg Compounds

As already mentioned, the first substrate material used for $Hg_{1-x}Cd_xTe$ was CdTe. Cadmium telluride has a zinc-blende structure which belongs to the cubic noncentrosymmetric space group $F\bar{4}3m$. The characteristic tetrahedral bonding of this compound is caused by sp^3 hybridization. $Hg_{1-x}Cd_xTe$ belongs to the same group as CdTe except that a fraction $(1-x)$ of the Cd is randomly replaced by Hg. Since $Hg_{1-x}Cd_xTe$, HgTe-CdTe superlattices, and other Hg-based semiconductors are of great importance for infrared technology, it is necessary to understand the role of the substrate in order to grow high-quality material with MBE [7.216–218].

Most $Hg_{1-x}Cd_xTe$ epilayers grown by MBE have been crystallized on one of the three low-index crystallographic orientations of CdTe, namely, (111)B, (100) or (110) [7.218], with (111)B being preferred [7.188]. These low-index crystallographic orientations are the most important because the physical properties of other high-index planes fall in between the primary low-index planes.

The (111) and (100) faces in the zinc-blende structure are called polar faces. This polarity leads to the two types of face in each case. A (111)A face is ideally terminated by a triply bonded Cd atom or a singly bonded Te (Fig. 7.39). Hence,

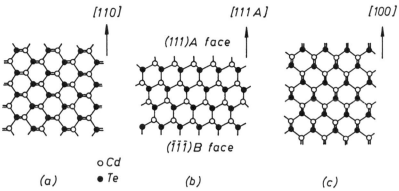

[110]   [111 A]   [100]

(111)A face

(ī̄ī)B face

○ Cd
(a)   ● Te   (b)   (c)

Fig. 7.39. Models of ideal CdTe crystal structure indicating the following faces: (a) (110), (b) (111), and (c) (100) [7.218]

it is called the Cd face, since the stable configuration has the Cd, or possibly Hg atom in the case of $Hg_{1-x}Cd_x Te$ at the surface. The other possible face, the (111)B face, is called the Te face, since the stable configuration is terminated by triply bonded Te. The (100) faces could be terminated with either a doubly bonded Te or a doubly bonded Cd. For the (100) face there is no difference in the bonding of Cd or Te at the surface, but for (111)A and (111)B either the Te or Cd atom is triply bonded while the other atom is slightly bonded. For (100), Cd and Te are equivalent but the bonding direction is rotated by 90°. An important difference, in terms of polarity, is observed between (111) and (100). For the first, two kinds of bonds are found for each element while only one kind of bond exists for the second.

Mercury, and to a lesser extent cadmium, is more volatile than tellurium. Because of the difference in bonding between the (111)A face, the (111)B face, and the (100) face, one can expect an influence of the crystal orientation on the growth conditions of $Hg_{1-x}Cd_x Te$ and CdTe (see the considerations presented in Sect. 6.2.2). This influence has been evidenced experimentally by *Sivananthan* et al. [7.216, 217], who investigated the condensations coefficients of Hg, Cd, and Te during MBE growth on (111) and (100) CdTe surfaces. Other material characteristics of $Hg_{1-x}Cd_x Te$ grown by MBE on these polar surfaces of CdTe also exhibit an orientation dependence [7.218]. For instance, as-grown HgCdTe (100) epilayers produce high-quality n-type conductivity at 77 K, while as-grown (111) epilayers may grow either n- or p-type, depending on the Te ratio and/or growth temperature. It has also been reported that (100) growth is easier to control than (111)B growth mainly because twinning [7.204], caused by excess Hg, is not observed in the (100) plane. The Hg flux needed to maintain HgCdTe single crystal growth shows orientation dependence, which has also been demonstrated for the polar (100) and (111) surfaces [7.216]. Among the low-index surfaces of CdTe the nonpolar (110) surface seems to be especially interesting. This surface is in an ideal truncated bulk case terminated by Cd and Te atoms, each with a single dangling bond at the surface where nucleation and growth occurs. It has been established [7.219] by low-energy electron diffraction, electron energy loss spectroscopy, and theoretical calculations that the (110) surface of CdTe undergoes a (1×1) reconstruction in which the Te atoms at the surface move out by 0.18 Å and the Cd atoms move in by 0.64 Å, giving a total displacement of 0.82 Å for atoms on this face. This (1×1) reconstruction is associated with a charge transfer from the Cd to the Te atom that creates a doubly occupied dangling bond at the surface, leading to a strong chemical preference for nucleation and growth.

These advantageous properties of the (110)-oriented CdTe surface for growing $Hg_{1-x}Cd_xTe$ $(0.16 < x < 0.41)$ by MBE have been confirmed experimentally. The grown epilayers [7.218] were twin free and exhibited high structural perfection. They also always exhibited p-type conductivity at 77 K with composition-dependent hole concentration ($p = 7.7 \times 10^{16}$ cm^{-3} for $x = 0.41$, and $p = 9.6 \times 10^{18}$ cm^{-3} for $x = 0.16$).

It has already been pointed out that from the economic point of view, the most attractive substrate for MBE of Hg compounds is the composite CdTe buffer layer on a GaAs (100) substrate [7.188]. Much discussion has been devoted to the epitaxy of CdTe on GaAs since both (100) CdTe $\|$ (100)GaAs [7.220, 221] and (111) CdTe $\|$ (100)GaAs [7.222, 223] orientations have been reported. The orientation can be controlled by the preheating temperature [7.224, 225]. Cadmium telluride grows in the (100) $\|$ (100) orientation when the preheating temperature is 480° C or less and in the (111) $\|$ (100) orientation when the temperature is 580° C.

It has also been shown by *Srinivasa* et al. [7.226] that the orientation of CdTe grown epitaxially on clean GaAs(100) can be predetermined by the GaAs precursor surface reconstruction which is present when the CdTe growth is initiated. A Ga-stabilized GaAs surface yields CdTe(111), and an As-stabilized GaAs surface yields CdTe(100).

An area of concern, when growing a CdTe buffer layer on GaAs(100) substrate in the (111) orientation, is the polarity (A or B) of the CdTe surface. This polarity can change the growth conditions of any Hg-containing compound grown by MBE on composite CdTe(111)/GaAs(100) substrates. The identification of the polarity has been investigated by *Hsu* et al. [7.227] using x-ray photoelectron spectroscopy, chemical etching and in situ electron diffraction. The results from all the used experimental techniques have shown that, when CdTe(111) is grown on GaAs(100) substrates, the terminating atomic plane in the CdTe(111) surface is the Te plane, which means that the composite CdTe(111)/GaAs(100) substrate exhibits the B polarity of its epitaxial surface.

Observations by TEM of the CdTe(111) epilayers grown on GaAs(100) substrates show twins and stacking faults parallel to the (111) epitaxial plane. A high density of dislocations located in the interfacial (CdTe/GaAs) plane has also been revealed with TEM [7.208]. The dislocation density in the CdTe epilayer may, however, be drastically reduced, when a ZnTe layer 8 % mismatched with

335

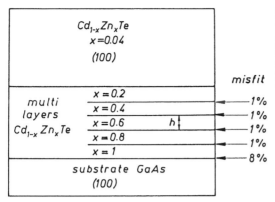

Fig. 7.40. Schematic illustration of the layer sequence in the multistepped composition buffer layer [7.208]. The thickness h of the layers is less than the critical thickness for the generation of misfit dislocations ($h = 50$ nm)

GaAs and 6.4 % mismatched with CdTe is grown between the GaAs substrate and the CdTe buffer layer. The majority of the dislocations generated by lattice misfit to GaAs are then bent at the ZnTe-CdTe interface due to the stress induced by the lattice mismatch between these two materials.

In an attempt to optimize this effect and to confine the misfit dislocations to the vicinity of the heterointerface, $Cd_x Zn_{1-x} Te$ buffer layers ($0 < x < 0.1$) have been grown on GaAs(100) or on single crystalline $Cd_{0.96} Zn_{0.04} Te$ substrates [7.228]. However, the best results have been obtained with a multistepped composition buffer layer (MCBL) [7.208]. The structure of this buffer layer is shown in Fig. 7.40. It is composed of a stack of $Cd_x Zn_{1-x} Te$ layers with the same thickness, but with different compositions. The first layer is the ZnTe binary compound, and then CdZnTe layers are deposited with composition steps of $\Delta x = 0.2$ to gradually adapt the lattice parameter from ZnTe to CdTe. With a composition step of 0.2 between each layer, the misfit variation from one layer to the next is about 1 %. It is well known that, when epilayers are not lattice matched with the substrate or the underlayer, misfit dislocations are generated when a critical thickness h_c is reached, releasing the stress (Sect. 6.2.1). The critical thickness variation of the CdTe layer in terms of the lattice mismatch have been determined by *Fontaine* et al. [7.229]. The critical thickness of ZnTe on GaAs is estimated as 9 nm. For the CdZnTe layers of the MCBL, with a misfit of 1 %, the critical thickness for each layer is about 70 nm. Therefore *Million* et al. [7.208] have grown two types of MCBL. In the first one, all the individual layer thicknesses are 200 nm, for the second they are 50 nm. These two thickness values have been chosen to be respectively above and below the critical thickness of a ZnCdTe layer of the MCBL. It has been shown with TEM that a MCBL with individual layer thickness of 200 nm does not improve the quality of the CdTe layer. The different interfaces are identified by the dislocations bent along each interfacial plane, but a high number of them still propagate through the CdTe layer. On the other hand, when the individual CdZnTe layer thickness is below the critical thickness, the dislocations are blocked in the MCBL. Observation by TEM also shows that the dislocation density is reduced from one layer to the next. Each interface is marked by dislocations bent along the interfacial plane. Moreover, the stacking faults, probably generated by thermal stress at the GaAs interface, are prevented from propagating to the film surface.

7.3.2 Hg-Compound Heterostructures Grown by MBE

Despite the fact that $Hg_{1-x}Cd_xTe$ is still the most important Hg-containing compound grown by MBE with high quality (the MBE-grown epilayers can be compared with the best $Hg_{1-x}Cd_xTe$ material grown by other techniques [7.230]), many other Hg-containing compounds and heterostructures have been grown. Because Hg has both a high vapor pressure and low sticking coefficient, at the usually used growth temperature of about 200° C special Hg MBE sources are required (see the design shown in Figs. 2.19 and 3.24). Also, the application of Hg ionization sources, as well as the MBE growth mode with substrate electrically biased, may be required in special cases. We will give here just some examples of the more interesting structures grown by MBE, indicating the original papers where the presented heterostructures has been described.

In $HgTe-Hg_{1-x}Cd_xTe$ heterostructures Hall mobility enhancement has been observed. Superlattices and heterojunctions can exhibit hole mobilities in p-type layers several times greater than in the alloys [7.195].

$Hg_{1-x}Cd_xTe$ epilayers of high quality, of both n-type and p-type conductivity, have been grown on CdTe(111) substrates as well as on GaAs(100) substrates. The electrical and optical investigations performed on this ternary compound and on $Hg_{1-x}Mn_xTe$-CdTe superlattices [7.198] provide evidence that it may be an interesting material for infrared detector device applications [7.231].

$Hg_{1-x}Zn_xTe$ epilayers seem to be becoming candidates for replacing MBE-grown $Hg_{1-x}Cd_xTe$ in infrared technology. This compound of high quality was first grown by MBE in 1985 [7.232]. It has been predicted that Cd destabilizes the Hg-Te bond and that Zn will be less harmful [7.233, 234]. In addition the fact that the bond length of ZnTe is 7 % shorter than that of CdTe is supposed to harden the structure. Thus the dislocation density is expected to be smaller in HgZnTe than in HgCdTe.

Since the very small Hg condensation coefficient is due to the weakness of the Hg-Te bond it is expected that if Cd or Zn changes the bond energy, the condensation coefficient of Hg should experience a change too [7.188].

Of all the Hg-containing heterostructures, the HgTe-CdTe superlattices belong to the most important from the point of view of the infrared detection technique [7.190, 191, 235–240]. This results mainly from the dependence of the cutoff wavelength λ_c of the superlattice on the HgTe layer thickness d ($\lambda_c = 0.1184d\,[\text{Å}] + 0.78$, [7.191]). A comparison of the λ_c for a $Hg_{1-x}Cd_xTe$ compound as a function of the compound composition (x), and for the HgTe-CdTe superlattice is shown in Fig. 7.41a. Figure 7.41b shows the room temperature λ_c of HgTe-CdTe superlattices with $d_{CdTe} \geq 35\,\text{Å}$ as a function of the HgTe layer thickness [7.191, 241]. One may easily recognize the advantage of the superlattice structure over the ternary compound.

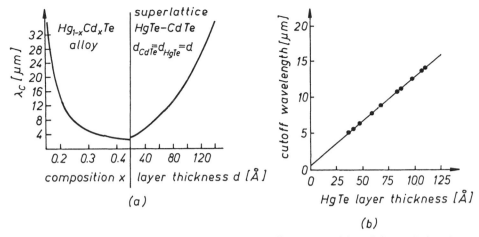

Fig. 7.41. (a) Cutoff wavelength as a function of alloy composition *(left panel)* for the $Hg_{1-x}Cd_xTe$ alloy and as a function of layer thickness *(right panel)* for the HgTe-CdTe superlattice with equally thick HgTe and CdTe layers [7.192]. **(b)** The room temperature cutoff wavelength of HgTe-CdTe superlattices with $d_{CdTe} \leq 35$ Å as a function of the HgTe layer thickness. The experimental data are given by the circles and the solid line is the linear fit. The experimental technique shows no variation in λ_c above 14 μm [7.191]

Following the tendency of substituting Cd with Zn, also in all-binary superlattices this procedure has been performed [7.242], by growing strained-layer HgTe-ZnTe superlattices. Three superlattices grown at 185° C have been characterized by electron and x-ray diffraction, infrared adsorption, and Hall measurements. The presence of satellite peaks in the x-ray spectra shows that the superlattices are of good quality despite the large lattice mismatch between HgTe and ZnTe ($\Delta a/a = 6.5\%$). These superlattices are p-type and the hole mobilities are very high compared to those of the corresponding alloy. Such a phenomenon has already been reported for HgTe-CdTe superlattices. Infrared transmission spectra show that HgTe-ZnTe superlattices have narrower band gaps than equivalent HgZnTe alloys [7.242].

7.3.3 Device Structures

Photovoltaic and photoconductive effects in semiconductors are widely used for detecting infrared radiation. When cooled down to low temperatures these detectors offer very high detectivities. Thermal imaging, space surveillance, detection of high-bit-rate optical fiber communication signals, and other commercial, medical and military applications have become possible with the $Hg_{1-x}Cd_xTe$ material in a broad wavelength region from 0.9 to 30 μm and beyond. Owing to the many interesting properties of $Hg_{1-x}Cd_xTe$, a variety of detectors can be fabricated for operation in the photovoltaic and photoconductive modes. Also, several types of infrared-sensitive metal-insulator-semiconductor (MIS) devices, including charge coupled devices (CCD), charge injection devices (CIDs), and metal-oxide semiconductor field effect transistors (MOSFETs) have been fabricated using $Hg_{1-x}Cd_xTe$ [7.192].

Significant advances in device performance have been achieved by incorporating the p-n homojunction consisting of $Hg_{1-x}Cd_xTe$ into a heterostructure configuration formed in contact with CdTe. This configuration has made it possible to use the backside illumination mode of photodiodes with improved quantum efficiency and reduction in noise.

A variety of methods have been used to construct the p-n junctions in $Hg_{1-x}Cd_xTe$. Both n-on-p and p-on-n junctions have been formed. Junction formation techniques include ion implantation of donors and acceptors, high-energy proton bombardment, diffusion of donors and acceptors, in-diffusion and out-diffusion of Hg, creation of p-type surface layers on n-type material by intense pulsed laser radiation, and in situ cosputtering of donors and acceptors. It appears that ion implantation and Hg in-diffusion are the most significant techniques for preparing high-performance infrared photodiodes.

One aim in designing an infrared detector is to provide efficient illumination of the absorbing material by only those photons which have their energy within the desired spectral band. Frontside illumination of the detector is a relatively simple conventional method. In frontside illumination the photon absorption takes place near the surface layer of the p-n junction (Fig. 7.42a). Unfortunately, recombination of carriers at the surface appreciably reduces the quantum efficiency of this structure, in particular, at high photon energies with large absorption coefficients. Reliability and efficiency of device performance are improved if a wide-gap semiconductor film (e.g., CdTe) is deposited onto the bulk photoabsorbing material (Fig. 7.42b). Now absorption of the desired radiation begins in the bulk. If the recombination rate at the heterointerface is lower than that in the surface of the bulk material, the device efficiency is increased relative to the p-n junction structure (Fig. 7.42a) mentioned above. A backside illumination configuration (Fig. 7.42c) is a further improvement in design. The optically active area is now located behind a wide-gap semiconductor substrate which acts as spectral filter for incident radiation. In this case, filtering is independent of angle of incidence and can be performed at the detector temperature. A reduction in the volume of the absorbing narrow-gap semiconductor leads to reduction in the noise of the detector. A sophisticated heterostructure photodetector is illustrated in Fig. 7.42d. Filtering is performed by a set of step-graded buffer layers and a thick filter layer which forms an interface with the absorbing active layer of the p-n junction. The main role of the buffer layers is the matching of the lattice parameters between the substrate and the filter layer to diminish the interface recombination processes. It is also possible in this detector to place the heterointerface directly in the p-n junction region. This configuration would maximize the quantum efficiency of the diode at high modulation frequencies because photon detection may take place, at sufficiently high reverse bias, within the depletion region of the heterojunction.

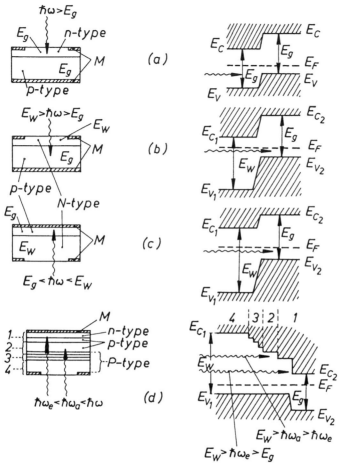

Fig. 7.42. Schematic illustrations of different photodiode structures designed to achieve optical illumination of the active layer with energy gap $E_g < h\nu$. (a) Conventional frontside illumination. (b) Frontside illumination through a wide band-gap layer with $E_W > E_g$. (c) Backside illumination through a wide band-gap substrate. (d) Backside illumination with spectral filtering in a multilayer heterostructure [7.192]

In principle, all the discussed heterostructure detectors may be produced with MBE, because this technique is already well advanced for the considered materials. Moreover, in-growth doping [7.243–246] has also recently been reported. Unfortunately only a few reports exist on device technology implemented with MBE [7.247, 248].

8. Outlook

In the preceding chapters we have presented a consistent picture of the current status of MBE. Emphasis has been put on technological equipment, characterization methods and growth processes. This presentation exhibits some unavoidable shortcomings, especially concerning completeness of the list of references, the background to the physical picture of the MBE growth process, and the detailed descriptions of the technological parameters used when growing specific materials by MBE.

It is very difficult to avoid such incompleteness in presenting a subject so intensively investigated, and so quickly developing as MBE. Also, this book serves the Springer Series in Material Science as an extended introduction to MBE and is intended to be followed by more specialized monographs dealing with MBE of selected systems of materials [8.1].

For the sake of completeness we will, however, briefly review the most interesting materials which are grown by MBE but are different from those discussed in Chap. 7. This is followed by a survey of MBE-related growth techniques, such as mass separated ion beam epitaxy (MIBE) [8.2–5], ionized cluster beam deposition (ICBD) [8.6–8], and molecular stream epitaxy (MSE) [8.9–11]. A short forecast of the development trends in MBE will conclude this chapter and the book.

8.1 Miscellaneous Material Systems Grown by MBE

The most important III–V material system grown by MBE which has not been discussed so far is the GaInAs(P)/InP system [8.12–21]. As the first representative of this system $Ga_{0.47}In_{0.53}As$ lattice-matched to InP may be considered. This heterojunction material pair has emerged as a very important semiconductor material [8.15]. High electron mobility and peak velocity are attractive for ultra-high speed devices. The band gap of 0.75 eV is ideal for photodetectors in optical communication systems in the optimum wavelength range 1.3–1.6 μm. Furthermore, heterojunction lasers with quantum well structures (QWSs) made of this pair of materials allow the emission wavelength to be shifted from 1.65 μm to the 1.3–1.55 μm region by making use of the quantum size effect in the QWS, that is, by having different well thicknesses in the heterostructure. This not only circumvents the necessity of using homogeneous GaInAsP alloys as the active medium, but also produces heterojunction lasers with significantly improved device performances not achievable by the bulk counterpart [8.21].

Extremely high quality $Ga_{0.47}In_{0.53}As/InP$ QWSs have been grown by MO MBE by *Tsang* and *Schubert* [8.15, 21]. Arsine (AsH_3) and phosphine (PH_3) were used as sources of As and P, while TEGa and TMIn gases were combined to form a single emerging beam as a source of Ga and In. This single beam nature of group III element reactants has guaranteed lateral spatial composition uniformity of the grown epilayers [8.22]. During the growth, thermal pyrolysis of the reactants occurred entirely on the substrate surface. Continuous growth was employed at the interfaces by switching out and in the appropriate gas components. $Ga_{0.47}In_{0.53}As/InP$ p-i-n photodiodes fabricated from wafers grown by MO MBE [8.23] have exhibited a very low dark current (less than 1 nA at 10 V bias), a quantum efficiency of 70 % (without antireflection coatings), and a transit-time-limited pulse response. These results attest to the high quality of heterojunction interfaces grown by MO MBE. They are among the best reported p-i-n photodiodes grown by other techniques [8.21]. State of the art avalanche photodiodes (APDs) and very low threshold double-heterojunction and multi-quantum-well lasers have also been fabricated from $Ga_{0.47}In_{0.53}As/InP$ wafers grown by MO MBE [8.21].

High quality QWSs and modulation doped heterostructures of this material system may also be grown by CPS MBE. This has been done by using as group V element sources cracker effusion cells equipped with a LN_2-cooled dewar located between the cracking zone and the mechanical shutter of the cell [8.18]. This allowed a very good control of the As_2/P_2 ratio during the growth of the heterostructures, thus enabling multiple quantum wells with heterointerface fluctuations of only 1 monolayer to be grown. A modulation-doped heterostructure of $Ga_{0.47}In_{0.53}As/InP$ with a 100 Å spacer layer gave a mobility of $8.2 \times 10^4 \, cm^2 \, V^{-1}s^{-1}$ for a charge density of $7.4 \times 10^{11} \, cm^{-2}$ when measured at 2 K [8.18].

The heterojunction pair $Ga_xIn_{1-x}As_yP_{1-y}/InP$ may be considered as the next representative of the GaInAs(P)/InP materials system [8.12, 13]. This heterojunction pair is usually grown by GS MBE because only this growth technique makes possible, in a fairly simple way, the well-controlled generation of P and P plus As beams for the growth of the quaternary compound. The detailed investigations on the use of GS MBE for the growth of $Ga_xIn_{1-x}As_yP_{1-y}/InP$ heterostructures have been extended to the study of a variety of heterostructures.

Single QWSs as thin as 5 Å have been grown with near monolayer abruptness. Superlattices have been reported with the observation of exciton structure in the absorption and photoresponse spectra [8.24]. These were 20-period structures with each period consisting of an 80 Å quantum well of $Ga_xIn_{1-x}As_yP_{1-y}$ and a 150 Å InP barrier. Superlattice APDs with high gain and low dark current have also been demonstrated, as have heterostructure bipolar transistors with direct current properties superior to any previously reported, and exhibiting potential for superior high-frequency performance [8.13]. Recently also, quantum well light modulators made of InGaAsP/InP rib waveguide structures grown by GS MBE have been demonstrated [8.20]. Using 100 Å wide wells of GaInAsP of 1.55 and 1.3 μm composition in a superlattice structure, the quantum confined Stark effect [8.25] on which the light modulators are based has been studied, and applied for modulators operating at a wavelength as short as 1.3 μm [8.20].

Avalanche photodiodes with separate absorption, grading and multiplication regions (SAGM-APD) have been grown by MO MBE using the system InP/GaInAsP/GaInAs [8.19]. These APDs exhibit low dark current (less than 25 nA at 90 % of breakdown voltage), low capacitance (about 0.2 pF), and good responsivity (0.75 A/W at 1.3 μm). The pulse response, being relatively independent of avalanche gain, is characterized by rise and fall times of approximately 1.4 ns.

The last representative of the considered materials system which will be mentioned here is the strained layer $Ga_x In_{1-x}As$-based heterostructure system [8.26–37]. Strained layer heterostructures have received increasing attention in the past few years [8.29]. These are high quality multilayered structures grown from lattice-mismatched materials. The large (> 0.1 %) mismatch is totally accommodated by uniform elastic strains in the heterostructure layers, because the layer thicknesses are kept below the critical value for generation of misfit dislocations (Sect. 6.2.1). The fairly easy control of epilayer thickness offered by MBE makes this growth technique especially useful for fabrication of strained-layer heterostructures (SLHs), with the one extreme of strained-layer superlattices (SLSs) [8.29] and the other extreme of double barrier quantum well structures [8.37].

Defect-free strained layer epitaxy opens up possibilities for further improvement of quantum well, two-dimensional electron gas, modulation-doped field effect transistor (MODFET) structures [8.33]. In the case of the GaInAs/AlInAs on InP system, for example [8.36], there exists an increased freedom with layer composition which allows certain properties of the SLHs in MODFET devices to be optimized. Among these properties are the conduction band edge discontinuity, which controls the maximum sheet concentration of electrons (n_s), and the electron effective mass, which influences the speed of the device. Enhancement of the device speed can be attained by increasing the Al concentration in the AlInAs layers and/or by decreasing the Ga concentration in the GaInAs. The maximum amount of strain which can be incorporated in the coherently grown layer structure sets, however, the upper limit on the compositional tolerances.

It has been shown [8.36] that with the $Ga_{0.47}In_{0.53}As/Al_{0.48}In_{0.52}As$ on InP system grown by MBE in the MODFET structure, high power microwave, low noise millimeter-wave, and high speed digital devices may be fabricated, with improved performance compared to the GaAs/AlGaAs MODFET technology. Moreover, the possibility exists of integration of the GaInAs/AlInAs SLH MODFETs with optoelectronic devices like lasers or photodiodes, as the $Ga_{0.47}In_{0.53}As$ bandgap is compatible with a minimum-loss transmission window for optical fibers (1.3–1.55 μm).

Another III–V materials system of interest for device applications is the InGaP/InGaAlP on GaAs system. Using CPS MBE with tetrameric group-V element beam sources and Be and Si as p- and n-type dopants, respectively, double heterostructure and multi-quantum-well laser diodes of this pair of materials have been grown [8.17]. Room-temperature continuous-wave operation of laser diodes operating in the visible (600 nm) with a threshold current of 110 mA has been achieved. This result was obtained by the introduction of H_2 into the growth chamber during MBE growth, and by growing a GaAs buffer layer on the Cr-doped semi-insulating and Si-doped n-type GaAs(100) substrates. It was found that the MQW lasers were stable in their operation under thermal treatment at temperatures as high as 750° C.

Among II–VI semiconductors which have attracted interest as being suitable for growth by MBE, the Zn-containing compounds should be mentioned. ZnS, ZnSe, ZnTe [8.38–45] and their ternaries, $ZnS_x Se_{1-x}$ and $ZnTe_x Se_{1-x}$ [8.45–47] have received considerable attention as potentially useful materials for optoelectronic applications in such devices as direct current thin film electroluminescent panels [8.48, 49], or light-emitting diodes and lasers operating in the blue spectral region.

Advantages offered by SLS structures, namely, pseudomorphic growth and modification of the band structure within the multiple-quantum-well region has also been demonstrated for this material system. Strained-layer superlattices of ZnSe-ZnTe have been grown by CPS MBE [8.50–52] on InP(001) substrates, or on GaAs(001) substrates covered with a 5000 Å thick buffer layer of ZnSe, or on InAs(001) with a ZnTe buffer layer. The strain and lattice dynamics in these structures have been investigated with transmission electron microscopy and Raman scattering [8.53]. It has been shown that the best results in terms of structural quality of the ZnSe-ZnTe SLSs are obtained on substrates with thick ZnSe buffer layers. The lattice mismatch in SLSs was completely accommodated by the tetragonal distortion of successive layers in the SL system. High resolution TEM imaging has shown the defect-free nature of the interfaces and an angular change between the (111) planes of the two SL component layers. Photoluminescence measurements [8.52] have indicated that ZnSe-ZnTe SLSs exhibit typical quantum-size effects. The luminescence intensity became stronger and the broadened tail of the spectrum was decreased when the SLS was grown on a buffer layer, that is, on a substrate with improved surface quality.

Using MO MBE, *Taike* et al. [8.54] have succeeded in growing ZnSe-ZnS SLSs. In their growth procedure DEZn, DESe and DES were used as source gases for Zn, Se and S, respectively. Hydrogen was used as the carrier gas for the MO reactants. The pyrolysis of DESe and DES was carried out in cracking cells at the outlet of the tube. For growth of ZnSe-ZnS SLSs, the DEZn beam was kept on during the entire growth, while the DESe and DES beams were alternated to produce the ZnSe and ZnS layers, respectively. Bypass lines have been added to the apparatus to stabilize the gas flow rate: DES (DESe) was ventilated while DESe (DES) was incident on the substrate for growth of the ZnSe (ZnS) layers.

The switching of DESe and DES flows was operated by on/off operation of the valves, thus, no shutters between the cracking cells and the substrate were used. The DESe and DES flow were switched at 30 s intervals, which resulted in abrupt SLS interfaces. The SL structure of the grown epilayers has been confirmed by x-ray diffraction measurements and by 4.2 K photoluminescence spectra.

A modification of ZnSe-based SLSs are the $ZnSe/Zn_{1-x}Mn_xSe$ $(0 < x \leq 1)$ superlattices [8.55–59]. The motivation for growing this interesting system in the form of both films and superlattices originated from potential applications as efficient blue light emitters and display devices [8.55]. The challenge involved in growing thin films and SLs of $Zn_{1-x}Mn_xSe$ stemmed from the tendency for bulk crystals of this ternary compound to exhibit wurtzite phases for Mn mole fractions in excess of 0.09, while ZnSe has zinc-blende structure.

The introduction of Mn into ZnSe produces not only a variation in band gap and lattice constant [8.60] but also interesting and potentially useful magnetic properties [8.59]. For low concentrations of Mn in $Zn_{1-x}Mn_xSe$ epilayers grown by MBE, this compound is strongly paramagnetic. Antiferromagnetic ordering becomes significant with increasing Mn fraction. The occurrence of antiferromagnetic ordering as the Mn fraction increases in MBE-grown zinc-blende epilayers is not surprising since rock-salt bulk crystals of MnSe are known to be antiferromagnetic. The unique feature of the MBE-grown MnSe/ZnSe SLS [8.58, 59, 61] is the occurrence of the heretofore "hypothetical" zinc-blende MnSe. Epilayers of this new phase of MnSe alternating with ZnSe have been incorporated into a variety of superlattice structures. Although MnSe is antiferromagnetic, ultrathin layers (down to one monolayer in thickness) of MnSe have been found to exhibit paramagnetic behavior [8.59]. Using RHEED intensity oscillations observed in ZnSe and MnSe during the MBE growth, it was possible to control precisely the thickness of the MnSe layers.

The considered structures were grown by CPS MBE on GaAs(100) substrates at 400° C using three effusion cells containing elemental Zn, Se and Mn of 6 N purity. The possibility of growing by MBE a crystalline phase of MnSe not existing in the bulk form indicates the great potential of MBE for fundamental investigations on crystal growth phenomena.

The MBE growth technique is currently used for growing many other materials in the form of thin layers or multilayer structures. For the sake of brevity, we will limit our treatment of this subject just to listing, without discussion, the most interesting examples.

The group of Pb-based IV–VI compounds providing tunable lasers operating in the far infrared (2.5–34 μm), and photovoltaic detectors for this spectral region, is usually grown by CPS MBE. Effusion cells used for the growth are charged with binary constituent compounds (they evaporate congruently) or with elemental constituents [8.62–66]. The standard RHEED oscillation technique for control of growth may also be used in MBE of these compounds [8.67].

Insulating fluorides, $(Ca,Sr)F_2$, grown by CPS MBE on Si [8.68] and on III–V compounds [8.69] seems to be promising for future application in very large scale integration (VLSI) technology.

Recently, high-temperature superconductor films, $DyBa_2Cu_3O_{7-x}$, have also been demonstrated as suitable for being grown by CPS MBE [8.70].

8.2 MBE-Related Growth Techniques

The observed constant interplay between materials advancement and device concepts fosters the progress in application of new thin film structures to new device designs. Molecular beam epitaxy has played a key role in the demonstration of many of the new concepts. The inherent features of this technology (Sect. 1.1.2) make it well suited to the research environment and to the demonstration of new concepts employing thin layers and abrupt interfaces. However, metalorganic chemical vapor deposition (MO CVD) is also employed currently with success to fabricate sophisticated multilayer devices with thin layers and abrupt interfaces. *Dapkus* [8.71] has compared MBE and MO CVD, taking into consideration the fundamental processes and results obtained in the fields of material properties and device performances. The main disadvantage in MO CVD is the occurrence of the diffusion boundary layer near the substrate surface, which results from the flow viscosity effects in the growth reactor. Therefore, the growth by MO CVD is to a large extent diffusion controlled, which has the consequence that the thickness and the abruptness of the epilayer structures are hard to control precisely. Lowering the pressure in the reactor, and using a vertical reactor geometry [8.72] diminishes this disadvantage of the MO CVD technique.

The competition and interplay of concepts on which MBE and MO CVD are based have led to a series of intermediate growth techniques. From the MBE side, the MO MBE modification of GS MBE may serve as an example, while the low pressure MO CVD [8.73] is the counter example from the MO CVD side. Some other hybrid processes that combine the principal advantages of these two growth techniques have also been proposed. These have been introduced under the names vacuum chemical epitaxy (VCE) [8.10, 11] and molecular stream epitaxy (MSE) [8.9].

Another direction of the development of MBE-related growth techniques is the application of ionized species during growth (compare the considerations in Sects. 6.5.4 and 7.3.2). In conventional MBE, surface atom migration can only be controlled by the substrate temperature (T_s) and growth rate (r_g). The migration increases when T_s increases and/or r_g decreases. This implies that where large surface atom migration is required, layers should be grown at extremely high T_s or extremely low r_g. However, for growth on substrates partially masked by a metal or an insulator, or on substrates grooved, pitted and mesa-etched, the enhancement of surface atom migration irrespective of T_s and r_g is desirable. This may be achieved by low energy ion irradiation during MBE growth [8.74–76]. By applying Ga^+ or H_2^+ ion irradiation, GaAs lateral single crystalline growth over a tungsten grating has been demonstrated [8.74] at a low substrate temperature where polycrystalline growth takes place without ion irradiation. Monte Carlo simulation studies of MBE growth [8.75, 76] in the presence of a low-energy ion beam have proved that smooth morphology can be realized at temperatures significantly lower than normal MBE growth temperatures. It was demonstrated by these studies that the smoothness of the growth front in ion-assisted MBE can be controlled by an appropriate choice of ion type, its energy, angle of incidence and flux.

In order to overcome the difficulties with the noncongruency of the III–V compounds in MBE growth (a large flux ratio of group-V element to group-III element is required in the growth of III–V compound semiconductors, because of the low sticking coefficient of group-V elements), so-called mass-separated ion beam epitaxy (MIBE) has been proposed by the ULVAC Corporation research team [8.2–5]. In this technique the group-III elements are supplied as molecular beams to the substrate from conventional effusion sources, while the group-V elements are supplied simultaneously as mass-separated high-purity ion beams and are implanted at low energy to increase the sticking coefficients. Semiconductor films of III–V compounds are grown on the substrate by reaction of the neutral group-III atoms and ion beams of the group-V elements. Therefore, the ion beam equipment should conform to the following requirements: (1) A stable supply of intense $(10^{13}–10^{14}\,\mathrm{cm}^{-2}\mathrm{s}^{-1})$ ion beams of various group-V elements; (2) a mass-separation facility for beam purification; (3) negligible contamination of the beam by thermal-energy neutrals effusing from the source.

Single crystal GaAs films are obtained with the MIBE method for growth temperatures between 220° C and 550° C, even with a flux ratio of unity, by implanting As$^+$ ions of 100 and 200 eV. High-quality InP layers have also been obtained between 210° C and 420° C with a flux ratio of unity by implanting P$^+$ ions of 100 and 200 eV. These results indicate that the controllability of the group-V element in MBE can be greatly improved by implanting mass-separated, high-purity group-V ions at low energies.

To enhance crystal growth and to promote chemical reactions, the energy of an ion beam should range from above thermal to a few tens of eV per atom. This is easily understood when we recognize that kinetic energies corresponding to thermal energies are approximately 0.01–0.1 eV, while the energy required to displace surface atoms is a few tens of eV. The latter energy is higher than the binding energy of atoms in a solid, and therefore at these energies different kinds of activation can be expected, which may promote the nucleation and growth processes vital to film formation [8.6]. However, acceleration of individual atoms to energies of tens of eV at sufficient beam intensities is very difficult because of space charge effects.

Ionized cluster beam deposition (ICBD) offers the capability of introducing such useful low energy at equivalent high current into the film formation process without the problems caused by space charge effects. This method was developed at Kyoto University [8.6].

In ICBD, material is placed in a crucible which can be heated to a very high temperature. The material inside the crucible is vaporized and is ejected through a small nozzle into a high vacuum region. Clusters of atoms are formed by nucleation in the beam. The cluster beam is then partially ionized by electron beam impact outside of the crucible and is accelerated toward the substrate.

The two most important advantages of ICBD which enable low temperature epitaxy are (1) the low charge-to-mass ratio of the ionized clusters, which can transport a high intensity beam at a suitable ion energy, and (2) control of the kinetic energy of the ionized clusters and the efficient transfer of this energy to the film formation process. A third and more subtle feature is the transfer of energy to migrating adatoms as a weakly bound cluster which breaks up at impact upon the substrate. Since ICBD provides tight control of the kinetic energy and ion content of the beam, this technique distinguishes itself from other deposition methods.

Capabilities of the ICBD have been demonstrated by growth of epitaxial Al films on various kinds of substrates, very large area deposition of GaAs, composition-controlled $Cd_{1-x}Mn_xTe$ films and CdTe/PbTe multilayer epitaxial films [8.6, 7]. The molecular dynamics simulation studies on ICBD of thin films performed by *Müller* [8.8] have confirmed the advantages of this method.

A schematic illustration showing the relation of the considered growth techniques to MBE, especially to its two modes CPS MBE and GS MBE, is shown in Fig. 8.1.

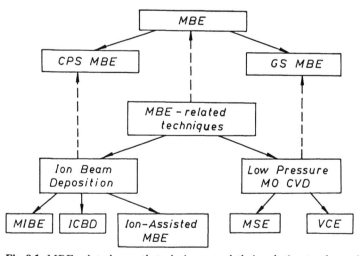

Fig. 8.1. MBE-related growth techniques and their relation to the main modes of MBE

8.3 Development Trends of the MBE Technique

Let us conclude the considerations on MBE presented in this book with a statement published by *Joyce* [8.77], one of the leading scientists in the field:

MBE has begun to make vital contributions in three fundamental areas. The first is crystal growth from the vapor phase, in which the UHV environment and in situ probes enable the growth process to be studied at the atomic level in real time. The second is in surface physics, where the preparation of surfaces with specified orientation, reconstruction and composition, which are also effectively damage-free, can be achieved with reasonable facility. The third, and by far the most widely exploited, area is the preparation of heterojunctions and quantum well structures, with their physics developing into the major new field of solid-state experimental and theoretical activity.

The next decade is sure to see all of these topics exploited further, with MBE playing a dominant role in research.

Figure 8.2 indicates some probable trends in the further development of MBE, and research activities in this field. One may distinguish four activity areas, namely, theory of MBE, experimental crystal growth, surface science for MBE applications, and device technology using MBE and combined MBE-UHV processing technology.

In the area of theory of MBE growth processes and of the physical phenomena determining the quality of the grown structures, computer simulation

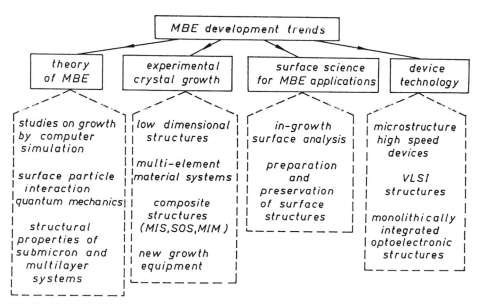

Fig. 8.2. Probable development trends of MBE and research activities in this field

techniques have already started to play an important role [8.78] and most probably will also be widely applied in the future. Here an understanding of the growth mechanism will be important, e.g., the transition from the two-dimensional nucleation mode to the step-flow mode [8.79], as well as an understanding of the influence which the beam parameters exert on the growth process [8.80].

The surface–particles interaction theory is the next important area of investigation in MBE theory [8.81, 82].

Very important theoretical activity will be connected with structural perfection of the structures grown by MBE. Dislocations, strains, and equilibrium atomic arrangements in submicrometer structures, analyzed from the point of view of so-called physical first principles, will probably be the subject of these investigations.

Experimental crystal growth concerns the optimization of growth parameters for low-dimensional structures, quantum wires and quantum boxes [8.83], as well as multielement material systems [8.70]. Composite structures, involving the growth of metals on insulators, or other combinations which are used in semiconductor devices are the next target of activity in this field. Metal-insulator-semiconductor (MIS), semiconductor-oxide-semiconductor (SOS), and metal-insulator-metal (MIM) structures are only the most important examples.

An interesting direction of the research on MBE growth is the area of equipment. Here some interesting results have already been obtained by replacing stainless steel with aluminum [8.84–86].

Molecular beam epitaxy is closely connected with surface physics, because surface kinetic phenomena play a decisive role in the growth by this technique. Therefore, it is obvious that in-growth surface analysis techniques like RHEED, ellipsometry or UPS will be intensively investigated in relation to MBE in the near future. This will serve for elaboration of new preparation and preservation techniques concerning the substrate surface structures. Picosecond time resolution RHEED may be mentioned as an interesting example [8.87].

The final, but most-needed, area of development and investigation is related to device technology realized with MBE. This is important from the commercial point of view. Here, the competition between MBE and other growth techniques is especially fierce. If MBE, which has very high equipment investment costs, is to survive in the future, it must offer new features in device fabrication.

It is possible that MBE will be introduced in the future to mass production lines, especially for fabrication of the most sophisticated devices, which are based on quantum-mechanical effects in solid-state structures. This tendency seems to be well confirmed by many contributions to the recent international conference on MBE (MBE-V,1988) held in Sapporo, Japan [8.88].

References

Chapter 1

1.1   L.L. Chang, K. Ploog (eds.): *Molecular Beam Epitaxy and Heterostructures*, NATO ASI Ser., Ser. E, no. 87 (Martinus Nijhoff, Dodrecht 1985)
1.2   E.H.C. Parker (ed.): *The Technology and Physics of Molecular Beam Epitaxy* (Plenum, New York 1985)
1.3   H. Morkoç (ed.): *Molecular Beam Epitaxy Workshop 1982*, Proc. 4th USA MBE Workshop; J. Vac. Sci. Technol. B 1, no. 2 (1983)
1.4   J.R. Arthur (ed.): *Molecular Beam Epitaxy Workshop 1983*, Proc. 5th USA MBE Workshop; J. Vac. Sci. Technol. B 2, no. 2 (1984)
1.5   G.Y. Robinson (ed.): *Molecular Beam Epitaxy Workshop 1985*, Proc. 6th USA MBE Workshop; J. Vac. Sci. Technol. B 4, no. 2 (1986)
1.6   D.L. Miller (ed.): *Molecular Beam Epitaxy Workshop 1986*, Proc. 7th USA MBE Workshop; J. Vac. Sci. Technol. B 5, no. 3 (1987)
1.7   J.R. Arthur (ed.): *Molecular Beam Epitaxy 1984*, Proc. 3rd Int. Conf.; J. Vac. Sci. Technol. B 3, no. 2 (1985)
1.8   C.T. Foxon, J.J. Harris (eds.): *Molecular Beam Epitaxy 1986*, Proc. 4th Int. Conf.; J. Cryst. Growth 81, nos. 1–4 (1984)
1.9   S. Ino: Jpn. J. Appl. Phys. 16, 891 (1977)
1.10   P.A. Maksym, J.L. Beeby: Surf. Sci. 110, 423 (1981)
1.11   B.A. Joyce, J.H. Neave, P.J. Dobson, P.K. Larsen: Phys. Rev. B 29, 814 (1984)
1.12   C.S. Lent, P.I. Cohen: Surf. Sci. 139, 121 (1984)
1.13   P.R. Pukite, C.S. Lent, P.I. Cohen: Surf. Sci. 161, 39 (1985)
1.14   T. Kawamura, P.A. Maksym: Surf. Sci. 161, 12 (1985)
1.15   P.K. Larsen, P.J. Dobson, J.H. Neave, B.A. Joyce, B. Boelger, J. Zhang: Surf. Sci. 169, 176 (1986)
1.16   O.K.T. Wu, E.M. Butler: J. Vac. Sci. Technol. 20, 453 (1982)
1.17   R. Memeo, F. Ciccacci, C. Mariani, S. Ossicini: Thin Solid Films 109, 159 (1983)
1.18   M.P. Seah: Vacuum 34, 453 (1984)
1.19   S.P. Svensson, P.O. Nilsson, T.G. Andersson: Phys. Rev. B 31, 5272 (1985)
1.20   A.V. Archipenko, Yu.A. Blyumkina, M.A. Lamin, O.P. Pcheljakov, L.V. Sokolov, S.I. Stenin, N.I. Kozlov, A.A. Kroshkov, A.V. Rzhanov: Poverkhnost 1, 93 (1985)
1.21   A.V. Rzhanov, L.A. Iljina (eds.): *Ellipsometry: Theory, Methods, Applications* (Nauka, Novosibirsk 1987) (in Russian)
1.22   D.A. Aspnes, A.A. Studna: Appl. Opt. 14, 220 (1975)
1.23   B.O. Seraphin (ed.): *Optical Properties of Solids: New Developments* (North-Holland, Amsterdam 1976) Chap. 15
1.24   Y. Demay, J.P. Gailliard, P. Medina: J. Cryst. Growth 81, 97 (1987); J. Vac. Sci. Technol. A 5, 3139 (1987)
1.25   L.I. Maissel, R. Glang (eds.): *Handbook of Thin Film Technology* (McGraw-Hill, New York 1970)
1.26   S.Dushman: *Scientific Foundations of Vacuum Technique* (Wiley, New York 1962)
1.27   L.B. Loeb: *The Kinetic Theory of Gases*, 2nd ed. (McGraw-Hill, New York 1934)
1.28   D.E. Gray (ed.): *American Institute of Physics Handbook*, 3rd ed. (McGraw-Hill, New York 1972) pp. 7–6
1.29   K. Ploog: "Molecular Beam Epitaxy of III–V Compounds", in *Crystals–Growth, Properties and Applications*, Vol. 3, ed. by H.C. Freyhardt (Springer, Berlin, Heidelberg 1980) p. 73

1.30 P.E. Luscher, D.M. Collins: Prog. Cryst. Growth Charact. **2**, 15 (1979)
1.31 G.J. Davies, D. Williams: "III–V MBE Growth Systems", in [1.2; p. 15]
1.32 C.W. Farley, G.E.Crook, V.P. Kesan, T.R. Block, H.A. Stevens, T.J. Mattord, D.P. Neikirk, B.G. Streetman: J. Vac. Sci. Technol. B **5**, 1374 (1987)
1.33 M.A. Herman: Vacuum **32**, 555 (1982)
1.34 J.A. Curless: J. Vac. Sci. Technol. B **3**, 531 (1985)
1.35 S.P. Svensson, P.G. Andersson: J. Phys. E **14**, 1076 (1981)
1.36 A.Y. Cho, K.Y. Cheng: Appl. Phys. Lett. **38**, 360 (1981)
1.37 A. Madhukar: Surf. Sci. **132**, 344 (1983)
1.38 M.A. Herman: *Semiconductor Superlattices* (Akademie-Verlag, Berlin 1986)
1.39 B. Lewis, J.C. Anderson: *Nucleation and Growth of Thin Films* (Academic, New York 1978)
1.40 L.D. Schmidt: "Condensation Kinetics and Mechanisms", in *The Physical Basis for Hetergeneous Catalysis,* ed. by E. Drauglis, R.I. Jaffee (Plenum, New York 1975) p. 451
1.41 G. Ehrlich: J. Phys. Chem. Solids **1**, 3 (1956); ibid. **5**, 47 (1958)
1.42 P. Kisliuk: J. Phys. Chem. Solids **3**, 95 (1957); ibid. **5**, 78 (1958)
1.43 E. Bauer: Z. Kristallogr. **110**, 372 (1958)
1.44 J.A. Venables, G.D.T. Spiller, M. Hanbuecken: Rep. Prog. Phys. **47**, 399 (1984)
1.45 S.I. Stenin: Vacuum **36**, 419 (1986)
1.46 K. Ploog: "Retrospect and Prospects of MBE", in [1.2; p. 647]
1.47 K. Ploog: J. Cryst. Growth **79**, 887 (1986)
1.48 A.Y. Cho: Thin Solid Films **100**, 291 (1983)
1.49 A.Y. Cho, J.R. Arthur: Prog. Solid State Chem. **10**, 157 (1975)
1.50 B.A. Joyce: Rep. Prog. Phys. **48**, 1637 (1985)
1.51 T. Yao: "MBE of II–VI Compounds", in [1.2; p. 313]
1.52 R.F.C. Farrow: "MBE Growth of II–VI and IV–VI Compounds and Alloys", in [1.1; p. 227]
1.53 R.F.C. Farrow: "Metallization by MBE", in [1.2; p. 387]
1.54 H.C. Casey, Jr.; A.Y. Cho: "Insulating Layers by MBE" in [1.2; p. 413]
1.55 J.C. Bean: J. Cryst. Growth **81**, 411 (1987)
1.56 Y. Shiraki: "Silicon Molecular Beam Deposition", in [1.2; p. 345]
1.57 Y. Ota: Thin Solid Films **106**, 1 (1983)
1.58 E. Kasper, J.C. Bean (eds.): *Silicon Molecular Beam Epitaxy* (CRC, Boca Raton. FL 1988)
1.59 C.H.L. Goodman, M.V. Pessa: J. Appl. Phys. **60**, R 65 (1986)
1.60 K.G. Günther: Z. Naturforsch. **13A**, 1081 (1958)
1.61 J.R. Arthur: J. Appl. Phys. **39**, 4032 (1968)
1.62 J.E. Davey, T. Pankey: J. Appl. Phys. **39**, 1941 (1968)
1.63 A. Lopez-Otero: Thin Solid Films **49**, 3 (1978)
1.64 H. Clemens, E.J. Fantner, G. Bauer: Rev. Sci. Instrum. **54**, 685 (1983)
1.65 H. Fujiyasu, A. Ishida, H. Kuwabara, S. Shimomura, S. Takaoka, K. Murase: Surf. Sci. **142**, 579 (1984)
1.66 H. Sitter: "Hot-Wall Epitaxy of IV–VI Compounds" in *Two-Dimensional Systems, Heterostructures and Superlattices,* ed. by G. Bauer, F. Kuchar, H. Heinrich, Springer Ser. Solid-State Sci., Vol. 53 (Springer, Berlin, Heidelberg 1984) p. 157
1.67 D. Schikora, H. Sitter, J. Humenberger, K. Lischka: Appl. Phys. Lett. **48**, 1276 (1986)
1.68 J.R. Arthur, J.J. LePore: J. Vac. Sci. Technol. **6**, 545 (1969)
1.69 A.Y. Cho: J. Vac. Sci. Technol. **8**, S 31 (1971)
1.70 A.Y. Cho: J. Appl. Phys. **42**, 2074 (1971)
1.71 J.R. Arthur: Surf. Sci. **43**, 449 (1974)
1.72 C.T. Foxon, M.R. Boudry, B.A. Joyce: Surf. Sci. **44**, 69 (1974)
1.73 C.T. Foxon, B.A. Joyce: Surf. Sci. **50**, 434 (1975)
1.74 C.T. Foxon, B.A. Joyce: Surf. Sci. **64**, 293 (1977)
1.75 C.T. Foxon, J.A. Harvey, B.A. Joyce: J. Phys. Chem. Solids **34**, 1693 (1973)
1.76 R.F.C. Farrow: J. Phys. D **7**, L 121 (1974)
1.77 C.T. Foxon, B.A. Joyce: J. Cryst. Growth **44**, 75 (1978)

1.78　C.T. Foxon, B.A. Joyce, M.T. Norris: J. Cryst. Growth **49**, 132 (1980)

1.79　C.T. Foxon, B.A. Joyce: "Fundamental Aspects of MBE", in *Current Topics in Materials Science,* Vol. 7, ed. by E. Kaldis (North-Holland, Amsterdam 1981) p. 1

1.80　R. Heckingbottom, G.J. Davies, K.A. Prior: Surf. Sci. **132**, 375 (1983)

1.81　J.J. Harris, D.E. Ashenford, C.T. Foxon, P.J. Dobson, B.A. Joyce: Appl. Phys. A **33**, 87 (1984)

1.82　D.A. Andrews, M.Y. Kong, R. Heckingbottom, G.J. Davies: J. Appl. Phys. **55**, 841 (1984)

1.83　K.A. Prior, G.J. Davies, R. Heckingbottom: J. Cryst. Growth **66**, 55 (1984)

1.84　D.A. Andrews, R. Heckingbottom, G.J. Davies: J. Appl. Phys. **60**, 1009 (1986)

1.85　L. Gonzales, J.B. Clegg, D. Hilton, J.P. Gowers, C.T. Foxon, B.A. Joyce: Appl. Phys. A **41**, 237 (1986)

1.86　W.T. Tsang, B. Tell, J.A. Ditzenberger, A.H. Dayem: J. Appl. Phys. **60**, 4182 (1986)

1.87　E.F. Schubert, K. Ploog: Jpn. J. Appl. Phys. **25**, 966 (1986)

1.88　K. Ploog: J. Cryst. Growth **81**, 304 (1987)

1.89　M.B. Panish: Prog. Cryst. Growth Charact. **12**, 1 (1986)

1.90　M.B. Panish: J. Cryst. Growth **81**, 249 (1987)

1.91　J.H. Neave, B.A. Joyce, P.J. Dobson, N. Norton: Appl. Phys. A **31**, 1 (1983)

1.92　J.H. Neave, B.A. Joyce, P.J. Dobson: Appl. Phys. A **34**, 179 (1984)

1.93　J.M. Van Hove, C.S. Lent, P.R. Pukite, P.I. Cohen: J. Vac. Sci. Technol. B **1**, 741 (1983)

1.94　T. Sakamoto, H. Funabashi, K. Ohta, T. Nakagawa, J.J. Kaway, T. Kojima: Jpn. J. Appl. Phys. **23**, L 657 (1984)

1.95　T. Sakamoto, H. Funabashi, K. Ohta, T. Nakagawa, N.J. Kawai, T. Kojima, Y. Bando: Superlattices and Microstructures **1**, 347 (1985)

1.96　T. Sakamoto: "RHEED oscillations in MBE and their applications to prcisely controlled crystal growth", in Proc. NATO Summer School on MBE, Ile de Bendor, June 1987

1.97　J. Nishizawa, H. Abe, T. Kurabayashi: J. Electrochem. Soc. **132**, 1197 (1985)

1.98　K. Takahashi: "Future epitaxial growth process – Photo MO MBE", in *GaAs and Related Compounds 1985*; Inst. Phys. Conf. Ser. no. 79 (Hilger, Bristol 1986) p. 73

1.99　E. Miyauchi, H. Hashimoto: J. Vac. Sci. Technol. A **4**, 933 (1986)

1.100　J. Melngailis: J. Vac. Sci. Technol. B **5**, 469 (1987)

1.101　J.P. Harbison: J. Vac. Sci. Technol. A **4**, 1033 (1986)

1.102　M.B. Panish: J. Electrochem. Soc. **127**, 2729 (1980)

1.103　M.B. Panish, H. Temkin, S. Sumski: J. Vac. Sci. Technol. B **3**, 687 (1985)

1.104　M.B. Panish, H. Temkin, R.A. Hamm, S.N.G. Chu: Appl. Phys. Lett. **49**, 164 (1986)

1.105　W.T. Tsang. Appl. Phys. Lett. **45**, 1234 (1984)

1.106　W.T. Tsang: J. Cryst. Growth **81**, 261 (1987)

1.107　E. Tokumitsu, Y. Kudou, M. Konagai, K. Takahashi: J. Appl. Phys. **55**, 3163 (1984)

1.108　E. Tokumitsu, Y. Kudou, M. Konagai, K. Takahashi: Jpn. J. Appl. Phys. **24**, 1189 (1985)

1.109　E. Tokumitsu, T. Katoh, R. Kimura, M. Konagai, K. Takahashi: Jpn. J. Appl. Phys. **25**, 1211 (1986)

1.110　N. Vogjdani, A. Lamarchand, M. Paradan: J. de Phys. Colloq. **43**, C5-339 (1982)

1.111　H. Heinecke, K. Werner, M. Weyers, H. Lueth, P. Balk: J. Cryst. Growth **81**, 270 (1987)

1.112　M.B. Panish: Private communication

1.113　G.B. Stringfellow (ed.): *Metalorganic Vapor Phase Epitaxy 1986,* Proc. 3rd Int. Conf.; J. Crystal Growth **77**, nos. 1–3 (1986)

1.114　F.J. Morris, H. Fukui: J. Vac. Sci. Technol. **11**, 506 (1974)

1.115　A.R. Calawa: Appl. Phys. Lett. **38**, 701 (1981)

1.116　W.T. Tsang: J. Appl. Phys. **58**, 1415 (1985)

1.117　V. Bostanov, R. Roussinova, E. Budevski: J. Electrochem. Soc. **119**, 1346 (1972)

1.118　Y. Namba, R.W. Vook, S.S. Chao: Surf. Sci. **109**, 320 (1981)

1.119　K.D. Gronwald, M. Henzler: Surf. Sci. **117**, 180 (1982)

1.120　J.D. Weeks, G.H. Gilmer: Adv. Chem. Phys. **40**, 157 (1979)

1.121 B.A. Joyce, P.J. Dobson, J.H. Neave, K. Woodbridge, J. Zhang, P.K. Larsen, B. Boelger: Surf. Sci. **168**, 423 (1986)
1.122 P.J. Dobson, B.A. Joyce, J.H. Neave, J. Zhang: J. Cryst. Growth **81**, 1 (1987)
1.123 T. Sakamoto, N.J. Kawai, T. Nakagawa, K. Ohta, T. Kojima, G. Hashiguchi: Surf. Sci. **174**, 65 (1986)
1.124 T. Sakamoto, T. Kawamura, S. Nago, G. Hashiguchi, K. Sakamoto, K. Kuniyoshi: J. Cryst. Growth **81**, 59 (1987)
1.125 T. Suntola, J. Hyvaerinen: Annu. Rev. Mater. Sci. **15**, 177 (1985)
1.126 M. Ahonen, M. Pessa, T. Suntola: Thin Solid Films **65**, 301 (1980)
1.127 T.A. Pakkanen, V. Nevalainen, M. Lindblad, P. Makkonen: Surf. Sci. **188**, 456 (1987)
1.128 M. Kawabe, N. Matsuura, H. Inuzuka: Jpn. J. Appl. Phys. **21**, L 447 (1982)
1.129 M. Kawabe, M. Kondo, N. Matsuura, K. Yamamoto: Jpn. J. Appl. Phys. **22**, 64 (1983)
1.130 Y. Horikoshi, M. Kawashima, H. Yamaguchi: Jpn. J. Appl. Phys. **25**, L 868 (1986)
1.131 J. Nishizawa, H. Abe, T. Kurabayashi, N. Sakurai: J. Vac. Sci. Technol. A **4**, 706 (1986)
1.132 A. Takamori, E. Miyauchi, H. Arimoto, Y. Bamba, H. Hashimoto: Jpn. J. Appl. Phys. **23**, L 599 (1984)
1.133 K. Ploog, K. Graf: *Molecular Beam Epitaxy of III–V Compounds, A Comprehensive Bibliography 1958–1983* (Springer, Berlin, Heidelberg 1984)
1.134 A. Madhukar, S.V. Ghaisas: CRC Crit. Rev. Solid State Mater. Sci. **14**, 1 (1988)

Chapter 2

2.1 W.T. Tsang: "Semiconductor Lasers and Photodetectors by Molecular Beam Epitaxy" in [2.20, 21]
2.2 K. Alavi, A.Y. Cho, F. Capasso, J. Alam: J. Vac. Sci. Technol. B **5**, 802 (1987)
2.3 H.T. Griem, K.H. Hsieh, I.J. D'Haenens, M.J. Delaney, J.A. Henige, G.H. Wicks, A.S. Brown: J. Vac. Sci. Technol. B **5**, 785 (1987)
2.4 S. Hiyamizu, T. Fujii, S. Muto, T. Inata, Y. Nakata, Y. Sugiyama, S. Sasa: J. Cryst. Growth **81**, 349 (1987)
2.5 F. Alexandre, J.C. Harmand, J.L. Lievin, C. Dubon-Chevallier, D. Ankri, C. Minot, J.F. Palmier: J. Cryst. Growth **81**, 391 (1987)
2.6 L. Goldstein, J.P. Praseuth, M.C. Joncour, J. Primot, P. Henoc, J.L. Pelouard, P. Hesto: J. Cryst. Growth **81**, 396 (1987)
2.7 N. Koguchi, T. Kiyosawa, S. Takahashi: J. Cryst. Growth **81**, 400 (1987)
2.8 P.A. Maki, S.C. Palmateer, A.R. Calawa, B.R. Lee: J. Electrochem. Soc. **132**, 2813 (1985)
2.9 P.A. Maki, S.C. Palmateer, A.R. Calawa, B.R. Lee: J. Vac. Sci. Technol. B **4**, 564 (1986)
2.10 J.H. Neave, P.K. Larsen, J.F. van der Veen, P.J. Dobson, B.A. Joyce: Surf. Sci. **133**, 267 (1983)
2.11 B.A. Joyce: Rep. Prog. Phys. **48**, 1637 (1985)
2.12 R. Chow, Y.G. Chai: J. Vac. Sci. Technol. A **1**, 49 (1983)
2.13 M.B. Panish: Prog. Cryst. Growth Charact. **12**, 1 (1986)
2.14 J.T. Cheung: Appl. Phys. Lett. **51**, 1940 (1987)
2.15 L.I. Maissel, R. Glang (eds.): *Handbook of Thin Film Technology* (McGraw-Hill, New York 1970)
2.16 H. Hertz: Ann. Phys. **17**, 177 (1882)
2.17 M. Knudsen: Ann. Phys. (Leipzig) **47**, 697 (1915)
2.18 I. Langmuir: Phys. Z. **14**, 1273 (1913)
2.19 M. Knudsen: Ann. Phys. (Leipzig) **29**, 179 (1909)
2.20 L.L. Chang, K. Ploog (eds.): *Molecular Beam Epitaxy and Heterostructures,* NATO ASI Ser., Ser. E, no. 87 (Martinus Nijhoff, Dordrecht 1985)
2.21 E.H.C. Parker (ed.): *The Technology and Physics of Molecular Beam Epitaxy* (Plenum, New York 1985)

2.22 C.T. Foxon, J.A. Harvey, B.A. Joyce: J. Phys. Chem. Solids **34**, 1693 (1973)

2.23 R.F.C. Farrow: J. Phys. D**7**, 2436 (1974)

2.24 E. Rutner: In *Condensation and Evaporation of Solids* ed. by E. Rutner, P. Goldfinger, J.P. Hirth (Gordon and Breach, New York 1964) p. 149

2.25 K. Ploog: "Molecular Beam Epitaxy of III–V Compounds", in *Crystals–Growth, Properties and Applications*, Vol. 3, ed. by H.C. Freyhardt (Springer, Berlin, Heidelberg 1980) p. 73

2.26 J.A. Curless: J. Vac. Sci. Technol. B**3**, 531 (1985)

2.27 T. Yamashita, T. Tomita, T. Sakurai: Jpn. J. Appl. Phys. **26**, 1192 (1987)

2.28 P. Clausing: Z. Phys. **66**, 471 (1930)

2.29 P. Clausing: Ann. Phys. (Leipzig) **12**, 961 (1932)

2.30 W.L. Winterbottom, J.P. Hirth: J. Chem. Phys. **37**, 784 (1962)

2.31 V. Ruth, J.P. Hirth: "The Angular Distribution of Vapor Flowing from a Knudsen Cell", in *Condensation and Evaporation of Solids*, ed. by E. Rutner, P. Goldfinger, J.P. Hirth (Gordon and Breach, New York 1964) p. 99

2.32 K. Motzfeldt: J. Phys. Chem. **59**, 139 (1955)

2.33 B.B. Dayton: "Gas Glow Patterns at Entrance and Exit of Cylindrical Tubes", in *1956 National Symposium on Vacuum Technology Transactions*, ed. by E.S. Perry, J.H. Durant (Pergamon, Oxford 1957) p. 5

2.34 M.A. Herman: Vacuum **32**, 555 (1982)

2.35 P.E. Luscher, D.M. Collins: Prog. Cryst. Growth Charact. **2**, 15 (1979)

2.36 W.H. Bröhl, H. Hartmann: Vacuum **31**, 117 (1981)

2.37 N.F. Ramsey: *Molecular Beams* (Oxford University Press, Oxford 1956)

2.38 W. Steckelmacher: Rep. Prog. Phys. **49**, 1083 (1986)

2.39 P. Krasuski: J. Vac. Sci. Technol. A**5**, 2488 (1987)

2.40 J. Humenberger, H. Sitter: Proc. 7th Int. Conf. Thin Films, New Delhi 1987, to be published in Thin Solid Films (1988)

2.41 J.K. Haviland, M.L. Levin: Phys. Fluids **5**, 1399 (1962)

2.42 D.L. Miller, G.J. Sullivan: J. Vac. Sci. Technol. B**5**, 1377 (1987)

2.43 L.Y.L. Shen: J. Vac. Sci. Technol. **15**, 10 (1978)

2.44 G.J. Davies, D. Williams: "III–V MBE Growth Systems", in [2.21; p. 15]

2.45 R.A.A. Kubiak, E.H.C. Parker, S.S. Lyer: "Si-MBE Growth Systems – Technology and Practice", in *Silicon Molecular Beam Epitaxy*, ed. by E. Kasper, J.C. Bean (CRC, Boca Raton, FL 1988) Chap. 2

2.46 J.T. Cheung, J. Madden: J. Vac. Sci. Technol. B**5**, 705 (1987)

2.47 R.J. Malik: J. Vac. Sci. Technol. B**5**, 722 (1987)

2.48 R.L. Lee, W.J. Schaffer, Y.G. Chai, D. Liu, J.S. Harris: J. Vac. Sci. Technol. B**4**, 568 (1986)

2.49 B.S. Krusor, R.Z. Bachrach: J. Vac. Sci. Technol. B**1**, 138 (1983)

2.50 C.R. Stanley, R.F.C. Farrow, P.W. Sullivan: "MBE of InP and other P-containing Compounds", in [2.21; Chap. 9]

2.51 J.B. Clegg, F. Grainger, I.G. Gale: J. Mater. Sci. **15**, 747 (1980)

2.52 R.F.C. Farrow, G.M. Williams: Thin Solid Films **55**, 303 (1978)

2.53 V.A. Borodin, V.V. Sidorov, T.A. Steriopolo, V.A. Tatarchenko: J. Cryst. Growth **82**, 89 (1987)

2.54 C. Chatillon, M. Allibert, A. Pattoret: Adv. Mass Spectrosc. **7A**, 615 (1978)

2.55 T.H. Myers, J.F. Schetzina: J. Vac. Sci. Technol. **20**, 134 (1982)

2.56 S.C. Jackson, B.N. Baron, R.E. Rocheleau, T.W.F. Russell: J. Vac. Sci. Technol. A**3**, 1916 (1985)

2.57 J.P. Faurie: J. Cryst. Growth **81**, 483 (1987)

2.58 H.F. Schaake (ed.): Proc. 1986 U.S. Workshop on the Phys. Chem. Mercury Cadmium Telluride, in J. Vac. Sci. Technol. A**5**, no. 5 (1987)

2.59 M.A. Herman, M. Pessa: J. Appl. Phys. **57**, 2671 (1985)

2.60 R.F.C. Farrow, G.R. Jones, G.M. Williams, P.W. Sullivan, W.J.O. Boyle, J.T.M. Wotherspoon: J. Phys. D**12**, L117 (1979)

2.61 T. Yao: "MBE of II–VI Compounds", in [2.21; Chap. 10]

2.62 J.M. Arias, S.H. Shin, J.T. Cheung, J.S. Chen, S. Sivananthan, J. Reno, J.P. Faurie: J. Vac. Sci. Technol. A**5**, 3133 (1987)

2.63 K.A. Harris, S. Hwang, D.K. Blanks, J.W. Cook Jr., J.F. Schetzina, N. Otsuka: J. Vac. Sci. Technol. A **4**, 2061 (1986)

2.64 M.B. Panish, R.A. Hamm: J. Cryst. Growth **78**, 445 (1986)

2.65 R.F.C. Farrow, P.W. Sullivan, G.M. Williams, C.R. Stanley: Collected Papers of 2nd Int. Symp. MBE and Related Clean Surface Techniques, Tokyo 1982 (Jpn. Soc. Appl. Phys., Tokyo 1982) p. 169

2.66 T. Henderson, W. Kopp, R. Fischer, J. Klem, H. Morkoç, L.P. Erickson, P.W. Palmberg: Rev. Sci. Instrum. **55**, 11 (1984)

2.67 D. Huet, M. Lambert, D. Bonnevie, D. Dufresne: J. Vac. Sci. Technol. B **3**, 823 (1985)

2.68 L.P. Erickson, T.J. Mattord, P.W. Palmberg, R. Fischer, H. Morkoç: Electron. Lett. **19**, 632 (1983)

2.69 M.B. Panish, S. Sumski: J. Appl. Phys. **55**, 3517 (1984)

2.70 L.W. Kapitan, C.W. Litton, G.C. Clark, P.C. Colter: J. Vac. Sci. Technol. B **2**, 280 (1984)

2.71 J.C. Garcia, A. Barski, J.P. Contour, J. Massies: Appl. Phys. Lett. **51**, 593 (1987)

2.72 E. Kasper: Appl. Phys. A **28**, 129 (1982)

2.73 U. König, H. Kibbel, E. Kasper: J. Vac. Sci. Technol. **16**, 985 (1979)

2.74 Y. Ota: Thin Solid Films **106**, 3 (1983)

2.75 Y. Ota: J. Electrochem. Soc. **124**, 1795 (1977)

2.76 J.C. Bean, J.M. Poate: Appl. Phys. Lett. **37**, 643 (1980)

2.77 R.T. Tung, J.M. Poate, J.C. Bean, J.M. Gibson, D.C. Jacobson: Thin Solid Films **93**, 77 (1982)

2.78 M. Tacano, Y. Sugiyama, M. Ogura, M. Kawashima: Collected Papers of 2nd Int. Symp. MBE and Related Clean Surface Techniques, Tokyo 1982 (Jpn. Soc. Appl. Phys., Tokyo 1982) p. 125

2.79 L. Ramberg, E. Flemming, T.G. Andersson: J. Vac. Sci. Technol. A **4**, 141 (1986)

2.80 J.T. Cheung, G. Niizawa, J. Moyle, N.P. Ong, B.M. Paine, T. Vreeland, Jr.: J. Vac. Sci. Technol. A **4**, 2086 (1986)

2.81 F.J. Morris, H. Fukui: J. Vac. Sci. Technol. **11**, 506 (1974)

2.82 M.B. Panish: J. Electrochem. Soc. **127**, 2729 (1980)

2.83 W.T. Tsang. J. Eectron. Mater. **15**, 235 (1986)

2.84 H. Ando, A. Taike, R. Kimura, M. Konagai, K. Takahashi: Jpn. J. Appl. Phys. **25**, L279 (1986)

2.85 M.B. Panish: J. Cryst. Growth **81**, 249 (1987)

2.86 A.R. Calawa: Appl. Phys. Lett. **38**, 701 (1981)

2.87 L.M. Fraas, P.S. McLeod, R.E. Weiss, L.D. Partain, J.A. Cape: J. Appl. Phys. **62**, 299 (1987)

2.88 R.D. Dupuis: Science **226**, 623 (1984)

Chapter 3

3.1 G.J. Davies, D. Williams: "III–V MBE Growth Systems", in *The Technology and Physics of Molecular Beam Epitaxy*, ed. by E.H.C. Parker (Plenum, New York 1985)

3.2 *Products for the Semiconductor Industry,* a VG Instruments advertisement pamphlet (1985)

3.3 *MBE,* an ISA Riber advertisement pamphlet (1987)

3.4 D.E. Mars, J.N. Miller: J. Vac. Sci. Technol. B **4**, 571 (1986)

3.5 J. Massies, J.P. Contour: Jpn. J. Appl. Phys. **26**, L38 (1987)

3.6 W.M. Lau, R.N.S. Sodhi, S. Ingrey: Appl. Phys. Lett. **52**, 386 (1988)

3.7 *High Yield GaAs/AlGaAs MBE with the V80H,* a VG Semicon advertisement pamphlet (1986)

3.8 *Gas Source MBE in the V80H,* a VG Semicon advertisement pamphlet (1987); ISA Riber: Private communication

3.9 B. Bölger, P.K. Larsen: Rev. Sci. Instrum. **57**, 1363 (1986)

3.10 H. Marten, G. Meyer-Ehmsen: Surf. Sci. **151**, 570 (1985)

3.11 B.A. Joyce, J.H. Neave, P.J. Dobson, P.K. Larsen: Phys. Rev. B **29**, 814 (1984)

3.12 K. Heinz, K. Müller: In *Structural Studies of Surfaces,* Springer Tracts Mod. Phys. Vol. 91 (Springer, Berlin, Heidelberg 1982) p. 91
3.13 F. Jona, J.A. Strozier, W.S. Yang: Rep. Prog. Phys. **45**, 527 (1982)
3.14 *HWBE 2500*, an advertisement leaflet of TOPLAB a division of Hainzl Industriesysteme (1987)
3.15 A. Lopez-Otero: Thin Solid Films **49**, 3 (1978)
3.16 J. Humenberger: Z. Kristallogr., in press
3.17 J. Melngailis: J. Vac. Sci. Technol. B **5**, 469 (1987)
3.18 E. Miyauchi, H. Hashimoto: J. Vac. Sci. Technol. A **4**, 933 (1986)
3.19 E. Miyauchi, H. Arimoto, H. Hashimoto, T. Utsumi: J. Vac. Sci. Technol. B **1**, 1113 (1983)
3.20 R.L. Seliger: J. Appl. Phys. **43**, 2352 (1972)
3.21 W.T. Tsang, A.Y. Cho: Appl. Phys. Lett. **30**, 293 (1977)
3.22 Y.C. Lin, A.R. Neureuther, W.G. Oldham: J. Electrochem. Soc. **130**, 939 (1983)
3.23 M.A. Hasan, J. Knall, S.A. Barnett, A. Rockett, J.E. Sundgren, J.E. Greene: J. Vac. Sci. Technol. B **5**, 1332 (1987)
3.24 C. Lejeune, G. Gautherin: Vacuum **34**, 251 (1984)

Chapter 4

4.1 S. Ino: Jpn. J. Appl. Phys. **16**, 891 (1977)
4.2 P.A. Maksym, J.L. Beeby: Surf. Sci. **110**, 423 (1981)
4.3 J.M. Van Hove, P.I. Cohen: J. Vac. Sci. Technol. **20**, 726 (1982)
4.4 P.J. Dobson, J.H. Neave, B.A. Joyce: Surf. Sci. **119**, L339 (1982)
4.5 J.M. Van Hove, P.I. Cohen, C.S. Lent: J. Vac. Sci. Technol. A **1**, 546 (1983)
4.6 J.H. Neave, B.A. Joyce, P.J. Dobson, N. Norton: Appl. Phys. A **31**, 1 (1983)
4.7 J.M. Van Hove, C.S. Lent, P.R. Pukite, P.I. Cohen: J. Vac. Sci. Technol. B **1**, 741 (1983)
4.8 J.H. Neave, B.A. Joyce, P.J. Dobson, P.K. Larsen: Phys. Rev. B **29**, 814 (1984)
4.9 J.H. Neave, B.A. Joyce, P.J. Dobson: Appl. Phys. A **34**, 179 (1984)
4.10 L. Däweritz, R. Hey, H. Berger: Thin Solid Films **116**, 165 (1984)
4.11 C.S. Lent, P.I. Cohen: Surf. Sci. **139**, 121 (1984)
4.12 P.R. Pukite, C.S. Lent, P.I. Cohen: Surf. Sci. **161**, 39 (1985)
4.13 T. Kawamura, P.A. Maksym: Surf. Sci. **161**, 12 (1985)
4.14 P.K. Larsen, P.J. Dobson, J.H. Neave, B.A. Joyce, B. Bölger, J. Zhang: Surf. Sci. **169**, 176 (1986)
4.15 B.A. Joyce, P.J. Dobson, J.H. Neave, K. Woodbridge, J. Zhang, P.K. Larsen, B. Bölger: Surf. Sci. **168**, 423 (1986)
4.16 T. Kawamura, T. Sakamoto, K. Ohta: Surf. Sci. **171**, L409 (1986)
4.17 T. Sakamoto, N.J. Kawai, T. Nakagawa, K. Ohta, T. Kojima, G. Hashiguchi: Surf. Sci. **174**, 651 (1986)
4.18 T. Sakamoto, G. Hashiguchi: Jpn. J. Appl. Phys. **25**, L78 (1986)
4.19 T. Sakamoto, T. Kawamura, G. Hashiguchi: Appl. Phys. Lett. **48**, 1612 (1986)
4.20 P.K. Larsen, G. Meyer-Ehmsen, B. Bölger, A.J. Hoeven: J. Vac. Sci. Technol. A **5**, 611 (1987)
4.21 A.V. Rzhanov, L.A. Iljina (eds.): *Ellipsometry – Theory, Methods and Applications* (Nauka, Novosibirsk 1987) (in Russian)
4.22 Y. Demay, J.P. Gailliard, P. Medina: J. Cryst. Growth **81**, 97 (1987)
4.23 Y. Demay, D. Arnoult, J.P. Gailliard, P. Medina: J. Vac. Sci. Technol. A **5**, 3139 (1987)
4.24 R.W. Collins, J.M. Cavese: J. Appl. Phys. **62**, 4146 (1987)
4.25 S.I. Stenin: Vacuum **36**, 419 (1986)
4.26 J.N. Eckstein, C. Webb, S.L. Weng, K.A. Bertness: Appl. Phys. Lett. **51**, 1833 (1987)
4.27 R.H. Williams, G.P. Srivastava, I.T. McGovern: Rep. Prog. Phys. **43**, 1357 (1980)
4.28 R. Memeo, F. Ciccacci, C. Mariani, S. Ossicini: Thin Solid Films **109**, 159 (1983)
4.29 G.E. McGuire, P.H. Holloway: "Application of Auger Spectroscopy in Materials Analysis", in *Electron Spectroscopy – Theory, Techniques and Applications*, Vol. 4, ed. by C.R. Brundle, A.D. Baker (Academic, London 1981) p. 1

4.30 D. Briggs, M.P. Seah (eds.): *Practical Surface Analysis by Auger and X-ray Photoelectron Spectroscopy* (Wiley, Chichester 1983)
4.31 R.K. Wild: Vacuum **31**, 183 (1981)
4.32 M.P. Seah: Vacuum **34**, 453 (1984)
4.33 A.W. Czanderna (ed.): *Methods of Surface Analysis* (Elsevier, New York 1975)
4.34 A. Browh, J.C. Vickerman: "Seconary Ion Mass Spectrometry", in Secondary Ion Mass Spectrometry, SIMS V, ed. by A. Benninghoven, R.J. Colton, D.S. Simons, H.W. Werner, Springer Ser. Chem. Phys. Vol. 144 (Springer, Berlin, Heidelberg 1986) p. 222
4.35 D. Briggs (ed.): *Handbook of X-Ray and Ultraviolet Spectroscopy* (Heyden, London 1977)
4.36 G. Margaritondo: Surf. Sci. **132**, 468 (1983); ibid. **168**, 439 (1986)
4.37 A.D. Katnani, G. Margaritondo: J. Appl. Phys. **54**, 2522 (1983)
4.38 R.S. Bauer, P. Zurcher, H.W. Sang: Appl. Phys. Lett. **43**, 663 (1983)
4.39 E.A. Kraut, R.W. Grant, J.R. Waldrop, S.P. Kowalczyk: Phys. Rev. Lett. **44**, 1620 (1980)
4.40 S. Mader: "Determination of Structures in Films", in *Handbook of Thin Film Technology*, ed. by L.I. Maissel, R. Glang (McGraw-Hill, New York 1970) Chap. 9
4.41 C.W. Oatley, W.C. Nixon, R.F.W. Pease: Adv.Electron. Electron Phys. **21**, 181 (1965)
4.42 P.J. Dobson, B.A. Joyce, J.H. Neave, J. Zhang: J. Cryst. Growth **81**, 1 (1987)
4.43 E. Bauer: "Reflection Electron Diffraction", in *Techniques of Metals Research,* Vol. 2, ed. by R.F.Bunshah (Wiley-Interscience, New York 1969) Chap. 15
4.44 B.A. Joyce, P.J. Dobson, J.H. Neave, J. Zhang: Surf. Sci. **178**, 110 (1986)
4.45 B.K. Vainshtein: *Analysis by Electron Diffraction* (Pergamon, Oxford 1964)
4.46 A.Y. Cho: Thin Solid Films **100**, 291 (1983)
4.47 P.K. Larsen: "RHEED and Photoemission Studies of Semiconductors Grown in situ by MBE", in *Dynamical Phenomena at Surfaces, Interfaces and Superlattices,* ed. by F. Nizzoli, K.-H. Rieder, R.F. Willis, Springer Ser. Surf. Sci., Vol. 3 (Springer, Berlin, Heidelberg 1985) p. 196
4.48 K. Britze, G. Meyer-Ehmsen: Surf. Sci. **67**, 358 (1977)
4.49 H. Marten, G. Meyer-Ehmsen: Surf. Sci. **151**, 570 (1985)
4.50 A. Kahn: Surf. Sci. Rep. **3**, 193 (1983)
4.51 K.H. Rieder: "Structural Determination of Surface and Overlayers with Diffraction Methods", in *Dynamical Phenomena at Surfaces, Interfaces and Superlattices,* ed. by F. Nizzoli, K.-H. Rieder, R.F. Willis, Springer Ser. Surf. Sci., Vol. 3 (Springer, Berlin, Heidelberg 1985) p. 2
4.52 D.J. Chadi: Vacuum **33**, 613 (1983)
4.53 E.A. Wood: J. Appl. Phys. **35**, 1306 (1964)
4.54 J.B. Pendry: *Low Energy Electron Diffraction* (Academic, London 1980)
4.55 A.Y. Cho: J. Appl. Phys. **41**, 2780 (1970)
4.56 A.Y. Cho: J. Appl. Phys. **42**, 2074 (1971)
4.57 A.Y. Cho, I. Hayashi: J. Appl. Phys. **42**, 4422 (1971)
4.58 P.K. Larsen, J.H. Neave, J.F. van der Veen, P.J. Dobson, B.A. Joyce: Phys. Rev. B **27**, 4966 (1983)
4.59 J. Massies, P. Devoldere, N.T. Linh: J. Vac. Sci. Technol. **16**, 1244 (1979)
4.60 J. Massies, P. Etienne, F. Dezaly, N.T. Linh: Surf. Sci. **99**, 121 (1980)
4.61 J.H. Neave, B.A. Joyce: J. Cryst. Growth **44**, 387 (1978)
4.62 A.Y. Cho: J. Appl. Phys. **47**, 2841 (1976)
4.63 R. Ludeke: IBM J. Res. Dev. **22**, 304 (1978)
4.64 R. Ludeke, R.M. King, E.H.C. Parker: "MBE Surface and Interface Studies", in *The Technology and Physics of Molecular Beam Epitaxy,* ed. by E.H.C. Parker (Plenum, New York 1985) Chap. 16
4.65 B.A. Joyce: Rep. Prog. Phys. **48**, 1637 (1985)
4.66 K. Woodbridge, J.P. Gowers, P.F. Fewster, J.H. Neave, B.A. Joyce: J. Cryst. Growth **81**, 224 (1987)
4.67 C.T. Foxon, B.A. Joyce: J. Cryst. Growth **45**, 75 (1978)
4.68 B.T. Meggitt, E.H.C. Parker, R.M. King: Appl. Phys. Lett. **33**, 528 (1978)

4.69 B.T. Meggitt, E.H.C. Parker, R.M. King, J.D. Drange: J. Cryst. Growth **50**, 538 (1980)

4.70 R.F.C. Farrow: J. Phys. D **7**, L121 (1974)

4.71 C.R. Stanley, R.F.C. Farrow, P.W. Sullivan: "MBE of InP and Other P-containing Compounds", in *The Technology and Physics of Molecular Beam Epitaxy,* ed. by E.H.C. Parker (Plenum, New York 1985) Chap. 9

4.72 A.Y. Cho, J.R. Arthur: Prog. Solid State Chem. **10**, 157 (1975)

4.73 K. Britze, G. Meyer-Ehmsen: Surf. Sci. **77**, 131 (1978)

4.74 J. Zhang, J.H. Neave, P.J. Dobson, B.A. Joyce: Appl. Phys. A **42**, 317 (1987)

4.75 A. Ichimiya, K. Kambe, G. Lehmpfuhl: J. Phys. Soc. Jpn. **49**, 684 (1980)

4.76 T. Sakamoto, H. Funabashi, K. Ohta, T. Nakagawa, N.J. Kawai, T. Kojima: Jpn. J. Appl. Phys. **23**, L657 (1984)

4.77 J.H. Neave, P.J. Dobson, B.A. Joyce, J. Zhang: Appl. Phys. Lett. **47**, 400 (1985)

4.78 T. Nishinaga, K.I. Cho: Jpn. J. Appl. Phys. **27**, L12 (1988)

4.79 J.M. Van Hove, P.I. Cohen: J. Cryst. Growth **81**, 13 (1987)

4.80 P. Chen, A. Madhukar, J.Y. Kim, N.M. Cho: in Proc. 18th Int. Conf. Phys. Semicond., Stockholm 1986, ed. by O. Engström (World Scientific, Singapore 1987) p.109 ,

4.81 B.F. Lewis, T.C. Lee, F.J. Grunthaner, A. Madhukar, R. Fernandez, J. Maserjian: J. Vac. Sci. Technol. B **2**, 419 (1984)

4.82 B.F. Lewis, F.J. Grunthaner, A. Madhukar, T.C. Lee, R. Fernandez: J. Vac. Sci. Technol. B **3**, 1317 (1985)

4.83 P. Chen, A. Madhukar, J.Y. Kim, T.C. Lee: Appl. Phys. Lett. **48**, 650 (1986)

4.84 M.Y. Yen, T.C. Lee, P. Chen, A. Madhukar: J. Vac. Sci. Technol. B **4**, 590 (1986)

4.85 T.C. Lee, M.Y. Yen, P. Chen, A. Madhukar: J. Vac. Sci. Technol. A **4**, 884 (1986)

4.86 T.C. Lee, M.Y. Yen, P. Chen, A. Madhukar: Surf. Sci. **174**, 55 (1986)

4.87 P. Chen, J.Y. Kim, A. Madhukar, T.C. Cho: J. Vac. Sci. Technol. B **4**, 890 (1986)

4.88 N.M. Cho, P. Chen, A. Madhukar: Appl. Phys. Lett. **50**, 1909 (1987)

4.89 K. Sakamoto, T. Sakamoto, S. Nagao, G. Hashiguchi, K. Kuniyoshi, Y. Bando: Jpn. J. Appl. Phys. **26**, 666 (1987)

4.90 P.R. Berger, P.K. Bhattacharya, J. Singh: J. Appl. Phys. **61**, 2856 (1987)

4.91 J.M.Van Hove, P.R. Pukite, P.I. Cohen: J. Vac. Sci. Technol. B **3**, 563 (1985)

4.92 Y. Iimura, M.Kawabe: Jpn. J. Appl. Phys. **25**, L81 (1986)

4.93 H. Sakaki, M. Tanaka, J. Yoshino: Jpn. J. Appl. Phys. **24**, L417 (1985)

4.94 M. Tanaka, H. Sakaki, J. Yoshino, T. Furuta: Surf. Sci. **174**, 65 (1986)

4.95 M. Tanaka, H. Sakaki, J. Yoshino: Jpn. J. Appl. Phys. **25**, L155 (1986)

4.96 M. Tanaka, H. Sakaki: J. Cryst. Growth **81**, 153 (1987)

4.97 T. Hayakawa, T. Suyama, K. Takahashi, K. Kondo, S. Yamamoto, S. Yano, T. Hijikata: Appl. Phys. Lett. **47**, 952 (1985)

4.98 D. Bimberg, D. Mars, J.N. Miller, R. Bauer, D. Ortel, J. Christen: Superlattices and Microstructures **3**, 79 (1987)

4.99 C.W. Tu, R.C. Miller, B.A. Wilson, P.M. Petroff, T.D. Harris, R.F. Kopf, S.K. Sputz, M.G. Lamont: J. Cryst. Growth **81**, 159 (1987)

4.100 F.Y. Juang, P.K. Bhattacharya, J. Singh: Appl. Phys. Lett. **48**, 290 (1986)

4.101 F. Briones, D. Golmayo, L. Gonzalez, J.L. DeMiguel: Jpn. J. Appl. Phys. **24**, L478 (1985)

4.102 L. Däweritz: Surf. Sci. **118**, 585 (1982)

4.103 L. Däweritz: Surf. Sci. **160**, 171 (1985)

4.104 B.A. Joyce, P.J. Dobson, J.H. Neave, J. Zhang: In *Two-Dimensional Systems: Physics and New Devices,* ed. by G. Bauer, F. Kuchar, H. Heinrich, Springer Ser. Solid-State Sci., Vol. 67 (Springer, Berlin, Heidelberg 1986) p.42

4.105 T.Sakamoto, N.J. Kawai, T. Nakagawa, K. Ohta, T. Kojima: Appl. Phys. Lett. **47**, 167 (1985)

4.106 J. Aarts, W.M. Gerits, P.K. Larsen: Appl. Phys. Lett. **48**, 931 (1986)

4.107 P.K. Larsen, P.J. Dobson (eds.): *Reflection High Energy Electron Diffraction and Reflection Electron Imaging of Surfaces,* Proc. NATO-ARW (Plenum, New York 1988)

4.108 D.E.Aspnes: "Spectroscopic Ellipsometry of Solids", in *Optical Porperties of Solids – New Developments,* ed. by B.O. Seraphin (North-Holland, Amsterdam 1976) p.799

4.109 D.E. Aspnes: Proc. Soc. Photo-Opt. Instrum. Eng. **276**, 188 (1981)
4.110 F.L. McCrackin, E. Passaglia, R.R. Stromberg, H.L. Steinberg: J. Res. Natl. Bur. Stand. **67A**, 363 (1963)
4.111 P.J. McMarr, K. Vedam, J. Narayan: J. Appl. Phys. **59**, 694 (1986)
4.112 M. Oikkonen: J. Appl. Phys. **62**, 1385 (1987)
4.113 R.M.A. Azzam, N.M. Bashara: *Ellipsometry and Polarized Light* (North-Holland, Amsterdam 1977)
4.114 D.E. Aspnes, S.M. Kelso, R.A. Logan, R. Bhat: J. Appl. Phys. **60**, 754 (1986)
4.115 J.R. Meyer-Arendt: *Introduction to Classical and Modern Optics* (Prentice-Hall, Englewood Cliffs, NJ 1972) p. 228
4.116 A.V. Archipenko, Yu.A. Blyumkina, M.A. Lamin, O.P. Pcheljakov, L.V.Sokolov, S.I. Stenin, N.I. Kozlov, A.A. Kroshkov, A.V. Rzhanov: Poverkhnost, No. 1, 93 (1985)
4.117 B.Z. Olshanetsky, A.A. Shklyaev: Surf. Sci. **82**, 445 (1979)
4.118 O.P.Pcheljakov, Yu.A. Blyumkina, S.I. Stenin: Poverkhnost, No. 1, 147 (1982)
4.119 D.E. Aspnes, A.A. Studna: Appl. Opt. **14**, 220 (1975)

Chapter 5

5.1   J.C. Campbell, W.T. Tsang, G.J. Qua, J.E. Bowers: Appl. Phys. Lett. **51**, 1454 (1987)
5.2   T.C. Chong, C.G. Fonstad: Appl. Phys. Lett. **51**, 221 (1987)
5.3   M. Sze: *Physics of Semiconductor Devices* (Wiley-Interscience, New York 1969)
5.4   G.L. Miller, D.V. Lang, L.C. Kimerling: Annu. Rev. Mater. Sci. **7**, 377 (1977)
5.5   T.A. Carlson: *Photoelectron and Auger Spectroscopy* (Plenum, New York 1975)
5.6   J.I. Hanoka, R.O. Bell: Annu. Rev. Mater. Sci. **11**, 353 (1981)
5.7   C.R. Brundle, A.D. Baker (eds.): *Electron Spectroscopy: Theory, Techniques and Applications* (Academic, New York 1981)
5.8   D.E. Aspnes: "Spectroscopic Ellipsometry of Solids", in *Optical Properties of Solids – New Developments,* ed. by B.O. Seraphin (North-Holland, Amsterdam 1976) p. 799
5.9   H.K. Herglotz, L.S. Birks: *X-ray Spectrometry* (Marcel Dekker, New York 1978)
5.10   E.P. Bertin: "The Electron-Probe Microanalyzer", in *Principles and Practice of X-Ray Spectrometric Analysis* (Plenum, New York 1970) Chap. 21
5.11   K. Sangwal: *Etching of Crystals,* ed. by S. Amelinckx, J. Nihoul (North-Holland, Amsterdam 1987)
5.12   D. Briggs, M.P. Seah (eds.): *Practical Surface Analysis by Auger and X-ray Photoelectron Spectroscopy* (Wiley, New York 1983)
5.13   E.H. Putley: *The Hall Effect and Related Phenomena* (Butterworth, London 1960)
5.14   D. Williams: "Molecular Spectroscopy – Infrared Region", in *Methods of Experimental Physics*, Vol. 13B, Spectroscopy, ed. by D. Williams (Academic, New York 1976) p. 2ff
5.15   M.D. Lumb: *Luminescence Spectroscopy* (Academic, New York 1978)
5.16   E. Bauer: "Reflection Electron Diffraction", in *Techniques of Metals Research*, Vol. 2, Part 2, ed. by R.F. Bunshah (Interscience, New York 1969) Chap. 15
5.17   W.K. Chu, J.W. Mayer, M.A. Nicolet: *Backscattering Spectrometry* (Academic, New York 1978)
5.18   L.E. Murr: *Electron and Ion Microscopy and Microanalysis* (Marcel Dekker, New York 1982)
5.19   J.A. McHugh: "Secondary Ion Mass Spectrometry", in *Methods of Surface Analysis,* ed. by A.W. Czanderna (Elsevier, New York 1975) Chap. 6
5.20   R.D. Heidenreich: *Fundamentals of Transmission Electron Microscopy* (Wiley-Interscience, New York 1964)
5.21   D. Briggs (ed.): *Handbook of X-ray and Ultraviolet Photoelectron Spectroscopy* (Heydon, London 1977, 1978)
5.22   B.D. Cullity: *Elements of X-ray Diffraction* (Addison-Wesley, Reading, MA 1967)
5.23   M.P. Seah: Vacuum **34**, 453 (1984)
5.24   R.K. Wild: Vacuum **31**, 183 (1981)
5.25   K.D. Sevier: *Low Energy Electron Spectroscopy* (Interscience, New York 1972)

5.26 C.R. Worthington, S.G. Tomlin: Proc. Phys. Soc., London A 69, 401 (1956)
5.27 P.W. Palmberg, T.N. Rhodin: J. Appl. Phys. 39, 2425 (1968)
5.28 P.W. Palmberg, F.K. Bohr, J.C. Tracy: Appl. Phys. Lett. 15, 254 (1969)
5.29 M.P. Seah, W.A. Dench: Surf. Interface Anal. 1, 2 (1979)
5.30 D.R. Penn: J. Electron. Spectrosc. 9, 29 (1976)
5.31 S. Tanuma, C.J. Powell, D.R. Penn: Surf. Sci. 192, 849 (1987)
5.32 P.M. Hall, J.M. Morabito: Surf. Sci. 83, 391 (1979)
5.33 S. Ichimura, R. Shimizu: Surf. Sci. 112, 386 (1981)
5.34 L.E.Davies, N.C. McDonald, P.W. Palmberg, G.E. Riach, R.E. Weber: *Handbook of Auger Electron Spectroscopy*, 2nd ed. (Physical Electronics Industries, Eden Prairie 1976)
5.35 J. Kempf:Surf. Interface Anal. 4, 116 (1982)
5.36 R. Memeo, F. Ciccacci, C. Mariani, S. Ossicini: Thin Solid Films 109, 159 (1983)
5.37 P. Sigmund: Phys. Rev. 184, 383 (1969); ibid. 187, 768 (1969)
5.38 M.P. Seah, C. Lea: Thin Solid Films 81, 257 (1981)
5.39 M.P. Seah, C.P. Hunt: Suf. Interface Anal. 5, 33 (1983)
5.40 C.G. Pantano, T.E. Madey: Appl. Surf. Sci. 7, 115 (1981)
5.41 H.H. Andersen: Appl. Phys. 18, 131 (1979)
5.42 V. Littmark, W.O. Hofer: Nucl. Instrum. Methods 168, 329 (1980)
5.43 A. Benninghoven: Z. Phys. 230, 403 (1971)
5.44 M.P. Seah, J.M. Sanz, S. Hofmann: Thin Solid Films 81, 239 (1981)
5.45 R.Pepinsky, V. Vand: In *Handbook of Physics,* 2nd ed., ed. by E.U.Condon, H. Odishawleds (McGraw-Hill, New York 1967) Chap. 1
5.46 F.C. Brown: In *The Physics of Solids – Ionic Crystals, Lattice Vibrations and Imperfections* (Benjamin, New York 1967)
5.47 J. Hornstra, W.J. Bartels: J. Cryst. Growth 44, 513 (1978)
5.48 W.J. Bartels, W. Nijman: J. Cryst. Growth 44, 518 (1978)
5.49 J. Matsui, K. Onabe, T. Kamejima, I. Hayashi: J. Electrochem. Soc. 126, 664 (1979)
5.50 E.J. Fantner, H. Clemens, G. Bauer: Adv. X-Ray Anal. 27, 171 (1984)
5.51 E.J. Fantner, G. Bauer: In *Two-Dimensional Systems, Heterostructures, and Superlattices,* ed. by G. Bauer, F. Kuchar, H. Heinrich, Springer Ser. Solid-State Sci., Vol. 53 (Springer, Berlin, Heidelberg 1984) p. 207
5.52 W.J. Bartels: J. Vac. Sci. Technol. B 1, 338 (1983)
5.53 T. Matsushita, S. Kikuta, K. Kohra: J. Phys. Soc. Jpn. 30, 1136 (1971)
5.54 W.J. Bartels: Philips Tech. Rev. 41, 183 (1983)
5.55 J.M. Vandenberg, R.A. Hamm, M.B. Panish, H. Temkin: J. Appl. Phys. 62, 1278 (1987)
5.56 J.M. Vandenberg, S.N. Chu, R.A. Hamm, M.B. Panish, H. Temkin: Appl. Phys. Lett. 49, 1302 (1986)
5.57 J.M. Vandenber, R.A. Hamm, A.T. Macrander, M.B. Panish, H. Temkin: Appl. Phys. Lett. 48, 1153 (1986)
5.58 R.W. James: *The Optical Principles of the Diffraction of X-rays* (Bell, London 1967)
5.59 S. Takagi: J. Phys. Soc. Jpn. 26, 1239 (1969)
5.60 D. Taupin: Bull. Soc. Fr. Mineral. Cristallogr. 87, 469 (1964)
5.61 D.M. Vardanyan, H.M. Manoukyan, H.M. Petrosyan: Acta Crystallogr. A 41, 212 (1985)
5.62 V.S.Speriosu, T. Vreeland, Jr.: J. Appl. Phys. 56, 1591 (1984)
5.63 W.J. Bartels, J. Hornstra, D.J.W. Lobeck: Acta Crystallogr. A 42, 539 (1986)
5.64 A. Segmüller, A.E. Blakeslee: J. Appl. Crystallogr. 6, 19 (1973)
5.65 L. Tapfer, K. Ploog: Phys. Rev. B 33, 5565 (1986)
5.66 P.J. Dean: Prog. Cryst. Growth Charact. 5, 89 (1982)
5.67 D.J. As, L. Palmetshofer: J. Appl. Phys. 62, 369 (1987)
5.68 P.J. Dean, C.H. Henry, C.J. Frosch: Phys. Rev. 168, 812 (1968)
5.69 D.Schikora, H. Sitter, J. Humenberger, K. Lischka: Appl. Phys. Lett. 48, 1276 (1986)
5.70 D.M. Larsen: Phys. Rev. B 13, 1681 (1976)
5.71 D.G. Thomas, J.J. Hopfield: Phys. Rev. 128, 2135 (1962)
5.72 J.J. Hopfield: Proc. Int. Conf. Phys. Semicond., Paris, 1964 (Dunod, Paris 1964) p. 725

5.73   P.J. Dean, H. Venghaus, J.C. Pfisten, B. Schaub, J. Marine: J. Lumin. **16**, 363 (1978)

5.74   W. Schairer, M. Schmidt: Phys. Rev. B**10**, 2501 (1974)

5.75   D.J. Ashen, P.J. Dean, D.T.J. Hurle, J.B. Mullin, A.M. White, P.D. Green: J. Phys. Chem. Solids **36**, 1041 (1975)

5.76   P.J. Dean: *Progress in Solid State Chemistry*, Vol. 8, ed. by J.O. McCladin, G. Somorjai (Pergamon, Oxford 1973) p. 1

5.77   K. Era, S. Shionoya, Y. Washizawa: J. Phys. Chem. Solids **29**, 1827 (1968)

5.78   T. Kamiya, E. Wagner: J. Appl. Phys. **48**, 1928 (1977)

5.79   P.J. Dean: J. Lumin. **18/19**, 755 (1978)

5.80   U. Heim, P. Hiesinger: Phys. Status Solidi (b) **66**, 461 (1974)

5.81   H. Künzel, K. Ploog: In *Gallium Arsenide and Related Compounds 1980*, Inst. Phys. Conf. Ser. **56**, 519 (1981)

5.82   J.K. Abrokwah, T.N. Peck, R.A. Walterson, G.E. Stillman, T.S. Low, B. Skromme: J. Electron. Mater. **12**, 681 (1983)

5.83   M. Ilegems: In *The Technology and Physics of Molecular Beam Epitaxy*, ed. by E.H.C. Parker (Plenum, New York 1985) p. 83

5.84   J.M. Langer, C. Delerue, M. Lanoo, H. Heinrich: to be published in Phys. Rev. B v. 38 (1988)

5.85   R. Dingle, R.A. Logan, J.R. Arthur: in GaAs and Related Compounds 1977; Inst. Phys. Conf. Ser. A**33**, (Inst. Phys., London 1977) p. 210

5.86   H.C. Casey, M.B. Panish: *Heterostructure Lasers* (Academic, New York 1978) p. 192, and S. Adachi: J. Appl. Phys. 58, R1 (1985)

5.87   B. Lambert, J. Caulet, A. Regreny, M. Baudet, B. Devaud, A. Chomette: Semicond. Sci. Technol. **2**, 491 (1987)

5.88   T.F. Kuech, D.J. Wolford, R. Potemski, J.A. Bradley, K.H. Kelleher, D. Yan, J.P. Farrell, P.M.S. Lesser, F. Polak: Appl. Phys. Lett. **51**, 505 (1987)

5.89   G. Oelgart, R. Schwabe, M. Heider, B. Jacobs: Semicond. Sci. Technol. **2**, 469 (1987)

5.90   N.C. Miller, S. Zemon, G.P. Werber, W. Powazinik: J. Appl. Phys. **57**, 512 (1985)

5.91   E.F. Schubert, E.O. Göbel, Y. Horikoshi, K. Ploog, H.J. Queisser: Phys. Rev. B**30**, 813 (1984)

5.92   J.M. Langer: to be published in Semicon. Sci. Technol. (1988)

5.93   Landolt-Börnstein: *Numerical Data and Functional Relationships in Sciences and Technology*, Group 3, Vol. 17, Semiconductors, Parts a, b, ed. by O. Madelung (Springer, Berlin, Heidelberg 1982)

5.94   J. Raczyńska, K. Fronc, J.M. Langer, K. Lischka, A. Pesek: to be published in Appl. Phys. Lett. (1988)

5.95   H. Heinrich, J.M. Langer: In *Festkörperprobleme*, Vol. 26, ed. by P. Grosse (Vieweg, Braunschweig 1986) p. 251

5.96   M.A. Herman: Semiconductor Superlattices (Akademie-Verlag, Berlin 1986)

5.97   R. Dingle, W. Wiegmann, C.H. Henry: Phys. Rev. Lett. **33**, 827 (1974), and R. Dingle: In *Festkörperprobleme*, Vol. 15, ed. by H.J. Queisser (Vieweg, Braunschweig 1975) p. 21

5.98   E.E. Mendez, G. Bastard, L.L. Chang, L. Esaki, H. Morkoç, R. Fischer: Phys. Rev. B**26**, 7101 (1982)

5.99   C. Weisbuch, R. Dingle. A.C. Gossard, W. Wiegmann: Solid State Commun. **38**, 709 (1981)

5.100 B.Deveaud, J.Y. Emery, A. Chomette, B. Lambert, M. Bandet: Superlattices and Microstructures **1**, 205 (1985)

5.101 J. Christen, D. Bimberg, A. Steckenborn, G. Weimann: Appl. Phys. Lett. **44**, 84 (1984)

5.102 D. Bimberg, D. Mars, J.N. Miller, R. Bauer, D. Oertel: J. Vac. Sci. Technol. B**4**, 1014 (1986)

5.103 D. Bimberg, J. Christen, T. Fukunaga, H. Nakashima, D.E. Mars, J.N. Miller: J. Vac. Sci. Technol. B**5**, 1191 (1987)

5.104 H. Brooks: In *Advances in Electronics and Electron Physics*, ed. by L. Morton (Academic, New York 1955) p. 158

5.105 H. Ehrenreich: J. Appl. Phys. **32**, 2155 (1961)
5.106 H.J. Meijer, D. Polder: Physica **19**, 255 (1953)
5.107 W.A. Harrison: Phys. Rev. **101**, 903 (1956)
5.108 G.E. Stillman, C.M. Wolf: Thin Solid Films **31**, 69 (1976)
5.109 P. Blood, J.W. Onton: Rep. Prog. Phys. **41**, 157 (1978)
5.110 A.Chandra, C.E.C. Wood, D.W. Woodard, L.F. Eastman: Solid-State Electron. **22**, 645 (1979)
5.111 H.L. Störmer: Surf. Sci. **132**, 519 (1983)
5.112 G. Weimann, W. Schlapp: Appl. Phys. Lett. **46**, 411 (1985)
5.113 K. Ploog: Extended Abstracts of the 18th Int. Conf. Solid State Devices and Materials, Tokyo, 1986, p.585
5.114 D.V. Lang: J. Appl. Phys. **45**, 3014 (1974)
5.115 H. Lefevre, M. Schulz: Appl. Phys. **12**, 45 (1977)
5.116 D.V. Lang, R.A. Logan: J. Electron. Mater. **4**, 1053 (1974)
5.117 H. Sitter, H. Heinrich, K. Lischka, A. Lopez-Otero: J. Appl. Phys. **53**, 4948 (1982)
5.118 D.V. Lang, A.Y. Cho, A.C. Gossard, M. Ilegems, W. Wiegmann: J. Appl. Phys. **47**, 2558 (1976)
5.119 J.H. Neave, P. Blood, B.A. Joyce: Appl. Phys. Lett. **36**, 311 (1980)
5.120 P. Blood, J.J. Harris: J. Appl. Phys. **56**, 993 (1984)
5.121 R.Y. DeJule, M.A. Haase, G.E. Stillman, S.C. Palmateer, J.C.M. Hwang: J. Appl. Phys. **57**, 5287 (1985)
5.122 K. Yamanaka, S. Naritsuka, K. Kanamoto, M. Mihara, M. Ishii: J. Appl. Phys. **61**, 5062 (1987)
5.123 A. Kitagawa, A. Usami, T. Wada, Y. Tokuda, H. Kano: J. Appl. Phys. **61**, 1215 (1987)
5.124 X.D. Xu, T.G. Andersson, J.M. Westin: J. Appl. Phys. **62**, 2136 (1987)
5.125 P. Pongratz, H. Sitter: J. Cryst. Growth **80**, 73 (1987)
5.126 R.B. Marcus, T.T. Sheng: *Transmission Electron Microscopy of Silicon VLSI Circuits and Structures* (Wiley-Interscience, New York 1983)
5.127 M.Y. Yen, A. Madhukar, B.F. Lewis, R. Fernandez, L. Eng, F.J. Grunthaner: Surf. Sci. **174**, 606 (1986)
5.128 M. Tanaka, H. Ichinose, T. Furuta, Y. Ishida, H. Sakaki: in Collected Papers of the 3rd Int. Conf. on Modulated Semiconductor Structures, Montpellier, France, 1987, p.85
5.129 J.M. Gibson, H.J. Grossmann, J.C. Bean, R.T. Tung, L.C. Feldman: Phys. Rev. Lett. **56**, 355 (1986)
5.130 H.L. Tsai, J.W. Lee: Appl. Phys. Lett. **51**, 130 (1987)
5.131 F.W. Saris, W.K. Chu, C.A. Chang, R. Ludeke, L. Esaki: Appl. Phys. Lett. **37**, 931 (1980)
5.132 W.K. Chu, F.W. Saris, C.A. Chang, R. Ludeke, L. Esaki: Phys. Rev. B **26**, 1999 (1982)
5.133 W.K. Chu, C.K. Pan, C.A. Chang: Phys. Rev. B **28**, 4033 (1983)
5.134 C.A. Chang, W.K. Chu: Appl. Phys. Lett. **42**, 463 (1983)
5.135 A.C. Chami, E. Ligeon, J. Fontenille, R. Danielon: J. Appl. Phys. **62**, 3718 (1987)
5.136 Y. Hishida, K. Yoneda, N. Matsunami, N. Itoh: J. Appl. Phys. **62**, 4460 (1987)
5.137 L.C. Feldman, J.W. Mayer, S.T. Picraux: *Material Analysis by Ion Channeling* (Academic, New York 1982)
5.138 G.C. Osbourn: J. Appl. Phys. **53**, 1586 (1982)
5.139 Y. Kido, J. Kawamoto: J. Appl. Phys. **61**, 956 (1987)
5.140 E. Rauhala: J. Appl. Phys. **62**, 2140 (1987)
5.141 C.K. Pan, D.C. Zheng, T.G. Finstad, W.K. Chu, V.S. Speriosu, M.A. Nicolet, J.H. Barret: Phys. Rev. B **31**, 1270 (1985)

Chapter 6

6.1 C.T.Foxon, B.A. Joyce: "Fundamental Aspects of MBE", in *Current Topics in Materials Science,* Vol. 7, ed. by E. Kaldis, (North-Holland, Amsterdam 1981) Chap.1
6.2 C.T. Foxon: J. Vac. Sci. Technol. B **1**, 293 (1983)

6.3   B.A. Joyce: Rep. Prog. Phys. **48**, 1637 (1985)
6.4   B.F. Lewis, T.C. Lee, F.J. Grunthaner, A. Madhukar, R. Fernandez, J. Maserjian: J. Vac. Sci. Technol. B**2**, 419 (1984)
6.5   B.F. Lewis, F.J. Grunthaner, A. Madhukar, T.C. Lee, R. Fernandez: J. Vac. Sci. Technol. B**3**, 1317 (1985)
6.6   T.C. Lee, M.Y. Yen, P. Chen. A. Madhukar: Surf. Sci. **174**, 55 (1986)
6.7   B.A. Joyce, P.J.Dobson, J.H. Neave, J. Zhang: Surf. Sci. **174**, 1 (1986)
6.8   B.A. Joyce, P.J. Dobson, J.H. Neave, K. Woodbridge, J. Zhang, P.K. Larsen, B. Bölger: Surf. Sci. **168**, 423 (1986)
6.9   P. Chen, J.Y. Kim, A. Madhukar, N.M. Cho: J. Vac. Sci. Technol. B**4**, 890 (1986)
6.10   B.F. Lewis, R. Fernandez, A. Madhukar, F.J. Grunthaner: J. Vac. Sci. Technol. B**4**, 560 (1986)
6.11   P.J. Dobson, B.A. Joyce, J.H. Neave, J. Zhang: J. Cryst. Growth **81**, 1 (1987)
6.12   B.A. Joyce, J. Zhang, J.H. Neave, P.J. Dobson: Appl. Phys. A **45**, 255 (1988)
6.13   T. Sakamoto: "RHEED Oscillations in MBE and Their Application to Precisely Controlled Crystal Growth", in Proc. NATO Summer School on MBE, Ile de Bendor, June 1987
6.14   A. Madhukar: Surf. Sci. **132**, 344 (1983)
6.15   A. Madhukar, S.V. Ghaisas: CRC Crit. Rev. Solid State Mater. Sci. **14**, 1 (1988)
6.16   G.M. Panchenkov, V.P. Lebedev: *Chemical Kinetics and Catalysis* (Mir, Moscow 1976)
6.17   R. Heckingbottom, G.J. Davies, K.A. Prior: Surf. Sci. **132**, 375 (1983)
6.18   M.A. Herman: Superlattices and Microstructures **2**, 345 (1986); Cryst. Res. Technol. **21**, 1413 (1986)
6.19   C. Ebner, C. Rottman, M. Wortis: Phys. Rev. B**28**, 4186 (1983)
6.20   J.W. Cahn, R. Kikuchi: Phys. Rev. B**31**, 4300 (1985)
6.21   G.B. Stringfellow: Rep. Prog. Phys. **45**, 469 (1982)
6.22   A. Zur, T.C. McGill: J. Appl. Phys. **55**, 378 (1984)
6.23   C. Hsu, S. Sivananthan, X. Chu, J.P. Faurie: Appl. Phys. Lett. **48**, 908 (1986)
6.24   J.M. Ballingall, M.L. Wroge, D.J. Leopold: Appl. Phys. Lett. **48**, 1273 (1986)
6.25   G.W. Cullen: "The Preparation and Properties of Heteroepitaxial Silicon", in *Heteroepitaxial Semiconductors for Electronic Devices,* ed. by G.W. Cullen, C.C. Wang (Springer, Berlin, Heidelberg 1978) Chap. 2
6.26   T.H. Myers, Y. Cheng, R.N. Bicknell, J.F. Schetzina: Appl. Phys. Lett. **42**, 247 (1983)
6.27   J. Yoshino, H. Munekata, L.L. Chang: J. Vac. Sci. Technol. B**5**, 683 (1987)
6.28   M.W. Geis, B.Y. Tsaur, D.C. Flanders: Appl. Phys. Lett. **41**, 526 (1982)
6.29   K. Kushida, H. Takeuchi, T. Kobayashi, K. Takagi: Appl. Phys. Lett. **48**, 764 (1986)
6.30   M.J. Stowell: "Defects in Epitaxial Deposits", in *Epitaxial Growth,* Part B, ed. by J.W. Matthews (Academic, New York 1975) Chap. 5
6.31   J.H. van der Merwe, C.A.B. Ball: "Energy of Interfaces Between Crystals", in *Epitaxial Growth,* Part B, ed. by J.W. Matthews (Academic, New York 1975) Chap. 6
6.32   J.H. van der Merwe: CRC Crit. Rev. Solid State Mater. Sci. **7**, 209 (1978)
6.33   J.W. Matthews: "Coherent Interfaces and Misfit Dislocations", in *Epitaxial Growth,* Part B, ed. by J.W.Matthews (Academic, New York 1975) Chap. 8
6.34   F.R.N. Nabarro: *Theory of Crystal Dislocations* (Clarendon, Oxford 1967)
6.35   C.B. Duke: CRC Crit. Rev. Solid State Mater. Sci. **8**, 69 (1978)
6.36   F.D. Auret, J.H. van der Merwe: Thin Solid Films **23**, 257 (1974)
6.37   F.C. Frank, J.H. van der Merwe: Proc. R. Soc., London **198**, 205, 216 (1949)
6.38   B.L. Sharma, R.K. Purohit: *Semiconductor Heterojunctions* (Pergamon, Oxford 1974)
6.39   J.W. Matthews: "Misfit Dislocations", in *Dislocations in Solids,* Vol. 2, ed. by F.R.N. Nabarro (North-Holland, Amsterdam 1979) Chap. 7
6.40   A.I. Finch, A.G. Quarrell: Proc. Phys. Soc., London **48**, 148 (1934)
6.41   W.A. Jesser, D. Kehlmann-Wilsdorf: Phys. Status Solidi **19**, 95 (1967)
6.42   J.W. Matthews, A.E. Blakeslee: J. Cryst. Growth **27**, 118 (1974)
6.43   T.G. Andersson, Z.G. Chen, V.D. Kulakovskii, A. Uddin, J.T. Vallin: Appl. Phys. Lett. **51**, 752 (1987)
6.44   M. Kobayashi, M. Konagai, K. Takahashi, K. Urabe: J. Appl. Phys. **61**, 1015 (1987)
6.45   E. Kasper, H.J. Herzog, H. Daembkes, G. Abstreiter: Mat. Res. Soc. Symp. Proc. **56**, 347 (1986)

6.46   E. Kasper, H.J. Herzog, H. Jorke, G. Abstreiter: Superlattices and Microstructures **3**, 141 (1987)

6.47   E. Kasper: "Silicon Germanium – Heterostructures on Silicon Substrates", in *Festkörperprobleme* (Advances in Solid State Physics) Vol. 27, ed. by P. Grosse (Vieweg, Braunschweig 1987) p. 265

6.48   J. Blanc: "Misfit, Strain, and Dislocations in Epitaxial Structures: Si/Si, Ge/Si, Si/Al$_2$O$_3$", in *Heteroepitaxial Semiconductors for Electronic Devices,* ed. by G.W. Cullen, C.C. Wang (Springer, Berlin, Heidelberg 1978) Chap. 8

6.49   C.A.B. Ball, J.H. van der Merwe: "The Growth of Dislocation-Free Layers", in *Dislocations in Solids,* Vol. 6, ed. by F.R.N. Nabarro (North-Holland, Amsterdam 1983) Chap. 27

6.50   W.T. Read, Jr.: *Dislocations in Crystals* (McGraw-Hill, New York 1953)

6.51   G.H. Olsen: J. Cryst. Growth **31**, 223 (1975)

6.52   C.A.B. Ball: Phys. Status Solidi **42**, 352 (1970)

6.53   J.W.Matthews, J.L. Crawford: Thin Solid Films **5**, 187 (1970)

6.54   J.H. Basson, C.A.B. Ball: Phys. Status Solidi (a) **46**, 707 (1978)

6.55   R. People: J. Appl. Phys. **59**, 3296 (1986)

6.56   G.C.Osbourn, P.L. Gourley, I.J. Fritz, R.M. Biefold, L.R. Dawson, T.E. Zipperian: "Strained Layer Superlattices", in *Semiconductors and Semimetals,* Vol. 32, ed. by R.K. Willardson, A.C. Beer (Academic, New York 1987) Chap. 2

6.57   W.I. Wang, E.E. Mendez, T.S. Kuan, L. Esaki: Appl. Phys. Lett. **47**, 826 (1985)

6.58   S. Subbanna, H. Kroemer, J.L. Merz: J. Appl. Phys. **59**, 488 (1986)

6.59   P.N. Uppal, J.S. Ahearn, D.P. Musser: J. Appl. Phys. **62**, 3766 (1987)

6.60   T. Takamori, T. Fukunaga, J. Kobayashi, K. Ishida, H. Nakashima: Jpn. J. Appl. Phys. **26**, 1097 (1987)

6.61   W.T. Wang, R.F. Marks, L. Vina: J. Appl. Phys. **59**, 937 (1986)

6.62   L. Vina, W.I. Wang: Appl. Phys. Lett. **48**, 36 (1986)

6.63   J.M. Ballingall, C.E.C. Wood: Appl. Phys. Lett. **41**, 947 (1982)

6.64   H.L. Strömer: Surf. Sci. **132**, 519 (1983)

6.65   W.I. Wang, E.E. Mendez, Y. Iye, B. Lee, M.H. Kim, G.E.Stillman: J. Appl. Phys. **60**, 1834 (1986)

6.66   G.D. Kramer, R.K. Tsui, J.A. Curless, M.S. Peffley: "Properties of AlGaAs Grown by MBE on Lenticular Substrates", in *GaAs and Related Compounds 1986* Inst. Phys. Conf. Ser. **83**: (Inst. of Phys., Bristol 1987) p. 117

6.67   S. Sivananthan, X. Chu, J. Reno, J.P. Faurie: J. Appl. Phys. **60**, 1359 (1986)

6.68   S. Sivananthan, X. Chu, J.P. Faurie: J. Vac. Sci. Technol. B **5**, 694 (1987)

6.69   L.I. Maissel, R. Glang (eds.): *Handbook of Thin Film Technology* (McGraw-Hill, New York 1970)

6.70   H. Kroemer: J. Cryst. Growth **81**, 193 (1987)

6.71   H. Kroemer: Surf. Sci. **132**, 543 (1983)

6.72   P.N. Uppal, H. Kroemer: J. Appl. Phys. **58**, 2195 (1985)

6.73   W.A. Harrison, E.A. Kraut, J.R. Waldrop, R.W. Grant: Phys. Rev. B **18**, 4402 (1978)

6.74   S.L. Wright, H. Kroemer, M. Inada: J. Appl. Phys. **55**, 2916 (1984)

6.75   R.C. Pond, J.P. Gowers, B.A. Joyce: Surf. Sci. **152/153**, 1191 (1985)

6.76   R.K. Willardson, H.L. Goering (ed.): *Compound Semiconductors*, Vol. 1 (Reinhold, London 1962) p. 241

6.77   A.Y. Cho, M.B. Panish, I. Hayashi: In Proc. 3rd Int. Symp. GaAs and Related Compounds 1970 (Inst. of Phys. London 1970) p. 18

6.78   A.D. Katnani, D.J. Chadi: Phys. Rev. B **31**, 2554 (1985)

6.79   E.E. Mendez, W.I. Wang, L.L. Chang, L. Esaki: Phys. Rev. B **30**, 1087 (1984)

6.80   D.A. Broido, L.J. Sham: Phys. Rev. B **31**, 888 (1985)

6.81   T.S. Low, G.E. Stillman, A.Y. Cho, H. Morkoç, A. Calawa: Appl. Phys. Lett. **40**, 611 (1982)

6.82   T.S. Low, G.E. Stillman, D.M. Collins, S. Tiwari, L.F. Eastman: Appl. Phys. Lett. **40**, 1034 (1982)

6.83   G.E. Stillman, T.S. Low, B. Lee: Solid State Commun. **53**, 1041 (1985)

6.84   K.C. Rustagi, W. Weber: Solid State Commun. **18**, 673 (1976)

6.85   S.B. Ogale, M. Thomsen, A. Madhukar: Mater. Res. Soc. Symp. Proc. **94**, 83 (1987)
6.86   M.Y. Yen, A. Madhukar, B.F. Lewis, R. Fernandez, L. Eng, F.J. Grunthaner: Surf. Sci. **174**, 606 (1986)
6.87   H.C. Casey, Jr., M.B. Panish: *Heterostructure Lasers* (Academic, New York 1978) Part B, Chap. 7
6.88   W.I. Wang: J. Appl. Phys. **58**, 3244 (1985)
6.89   T.S. Kuan, T.F. Kuech, W.I. Wang, E.L. Wilkie: Phys. Rev. Lett. **54**, 201 (1985)
6.90   T.S. Kuan, W.I. Wang, E.L. Wilkie: Appl. Phys. Lett. **51**, 51 (1987)
6.91   P.M. Petroff, A.Y. Cho, F.K. Reinhart, A.C. Gossard, W. Weigmann: Phys. Rev. Lett. **48**, 170 (1982)
6.92   Y.E. Ihm, N. Otsuka, J. Klem, H. Morkoç: Appl. Phys. Lett. **51**, 2013 (1987)
6.93   S.V. Ghaisas, A. Madhukar: Phys.Rev. Lett. **56**, 1066 (1986)
6.94   A. Madhukar, S.V. Ghaisas: Appl. Phys. Lett. **47**, 247 (1985)
6.95   S.V. Ghaisas, A. Madhukar: J. Vac. Sci. Technol. B**3**, 540 (1985)
6.96   S.B. Ogale, M. Thomsen, A. Madhukar: Appl. Phys. Lett. **52**, 568 (1988)
6.97   D.J. Chadi: J. Vac. Sci. Technol. A**5**, 834 (1987)
6.98   D.J. Frankel, C. Yu, J.P. Harbison, H.H. Farrel: J. Vac. Sci. Technol. B**5**, 1113 (1987)
6.99   P.K. Larsen, D.J. Chadi: Phys. Rev. B **37**, 8282 (1988)
6.100   H.H. Farrell, J.P. Harbison, L.D. Peterson: J. Vac. Sci. Technol. B**5**, 1482 (1987)
6.101   M. Thomsen, A. Madhukar: J. Cryst. Growth **80**, 275 (1987)
6.102   M. Thomsen, S.V. Ghaisas, A. Madhukar: J. Cryst. Growth **84**, 79 (1987)
6.103   M. Thomsen, A. Madhukar: J. Cryst. Growth **84**, 98 (1987)
6.104   P.W. Atkins: *Physical Chemistry* (Oxford University Press, Oxford 1978)
6.105   J.M. Blakely: *Introduction to the Properties of Crystal Surfaces* (Pergamon, Oxford 1979)
6.106   J.P. Gaspard: "Physisorption and Chemisorption", in *Interfacial Aspects of Phase Transformations,* ed. by B. Mutaftschiev, NATO Adv. Stud. Inst. Ser. C, Vol. 87 (D. Reidel, Dordrecht 1982)
6.107   C.T. Foxon, M.R.Boudry, B.A. Joyce: Surf. Sci. **44**, 69 (1974)
6.108   J.R. Arthur: Surf. Sci. **43**, 449 (1974)
6.109   C.T. Foxon, B.A. Joyce: Surf. Sci. **50**, 434 (1975)
6.110   C.T. Foxon, B.A. Joyce: Surf. Sci. **64**, 293 (1977)
6.111   C.T.Foxon: Acta Electron. **21**, 139 (1978)
6.112   L.L. Chang, K. Ploog (eds.): *Molecular Beam Eiptaxy and Heterostructures,* NATO ASI Ser., Ser. E, no. 87 (Martinus Nijhoff, Dordrecht 1985)
6.113   E.H.C. Parker (ed.): *The Technology and Physics of Molecular Beam Epitaxy* (Plenum, New York 1985)
6.114   K. Ploog: "Molecular Beam Epitaxy of III–V Compounds", in *Crystals–Growth, Properties and Applications,* Vol. 3, ed. by H.C. Freyhardt (Springer, Berlin, Heidelberg 1980) p. 73
6.115   J. Singh, A. Madhukar: J. Vac. Sci. Technol. **20**, 716 (1982)
6.116   J.Singh, K.K. Bajaj: J. Vac. Sci. Technol. B**2**, 576 (1984)
6.117   J. Singh, K.K. Bajaj: J. Vac. Sci. Technol. B**2**, 276 (1984)
6.118   J. Singh, S. Dudley, K.K. Bajaj: J. Vac. Sci. Technol. B**4**, 878 (1986)
6.119   J. Singh, K.K. Bajaj: Superlattices and Microstructures **2**, 185 (1986)
6.120   S. Dudley, J. Singh, K.K. Bajaj: J. Vac. Sci. Technol. B**5**, 712 (1987)
6.121   P.R. Berger, P.K. Bhattacharya, J. Singh: J. Appl. Phys. **61**, 2856 (1987)
6.122   J.P. Faurie, M. Boukerche, J. Reno, S. Sivananthan, C. Hsu: J. Vac. Sci. Technol. A**3**, 55 (1985)
6.123   J.P. Faurie: J. Cryst. Growth **81**, 483 (1987)
6.124   J. Reno, R. Sporken, Y.J. Kim, C. Hsu, J.P. Faurie: Appl. Phys. Lett. **51**, 1545 (1987)
6.125   M. Pessa, P. Huttunen, M.A. Herman: J. Appl. Phys. **54**, 6047 (1983)
6.126   M.A.Herman, O. Jylhä, M. Pessa: J. Cryst. Growth **66**, 480 (1984)
6.127   M. Pessa, O. Jylhä: Appl. Phys. Lett. **45**, 646 (1984)
6.128   M.A. Herman; M. Vulli, M. Pessa: J. Cryst. Growth **73**, 403 (1985)
6.129   M.A. Herman, O. Jylhä, M. Pessa: Cryst. Res. Technol. **21**, 841 (1986)
6.130   M.A. Herman, O. Jylhä, M. Pessa: Cryst. Res. Technol. **21**, 969 (1986)

6.131 M.A. Herman, P. Juza, W. Faschinger, H. Sitter: Cryst. Res. Technol. **23**, 3 (1988), and P. Juza, H. Sitter, M.A. Herman: Appl. Phys. Lett. **53**, 1396 (1988)
6.132 C.H.L. Goodman, M. Pessa: J. Appl. Phys. **60**, R.65 (1986)
6.133 J.B. Taylor, I. Langmuir: Phys. Rev. **44**, 423 (1933)
6.134 D.A. King: CRC Crit. Rev. Solid State Mater. Sci. **7**, 167 (1978)
6.135 A. Cassuto, D.A. King: Surf. Sci. **102**, 388 (1981)
6.136 M. Alnot, A. Cassuto: Surf. Sci. **112**, 325 (1981)
6.137 R.S. Chambers, G. Ehrlich: Surf. Sci. **186**, L535 (1987)
6.138 P. Summerside, E. Sommer, R. Teshima, H.J. Kreuzer: Phys. Rev. B **25**, 6235 (1982)
6.139 H.J. Kreuzer, Z.W. Gortel: *Physisorption Kinetics,* Springer Ser. Surf. Sci., Vol. 1 (Springer, Berlin, Heidelberg 1986)
6.140 J.H. Neave, B.A. Joyce, P.J. Dobson, N. Norton: Appl. Phys. A **31**, 1 (1983)
6.141 B.A. Joyce, P.J. Dobson, J.H. Neave, J. Zhang: Surf. Sci. **178**, 110 (1986)
6.142 T.C. Lee, M.Y. Yen, P. Chen, A. Madhukar: J. Vac. Sci. Technol. A **4**, 884 (1986)
6.143 A. Madhukar: "The atomistic nature of compound semiconductor interfaces and the role of growth interruption", in Proc. MRS Symp. on Epitaxy of Semiconductor Layered Sturctures, Boston, Nov. 1987
6.144 M. Tanaka, H. Sakaki, J. Yoshino, T. Furuta: Surf. Sci. **174**, 65 (1986)
6.145 M. Tanaka, H. Sakaki: J. Cryst. Growth **81**, 153 (1987)
6.146 D.Bimberg, D. Mars, J.N. Miller, R. Bauer, D. Oertel: J. Vac. Sci. Technol. B **4**, 1014 (1986)
6.147 D.Bimberg, D. Mars, J.N. Miller, R. Bauer, D.Oertel, J. Christen: Superlattices and Microstructures **3**, 79 (1987)
6.148 D. Bimberg, J. Christen, T. Fukunaga, H. Nakashima, D.E. Mars, J.N. Miller: J. Vac. Sci. Technol. B **5**, 1191 (1987); and to be published in Acta Phys. Polon. (1989)
6.149 D. Bimberg, J. Christen, T. Fukunaga, H. Nakashima, D.E. Mars, J.N. Miller: to be published in Superlattices and Microstructures (1988)
6.150 H. Sakaki, M. Tanaka, J. Yoshino: Jpn. J. Appl. Phys. **24**, L417 (1985)
6.151 T. Fukunaga, K.L.I. Kobayashi, H. Nakashima: Jpn. J. Appl. Phys. **24**; L510 (1985)
6.152 M. Tanaka, H. Sakaki, J. Yoshino: Jpn. J. Appl. Phys. **25**, L155 (1986)
6.153 T. Fukunaga, H. Nakashima: Jpn. J. Appl. Phys. **25**, L856 (1986)
6.154 M. Tanaka, H. Ichinose, T. Furuta, Y. Ishida, H. Sakaki: In Proc. 3rd Int. Conf. Modulated Semiconductor Structures, Montpellier, France, July 1987, p.85
6.155 T. Sakamoto, H. Funabashi, K. Ohta, T. Nakagawa, J.J. Kawai, T. Kojima: Jpn J. Appl. Phys. **23**, L657 (1984)
6.156 T. Sakamoto, H. Funabashi, K. Ohta, T. Nakagawa, N.J. Kawai, T.Kojima, Y.Bando: Superlattices and Microstructures **1**, 347 (1985)
6.157 H. Watanabe, A. Usui: "Atomic Layer Epitaxy", in Inst. Phys. Conf. Ser. **83**, *GaAs and Related Compounds 1986* (Inst. of Phys., Bristol 1987) p.1
6.158 Y. Aoyagi, A. Doi, S. Iwai, S. Namba: J. Vac. Sci. Technol. B **5**, 1460 (1987)
6.159 B.T.McDermott, N.A. El-Masry, M.A. Tischler, S.M. Bedair: Appl. Phys. Lett. **51**, 1830 (1987)
6.160 S.P. DenBaars, C.A. Beyler, A. Hariz, P.D.Dapkus: Appl. Phys. Lett. **51**, 1530 (1987)
6.161 K. Mori, M. Yoshida, A. Usui, H. Ferao: Appl. Phys. Lett. **52**, 27 (1988)
6.162 T. Suntola, J. Hyvaerinen: Annu. Rev. Mater. Sci. **15**, 177 (1985)
6.163 J. Nishizawa, H. Abe, T. Kurabayashi: J. Electrochem. Soc. **132**, 1197 (1985)
6.164 J. Nishizawa, H. Abe, T. Kurabayashi, N. Sakurai: J. Vac. Sci. Technol. A **4**, 706 (1986)
6.165 G.Oya, M. Yoshida, Y. Sawada: Appl. Phys. Lett. **51**, 1143 (1987)
6.166 M.B. Panish: J. Electrochem. Soc. **127**, 2729 (1980)
6.167 M.B. Panish, H. Temkin, S. Sumski: J. Vac. Sci. Technol. B **3**, 687 (1985)
6.168 M.B. Panish, H. Temkin, R.A. Hamm, S.N.G. Chu: Appl. Phys. Lett. **49**, 164 (1986)
6.169 W.T. Tsang: Appl. Phys. Lett. **45**, 1234 (1984)
6.170 W.T. Tsang: J. Cryst. Growth **81**, 261 (1987)
6.171 E. Tokumitsu, Y. Kudou, M. Konagai, K. Takahashi: J. Appl. Phys. **55**, 3163 (1984)
6.172 E. Tokumitsu, Y. Kudou, M. Konagai, K. Takahashi: Jpn. J. Appl. Phys. **24**, 1189 (1985)

6.173 E. Tokumitsu, T. Katoh, R. Kimura, M. Konagai, K. Takahashi: Jpn. J. Appl. Phys. **25**, 1211 (1986)
6.174 M. Pessa, O. Jylhä, M.A. Herman: J. Cryst. Growth **67**, 255 (1984)
6.175 T. Yao: Jpn. J. Appl. Phys. **25**, L942 (1986)
6.176 T. Takeda, T. Yao, T. Kurosu, M. Iida: In Extend. Abstr. 17th Conf. Solid State Dev. Mater., Tokyo, 1985, p. 221
6.177 T. Yao: Jpn. J. Appl. Phys. **25**, L544 (1986)
6.178 T. Yao, T. Takeda: Appl. Phys. Lett. **48**, 160 (1986)
6.179 T. Yao, T. Takeda, R. Watanuki: Appl. Phys. Lett. **48**, 1615 (1986)
6.180 J.A. Venables, G.D.T. Spiller, M. Hanbuecken: Rep. Prog. Phys. **47**, 399 (1984)
6.181 Y. Horikoshi, M. Kawashima, H. Yamaguchi: Jpn. J. Appl. Phys. **25**, L 868 (1986)
6.182 A. Salokatve, J. Varrio, J. Lammasniemi, H. Asonen, M. Pessa: Appl. Phys. Lett. **51**, 1340 (1987)
6.183 J. Varrio, H. Asonen, A. Salokatve, M. Pessa, E. Rauhala, J. Keinonen: Appl. Phys. Lett. **51**, 1801 (1987)
6.184 J. Nishizawa, T. Kurabayashi, H. Abe, A. Nozoe: Surf. Sci. **185**, 249 (1987)
6.185 J. Nishizawa, H. Shimawaki, Y. Sakuma: J. Electrochem. Soc. **134**, 3155 (1987)
6.186 J. Nishizawa, T. Kurabayashi, H. Abe, N. Sakurai: J. Electrochem. Soc. **134**, 945 (1987)
6.187 J. Nishizawa, T. Kurabayashi, J. Hoshina: J. Electrochem. Soc. **134**, 502 (1987)
6.188 G.J. Davies, D. Williams: "III–V MBE Growth Systems", in [6.113, p. 15]
6.189 K.A. Prior, G.J. Davies, R. Heckingbottom: J. Cryst. Growth **66**, 52 (1984)
6.190 R. Heckingbottom: "The Application of Thermodynamics to Molecular Beam Epitaxy", in [6.112, p. 71]
6.191 C.E.C. Wood: "MBE III–V Compounds: Dopant Incorporation, Characteristics and Behavior", in [6.112,.p. 149]
6.192 C.E.C. Wood: "Dopant Incorporation, Characteristics and Behavior", in [6.113, p. 61]
6.193 K. Seeger: *Semiconductor Physics – An Introduction,* 3rd ed., Springer Ser. Solid-State Sci., Vol. 40 (Springer, Berlin, Heidelberg 1985)
6.194 R. Heckingbottom, C.J. Todd, G.J. Davies: J. Electrochem. Soc. **127**, 444 (1980)
6.195 D.L. Miller (ed.): *Molecular Beam Epitaxy Workshop 1986,* Proc. 7th USA MBE Workshop; J. Vac. Sci. Technol. B **5**, no. 3 (1987)
6.196 C.T. Foxon, J.J. Harris (eds.): *Molecular Beam Epitaxy 1986,* Proc. 4th Int. Conf.; J. Cryst. Growth **81**, nos. 1–4 (1987)
6.197 F. Rosenberger: *Fundamentals of Crystal Growth I,* Springer Series in Solid-State Sci., Vol. 5 (Springer, Berlin, Heidelberg 1979) Chap. 2
6.198 M. Heyen, H. Bruch, K.H. Bachem, P. Balk: J. Cryst. Growth **42**, 127 (1977)
6.199 H.C. Casey, M.B. Panish: J. Cryst. Growth **13/14**, 818 (1972)
6.200 D.T.J. Hurle: In Proc. 6th Int. Symp. on GaAs and Related Compounds, Inst. Phys. Conf. Ser. **33A** (Inst. of Phys., Bristol 1977) p. 113
6.201 J.B. Mullin: J. Cryst. Growth **42**, 77 (1977)
6.202 E. Venhoff, M. Maier, K.H. Bachem, P. Balk: J. Cryst. Growth **53**, 598 (1981)
6.203 M.A. Savva: J. Electrochem. Soc. **123**, 1498 (1976)
6.204 D.J. Ashen, P.J. Dean, D.T.J.Hurle, J.B. Mullin, A.M. White, P.D. Greene: J. Phys. Chem. Solids **36**, 1041 (1975)
6.205 G.J. Davies, D.A. Andrews, R. Heckingbottom: J. Appl. Phys. **52**, 7214 (1981)
6.206 O.M. Uy, D.W. Muenow, P.J. Ficalora, J.L. Margrave: Trans. Faraday Soc. **64**, 2998 (1968)
6.207 M.B. Panish: J. Electrochem. Soc. **127**, 2729 (1980)
6.208 D.A. Andrews, R. Heckingbottom, G.J. Davies: J. Appl. Phys. **60**, 1009 (1986)
6.209 K. Ploog: J. Cryst. Growth **81**, 304 (1987), and K. Ploog, M. Hauser, A. Fischer: Appl. Phys. A **45**, 233 (1988)
6.210 C.E.C. Wood, G. Metze, J. Berry, L.F. Eastman, J. Appl. Phys. **51**, 383 (1980)
6.211 K. Ploog, A. Fischer, H. Künzel: J. Electrochem. Soc. **127**, 400 (1981)
6.212 A. Zrenner, H. Reisinger, F. Koch, K. Ploog: In Proc. 17th Int. Conf. on Phys. of Semicond., San Francisco 1984, ed. by J.D. Chadi, W.A. Harrison (Springer, Berlin, Heidelberg 1985) p. 325

6.213 E.F. Schubert, Y. Horikoshi, K. Ploog: Phys. Rev. B **32**, 1085 (1985), and K. Ploog, A. Fischer, E.F. Schubert: Surf. Sci. **174**, 120 (1986)
6.214 E.F. Schubert, A. Fischer, K. Ploog: IEEE Trans. ED-**33**, 625 (1986)
6.215 K. Ploog. Annu. Rev. Mater Sci. **11**, 171 (1981)
6.216 E. Miyauchi, H. Hashimoto: J. Vac. Sci. Technol. A **4**, 933 (1986)
6.217 E. Miyauchi, H. Arimoto, H. Hashimoto, T. Utsumi: J. Vac. Sci. Technol. B **1**, 1113 (1983)
6.218 J.C. Bean: J. Cryst. Growth **81**, 411 (1987)
6.219 Y. Ota: J. Appl. Phys. **51**, 1102 (1980)
6.220 M. Naganuma, K. Takahashi: Appl. Phys. Lett. **27**, 342 (1975)
6.221 N. Matsunaga, T. Suzuki, K. Takahashi: J. Appl. Phys. **49**, 5710 (1978)
6.222 J.C. Bean, R. Dingle: Appl. Phys. Lett. **35**, 925 (1979)
6.223 Y. Matsushima, S.I. Gonda, Y. Makita, S. Mukai: J. Cryst. Growth **43**, 281 (1978)
6.224 H. Sugiura: J. Appl. Phys. **51**, 2630 (1980)
6.225 H. Jorke, H.J. Herzog, H. Kibbel: Appl. Phys. Lett. **47**, 511 (1985)
6.226 H.Jorke, H. Kibbel: J. Electrochem. Soc. **133**, 774 (1986)
6.227 H. Jorke, A. Casel, H. Kibbel, H.J. Herzog: In Proc. 2nd Int. Symp. on Si-MBE, Honolulu 1987
6.228 R.A. Metzger, F.G. Allen: J. Appl. Phys. **55**, 931 (1984)
6.229 M. Tabe, K. Kajiyama: Jpn. J. Appl. Phys. **22**, 423 (1983)
6.230 R.A.A. Kubiak, W.Y. Leong, E.H.C. Parker: Appl. Phys. Lett. **46**, 565 (1985)
6.231 R.A.A. Kubiak, W.Y. Leong, E.H.C. Parker: J. Electrochem. Soc. **132**, 2738 (1985)
6.232 R.A.A. Kubiak, W.Y. Leong, E.H.C. Parker: J. Vac. Sci. Technol. B **3**, 588 (1985)
6.233 W.Y. Leong, R.A.A. Kubiak, E.H.C. Parker: In *Silicon MBE,* ed. by J.C. Bean (Electrochemical Society, Pennington, NJ 1985) p. 140
6.234 C.T. Foxon, J.A. Harvey, B.A. Joyce: J. Phys. Chem. Solids **34**, 1693 (1973)

Chapter 7

7.1 E. Kasper: "Silicon Germanium Heterostructures on Silicon Substrates", in *Festkörperprobleme* (Advances in Solid State Physics), Vol. 27, ed. by P. Grosse (Vieweg, Braunschweig 1987) p. 265
7.2 S.S. Iyer, S.L. Delage, R.D. Thompson, B.A. Ek: In Proc. 7th. Intl. Conf. Thin Films, New Delhi 1987, to be published in Thin Solid Films (1988)
7.3 J.C. Bean: J. Cryst. Growth **81**, 411 (1987)
7.4 E. Kasper, J.C. Bean (eds.): *Silicon Molecular Beam Epitaxy* (CRC Press, Boca Raton 1988)
7.5 K. Sakamoto, T. Sakamoto, S. Nagao, G. Hashiguchi, K. Kuniyoshi, Y. Bando: Jpn J. Appl. Phys. **26**, 666 (1987)
7.6 T. Sakamoto: "RHEED Oscillations in MBE and Their Applications to Precisely Controlled Crystal Growth", in Proc. NATO Summer School on MBE, Ile de Bendor, June 1987
7.7 H.J. Gossmann: Surf. Sci. **179**, 453 (1987)
7.8 M. Zinke-Allmang, H.J. Gossmann, L.C. Feldman, G.J. Fisanick: In MRS Symposia Proceedings, Vol. 77, ed. by J.D. Dow, I.K. Schuwer (MRS, Pittsburgh 1987) p. 703
7.9 R.F.C. Farrow: J. Vac. Sci. Technol. B **1**, 222 (1983)
7.10 E. Kasper: Appl. Phys. A **28**, 129 (1982)
7.11 J.C. Bean: J. Cryst. Growth **70**, 444 (1984)
7.12 J.C. Bean, G.E. Becker, P.M. Petroff, T.E. Seidel: J. Appl. Phys. **48**, 907 (1977)
7.13 R.C. Henderson: J. Electrochem. Soc. **119**, 772 (1972)
7.14 G.E. Becker, J.C. Bean: J. Appl. Phys. **48**, 3395 (1977)
7.15 Y. Ota: J. Electrochem. Soc. **124**, 1795 (1977)
7.16 A. Ishizaka, K. Nakagawa, Y. Shiraki: In Proc. 2nd Intl. Symposium on MBE and Related Clean Surface Techniques (Japanese Soc. Appl. Phys., Tokyo 1982) p. 183
7.17 Y.H. Xie, Y.Y. Wu, K.L. Wang: Appl. Phys. Lett. **48**, 287 (1986)
7.18 R.Hull, J.C. Bean, D.C. Joy, M.E. Twing: Appl. Phys. Lett. **49**, 1714 (1986)
7.19 K. Kugimiya, Y. Shirafuji, N. Matsuo: Jpn. J. Appl. Phys. **24**, 564 (1985)
7.20 T. Tatsumi, N. Aizak, H. Tsuya: Jpn. J. Appl. Phys. **24**, 1227 (1985)

7.21 M. Tabe: Appl. Phys. Lett. **45**, 1073 (1984)
7.22 T. Sakamoto, N.J. Kawai, T. Nakagawa, K. Ohta, T. Kojima: Appl. Phys. Lett. **47**, 617 (1985)
7.23 T. Sakamoto, N.J. Kawai, T. Nakagawa, K. Ohta, T. Kojima, G. Hashiguchi: Surf. Sci. **174**, 651 (1986)
7.24 T. Sakamoto, G. Hashiguchi: Jpn. J. Appl. Phys. **25**, L78 (1986)
7.25 T. Sakamoto, T. Kawamura, S. Nago, G. Hashiguchi, K. Sakamoto, K. Kuniyoshi: J. Cryst. Growth **81**, 59 (1987)
7.26 T. Sakamoto, K. Sakamoto, G. Hashiguchi, N. Takahashi, S. Nago, K. Kuniyoshi, K. Miki: In Proc. 2nd Intl. Symposium on Si-MBE, Honolulu, 1987
7.27 V. Fuenzalida, I. Eisele: J. Cryst. Growth **74**, 597 (1986)
7.28 A. Ishizaka, Y. Shiraki: J. Electrochem. Soc. **133**, 666 (1986)
7.29 R. Kaplan: Surf. Sci. **93**, 145 (1980)
7.30 N. Aizaki, T. Tatsumi: Surf. Sci. **174**, 658 (1986)
7.31 T. Sakamoto, T. Kawamura, G. Hashiguchi: Appl. Phys. Lett. **48**, 1612 (1986)
7.32 T. Kawamura, P.A. Maksym: Surf. Sci. **161**, 12 (1985)
7.33 T. Kawamura, T. Sakamoto, K. Ohta: Surf. Sci. **171**, L409 (1986)
7.34 T. Kawamura, T. Natori, T. Sakamoto, P.A. Maksym: Surf. Sci. **181**, L171 (1987)
7.35 H.J. Gossmann, L.C. Feldman: Phys. Rev. B **32**, 6 (1985)
7.36 H.J. Gossmann, L.C. Feldman: J. Vac. Sci. Technol. B **3**, 1065 (1985)
7.37 H.J. Gossmann, L.C. Feldman: Appl. Phys. A **38**, 171 (1985)
7.38 M. Hansen: *Constitution of Binary Alloys,* 2nd ed. (McGraw-Hill, New York 1958) p.1268
7.39 M.H. Grabow, G.H. Gilmer: Surf. Sci. **194**, 333 (1988)
7.40 H.J. Gossmann, L.C. Feldman, W.M. Gibson: Surf. Sci. **155**, 413 (1985)
7.41 P. Chen, D. Bolmont, C.A. Sebenne: Thin Solid Films **111**, 367 (1984)
7.42 T. Narusawa, W.M. Gibson: Phys. Rev. Lett. **47**, 1459 (1981)
7.43 G.O. Krause: Phys. Status Solidi (a) **3**, 907 (1970)
7.44 A.G. Cullis, G.R. Booker: J. Cryst. Growth **9**, 132 (1971)
7.45 L.N. Aleksandrov, R.N. Lovyagin, O.P. Pchelyakov, S.I. Stenin: J. Cryst. Growth **24/25**, 289 (1974)
7.46 G.L. McVay, A.R. Ducharme: J. Appl. Phys. **44**, 1409 (1973)
7.47 L.C. Feldman, J. Bevk, B.A. Davidson, H.J. Gossmann, J.P. Mannaerts: Phys. Rev. Lett. **59**, 664 (1987)
7.48 H.J. Gossmann, J.C. Bean, L.C. Feldman, E.G. McRae, I.K. Robinson: Phys. Rev. Lett. **55**, 1106 (1985)
7.49 H.J. Gossmann, J.C. Bean, L.C. Feldman, E.G. McRae, I.K. Robinson: J. Vac. Sci. Technol. A **3**, 1633 (1985)
7.50 A.V. Rzhanov, S.I. Stenin, O.P. Pchelyakov, B.Z. Kanter: Thin Solid Films **139**, 169 (1986)
7.51 S.M. Pintus, S.I. Stenin, A.I. Toropov, E.M. Trukhanov, V.Yu. Karasyov: Thin Solid Films **151**, 275 (1987)
7.52 P. Sheldon, B.G. Yacobi, S.E. Asher, K.M. Jones, M.J. Hafich, G.Y.Robinson: J. Vac. Sci. Technol. **4**, 889 (1986)
7.53 M. Zinke-Allmang, H.J. Gossmann, L.C. Feldman, G.J. Fisanick: J. Vac. Sci. Technol. A **5**, 2030 (1987)
7.54 H.J. Gossmann, L.C. Feldman: Appl. Phys. Lett. **48**, 1141 (1986)
7.55 J.C. Bean, L.C. Feldman, A.T. Fiory, S. Nakahara, I.K. Robinson: J. Vac. Sci. Technol. A **2**, 436 (1984)
7.56 H. Jorke, H.J. Herzog: J. Electrochem. Soc. **133**, 998 (1986)
7.57 R. People, J.C. Bean, V.D. Lang: J. Vac. Sci. Technol. A **3**, 846 (1985)
7.58 H. Daembkes, H.J. Herzog, H. Jorke, H. Kibbel, E. Kasper: IEEE Trans. ED-33, 633 (1986)
7.59 T.P. Pearsall, J.C. Bean: IEEE Trans. EDL-7, 308 (1986)
7.60 E. Kasper, H.J. Herzog, H. Jorke, G. Abstreiter: Superlattices and Microstructures **3**, 141 (1987)
7.61 J.H. Van der Merwe: J. Appl. Phys. **34**, 123 (1962)
7.62 J.W. Matthews, A.E. Blakeslee: J. Cryst. Growth **27**, 118 (1974)

7.63 R. People, J.C. Bean: Appl. Phys. Lett. **47**, 322 (1986) [Erratum: **49**, 229 (1986)]
7.64 J.C. Bean: J. Vac. Sci. Technol. B **4**, 1427 (1986)
7.65 S. Luryi, T. Pearsall, H. Temkin, J.C. Bean: IEEE Trans. EDL-7, 104 (1986)
7.66 H. Temkin, T.P. Pearsall, J.C. Bean, R.A. Logan, S. Luryi: Appl. Phys. Lett. **48**, 963 (1986)
7.67 T.P. Pearsall, H. Temkin, J.C. Bean, S. Luryi: IEEE Trans. EDL-7, 330 (1986)
7.68 J.F. Luy, E. Kasper, W.Behr: In Proc. of the 17th European Microwave Conf., Rome, September 1987, p. 820
7.69 J.F. Luy, A. Casel, W. Behr, E. Kasper: IEEE Trans. ED-34, 1084 (1987)
7.70 J.F. Luy, H. Kibbel, E. Kasper: Intl. J. Infrared and Millimeter Waves 7, 305 (1986)
7.71 J. Buechler, E. Kasper, P. Russer, K.M. Strohm: IEEE Trans. MTT-34, 1516 (1986)
7.72 H. Daembkes: In Proc. 2nd Intl. Symp. on Si-MBE, Honolulu, 1987
7.73 E. Kasper, H.J. Herzog, K. Wörner: J. Cryst. Growth 81, 458 (1987)
7.74 R. People, J.C. Bean, D.V. Lang, A.M. Sergent, H.L. Störmer, K.W. Wecht, R.L. Lynch, K. Baldwin: Appl. Phys. Lett. **45**, 1231 (1984)
7.75 G. Abstreiter, H. Brugger, T. Wolff, H. Jorke, H.J. Herzog: Phys. Rev. lett. **54**, 2441 (1985)
7.76 H.L. Störmer: Surf. Sci. **132**, 519 (1983)
7.77 R. People, J.C. Bean: Appl. Phys. Lett. **48**, 538 (1986)
7.78 J.C.Bean, E.A. Sadowsky: J. Vac. Sci. Technol. **20**, 137 (1982)
7.79 E. Kasper, H.J. Herzog, H. Daembkes, G. Abstreiter: Mater. Res. Soc. Symp. Proc. **56**, 347 (1987)
7.80 E. Kasper, H.J. Herzog, H. Daembkes, T. Ricker: "Growth Mode and Interface Structure of MBE Grown Ge_xSi_{1-x}/Si Structures", in *Two-Dimensional Systems: Physics and New Devices,* ed. by G. Bauer, F. Kuchar, H. Heinrich, Springer Ser. Solid-State Sci., Vol. 53 (Springer, Berlin, Heidelberg 1984)
7.81 H.H. Farrell, J.P. Harbison, L.D. Peterson: J. Vac. Sci. Technol. B **5**, 1482 (1987)
7.82 A.Y. Cho: "Recent Advances in GaAs on Si", in Digest of the Int. Electron Device Meeting, Washington DC, December 1987
7.83 R.M. Fletcher, D.K. Wagner, J.M. Ballantyne: Appl. Phys. Lett. **44**, 967 (1984)
7.84 T. Soga, S. Hattori, S. Sakai, M. Takeyasu, M. Umeno: Electron. Lett. **20**, 916 (1984)
7.85 R. Fischer, D. Neumann, H. Zabel, H. Morkoç, C. Choi, N. Otsuka: Appl. Phys. Lett. **48**, 1223 (1986)
7.86 R. Fischer, H. Morkoç, D. Neumann, H. Zabel, C. Choi, N. Otsuka, M. Longerbone, L.P. Erickson: J. Appl. Phys. **60**, 1640 (1986)
7.87 M. Kawabe, T. Ueda: Jpn. J. Appl. Phys. **25**, L285 (1986)
7.88 T. Ueda, S. Nishi, Y. Kawarada, M.Akiyama, K. Kaminishi: Jpn. J. Appl. Phys. **25**, L789 (1986)
7.89 S. Nishi, H. Inomata, M. Akiyama, K. Kaminishi: Jpn. J. Appl. Phys. **24**, L391 (1985)
7.90 J. Varrio, H. Asonen, A. Salokatve, M. Pessa, E.Rauhala, J. Keinonen: Appl. Phys. Lett. **52**, 1801 (1987)
7.91 H. Noge, H. Kano, T. Kato, M. Hashimoto, I. Igarashi: J. Cryst. Growth 83, 431 (1987)
7.92 P.K. Larsen, J.H. Neave, J.F. Van der Veen, P.J. Dobson, B.A. Joyce: Phys. Rev. B **27**, 4966 (1983)
7.93 P.J. Dobson, J.H. Neave, B.A. Joyce: Surf. Sci. **119**, L339 (1982)
7.94 F. Briones, D. Golmayo, L. Gonzales, J.L. De Miguel: Jpn. J. Appl. Phys. **24**, L478 (1985)
7.95 P.K. Larsen, D.J. Chadi: Phys. Rev. B **37**, 8282 (1988)
7.96 K. Ploog: Annu. Rev. Mater. Sci. **11**, 171 (1981)
7.97 A.Y. Cho: Thin Solid Films **100**, 291 (1983)
7.98 D.E. Mars, J.N. Miller: J. Vac. Sci. Technol. B **4**, 571 (1986)
7.99 J. Massies, J.P. Contour: J. Appl. Phys. **58**, 806 (1985)
7.100 H. Fronius, A. Fischer, K. Ploog: Jpn. J. Appl. Phys. **25**, L137 (1986)
7.101 K. Ploog, A. Fischer: Appl. Phys. Lett. **48**, 1392 (1986)
7.102 K.Fujiwara, Y. Nishikawa, Y. Tokuda, T. Nakayama: Appl. Phys. Lett. **48**, 701 (1986)

7.103 W.T. Tsang: Appl. Phys. Lett. **46**, 1086 (1985)
7.104 N. Watanabe, T. Fukunaga, K.L.I. Kobayashi, H. Nakashima: Jpn. J. Appl. Phys. **24**, L498 (1985)
7.105 M. Bafleur, A. Munoz-Yaque, A. Rocher: J. Cryst. Growth **59**, 531 (1982)
7.106 Y. Suzuki, M.Seki, Y. Horikoshi, H. Okamoto: Jpn. J. Appl. Phys. **23**, 164 (1984)
7.107 S.L. Weng, C. Webb, Y.G. Chai, S.G. Bandy: Appl. Phys. Lett. **47**, 391 (1985)
7.108 G.M. Metze, A.R. Calawa, J.G. Mavroides: J. Vac. Sci. Technol. B **1**, 166 (1983)
7.109 Y.G. Chai, Y.C. Pao, T. Hierl: Appl. Phys. Lett. **47**, 1327 (1985)
7.110 J. Massies, J.P. Contour: Jpn. J. Appl. Phys. **26**, L38 (1987)
7.111 H. Fronius, A. Fischer, K. Ploog: J. Cryst. Growth **81**, 169 (1987)
7.112 A. Madhukar, S.V. Ghaisas: CRC Crit. Rev. Solid State Mater. Sci. **14**, 1 (1988)
7.113 D.J. Frankel, C. Yu, J.P. Harbison, H.H. Farrell: J. Vac. Sci. Technol. B **5**, 1113 (1987)
7.114 D.J. Chadi: J. Vac. Sci. Technol. A **5**, 834 (1987)
7.115 R.I.G. Uhrberg, R.D. Bringans, R.Z. Bachrach, J.E. Northrup: Phys. Rev. Lett. **56**, 520 (1986)
7.116 Y. Horikoshi, M. Kawashima, H. Yamaguchi: Jpn. J. Appl. Phys. **25**, L868 (1986)
7.117 R.N. Bicknell, N.C. Giles, J.F. Schetzina: J. Vac. Sci. Technol. B **5**, 701 (1987)
7.118 M. Tanaka, H. Sakaki: J. Cryst. Growth **81**, 153 (1987)
7.119 W.T. Tsang, R.C. Miller: Appl. Phys. Lett. **48**, 1288 (1986)
7.120 C.W. Tu, R.C. Miller, B.A. Wilson, P.M. Petroff, T.D. Harris, R.F. Kopf, S.K. Sputz, M.G. Lamont: J. Cryst. Growth **81**, 159 (1987)
7.121 T. Hayakawa, T. Suyama, K. Takahashi, M. Kondo, S. Yamamoto, S. Yano, T. Hijikata: Surf. Sci. **174**, 76 (1986)
7.122 M. Tanaka, H. Sakaki, J. Yoshino, T. Furuta: Surf. Sci. **174**, 65 (1986)
7.123 H. Sakaki, T. Noda, K. Hirakawa, M. Tanaka, T. Matsusue: Appl. Phys. Lett. **51**, 1934 (1987)
7.124 J. Singh, K.K. Bajaj: J. Appl. Phys. **57**, 5433 (1985)
7.125 D. Bimberg, D. Mars, J.N. Miller, R. Bauer, D. Oertel, J. Christen: Superlattices and Microstructures **3**, 79 (1987)
7.126 J.Y. Kim, P. Chen, F. Voillot, A. Madhukar: Appl. Phys. Lett. **50**, 739 (1987)
7.127 F. Voillot, J.Y. Kim, W.C. Tang, A. Madhukar, P. Chen: Superlattices and Microstructures **3**, 313 (1987)
7.128 S.B. Ogale, A. Madhukar, F. Voillot, M. Thomsen, W.C. Tang, T.C. Lee, J.Y. Kim, P. Chen: Phys. Rev. B **36**, 1662 (1987)
7.129 J. Singh, K.K. Bajaj, S. Chaudhari: Appl. Phys. Lett. **44**, 805 (1984)
7.130 J. Singh, K.K. Bajaj: Appl. Phys. Lett. **44**, 1075 (1984)
7.131 H. Sakaki, M. Tanaka, J. Yoshino: Jpn. J. Appl. Phys. **24**, L417 (1985)
7.132 M. Tanaka, H. Sakaki, J. Yoshino: Jpn. J. Appl. Phys. **25**, L155 (1986)
7.133 T. Sakamoto, H. Funabashi, K. Ohta, T. Nakagawa, N.J. Kawai, T. Kojima: Jpn. J. Appl. Phys. **23**, L657 (1984)
7.134 T. Kojima, N.J. Kawai, T. Nakagawa, K. Ohta, T. Sakamoto, M. Kawashima: Appl. Phys. Lett. **47**, 286 (1985)
7.135 J.M. Van Hove, P.I. Cohen: Appl. Phys. Lett. **47**, 726 (1985)
7.136 T. Sakamoto: "RHEED Oscillations in MBE and Their Applications to Precisely Controlled Crystal Growth", in Proc. NATO Summer School on MBE, Ile de Bendor, June 1987
7.137 R.C. Miller, A.C. Gossard, D.A. Kleinman, O. Munteanu: Phys. Rev. B **29**, 3740 (1984)
7.138 R.C. Miller, A.C. Gossard, D.A. Kleinman: Phys. Rev. B **32**, 5443 (1985)
7.139 R.C. Miller, A.C. Gossard, D.A. Kleinman: Phys. Rev. B **29**, 7085 (1984)
7.140 A.C. Gossard, R.C. Miller, W. Wiegmann: Surf. Sci. **174**, 131 (1986)
7.141 M. Kawabe, M. Kondo, N. Matsuura, K. Yamamoto: Jpn. J. Appl. Phys. **22**, L64 (1983)
7.142 J.C.C. Fan, J.M. Poate (eds.): *Heteroepitaxy on Silicon*, MRS Symp. Proc., Vol. 67 (MRS, Pittsburgh, PN 1986)
7.143 R.D. Bringans, M.A. Olmstead, R.I.G. Uhrberg, R.Z. Bachrach: Appl. Phys. Lett. **51**, 523 (1987)

7.144 R. Hull, A. Fischer-Colbrie, S.J. Rosner, S.M. Koch, J.S. Harris, Jr.:Appl. Phys. Lett. **51**, 1723 (1987)
7.145 T. Won, G. Munns, M.S. Unlu, J. Chyi, H. Morkoç: J. Appl. Phys. **62**, 3860 (1987)
7.146 H. Kroemer: J. Cryst. Growth **81**, 193 (1987)
7.147 P.N. Uppal, H. Kroemer: J. Appl. Phys. **58**, 2195 (1985)
7.148 R.J. Fischer, N.C. Chang, W.F. Kopp, H. Morkoç, L.P. Erickson, R. Youngman: Appl. Phys. Lett. **47**, 397 (1985)
7.149 R.J. Fischer, C.K. Peng, J. Klem, T. Henderson, H. Morkoç: Solid-State Electron. **29**, 269 (1986)
7.150 W.A. Harrison, E.A. Kraut, J.R. Waldrop, R.W. Grant: Phys. Rev. B18, 4402 (1978)
7.151 D.A. Neumann, H. Zabel, R.J. Fischer, H. Morkoç: J. Appl. Phys. **61**, 1023 (1987)
7.152 D.K. Biegelsen, F.A. Ponce, A.J. Smith, J.C. Tramontana: J. Appl. Phys. **61**, 1856 (1987)
7.153 S.M. Koch, S.J. Rosner, R. Hull, G.W. Yoffe, J.S. Harris, Jr.: J. Cryst. Growth **81**, 205 (1987)
7.154 H.L. Tsai, J.W. Lee: Appl. Phys. Lett. **51**, 130 (1987)
7.155 R.M. Lum, J.K. Klingert, B.A. Davidson, M.G. Lamont: Appl. Phys. Lett. **51**, 36 (1987)
7.156 N. Chand, R. People, F.A. Baiocchi, K.W. Wecht, A.Y. Cho: Appl. Phys. Lett. **49**, 815 (1986)
7.157 N. Chand, R. Fischer, A.M. Sergent, D.V. Lang, S.J. Pearton, A.Y. Cho: Appl. Phys. Lett. **51**, 1013 (1987)
7.158 M. Kawabe, T. Ueda, H. Takasugi: Jpn. J. Appl. Phys. **26**, L114 (1987)
7.159 H. Takasugi, M. Kawabe, Y. Bando: Jpn. J. Appl. Phys. **26**, L584 (1987)
7.160 M. Kawabe, T. Ueda: Jpn. J. Appl. Phys. **26**, L944 (1987)
7.161 M. Zinke-Allmang, L.C. Feldman, S. Nakahara: Appl. Phys. Lett. **52**, 144 (1988)
7.162 K. Ploog: J. Cryst. Growth **79**, 887 (1986)
7.163 W.T. Tsang, R.C. Miller: J. Cryst. Growth **77**, 55 (1986)
7.164 D. Fritzsche: Solid-State Electron. **30**, 1183 (1987)
7.165 D.L. Miller, P.M. Asbeck: J. Cryst. Growth **81**, 368 (1987)
7.166 F. Alexandre, J.C. Harmand, J.L. Lievin, C. Dubon-Chevalier, D. Ankri, C. Minot, J.F. Palmier: J. Cryst. Growth **81**, 391 (1987)
7.167 T. Hayakawa, T.Suyama, K. Takahashi, M. Kondo, S. Yamamoto, T. Hijikata: Appl. Phys. Lett. **52**, 339 (1988)
7.168 T. Hayakawa, T.Suyama, K. Takahashi, M. Kondo, S. Yamamoto, T. Hijikata: Appl. Phys. Lett. **51**, 707 (1987)
7.169 T. Hayakawa, K. Takahashi, T. Suyama, M. Kondo, S. Yamamoto, T. Hijikata: Appl. Phys. Lett. **52**, 252 (1988)
7.170 H.Z. Chen, A. Ghaffari, H. Wang, H. Morcoç, A. Yariv: Appl. Phys. Lett. **51**, 1320 (1987)
7.171 T. Yuasa, M. Mannoh, T. Yamada, S. Naritsuka, K. Shinozaki, M. Ishii: J. Appl. Phys. **62**, 764 (1987)
7.172 W.T. Tsang, A.Y. Cho: Appl. Phys. Lett. **30**, 293 (1977)
7.173 J.S. Smith, P.L. Perry, S. Margalit, A. Yariv: Appl. Phys. Lett. **47**, 712 (1985)
7.174 W.J. Grande, C.L. Tang: Appl. Phys. Lett. **51**, 1780 (1987)
7.175 C. Amano, H. Sugiura, A. Yamamoto, M. Yamaguchi: Appl. Phys. Lett. **51**, 1998 (1987)
7.176 C. Amano, H. Sugiura, K. Ando, M. Yamaguchi, A. Saletes: Appl. Phys. Lett. **51**, 1075 (1987)
7.177 Y. Horikoshi, A. Fischer, K. Ploog: Appl. Phys. Lett. **45**, 919 (1984)
7.178 Y. Horikoshi, K. Ploog: Appl. Phys. Lett. **37**, 47 (1985)
7.179 F. Capasso: Surf. Sci. **132**, 527 (1983)
7.180 F. Capasso: Surf. Sci. **142**, 513 (1984)
7.181 E.F. Schubert. A. Fischer, K. Ploog: Electron. Lett. **21**, 411 (1985)
7.182 E.F. Schubert. A. Fischer, K. Ploog: IEEE Trans. ED-33, 625 (1986)
7.183 F. Capasso, S.Sen, A.Y. Cho: Appl. Phys. Lett. **51**, 526 (1987)
7.184 S. Sen, F. Capasso, A.C. Gossard, R.A. Spah, A.L. Hutchinson, S.N.G. Chu: Appl. Phys. Lett. **51**, 1428 (1987)

7.185 V.J. Goldman, D.C. Tsui, J.E. Cunningham, W.T. Tsang: J. Appl. Phys. **61** 2693 (1987)
7.186 M. Heiblum, D.C. Thomas, C.M. Knoedler, M.I. Nathan:Surf. Sci. **174**, 478 (1986)
7.187 K. Imamura, S. Muto, N. Yokoyama, M. Sasa, H. Ohnishi, S. Hiyamizu, H. Nishi: Surf. Sci. **174**, 481 (1986)
7.188 J.P. Faurie: J. Cryst. Growth **81**, 483 (1987)
7.189 J.P. Faurie, J. Reno, S. Sivananthan, I.K. Sou, X. Chu, M. Boukerche, P.S. Wijewarnasuriya: J. Vac. Sci. Technol. B **4**, 585 (1986)
7.190 J.P. Faurie: IEEE J. QE-**22**, 1656 (1986)
7.191 J. Reno, I.K. Sou, J.P. Faurie, J.M. Berroir, Y. Guldner: J. Vac. Sci. Technol. A **5**, 3107 (1987)
7.192 M.A. Herman, M. Pessa: J. Appl. Phys. **57**, 2671 (1985)
7.193 K.A. Harris, S. Hwang, D.K. Blanks, J.W. Cook, Jr., J.F. Schetzina, N. Otsuka: J. Vac. Sci. Technol. A **4**, 2061 (1986)
7.194 J. Reno, I.K. Sou, P.S. Wijewarnasuriya, J.P. Faurie: Appl. Phys. Lett. **48**, 1069 (1986)
7.195 J.P. Faurie, I.K. Sou, P.S. Wijewarnasuriya, S. Rafol, K.C. Woo: Phys. Rev. B **34**, 6000 (1986)
7.196 M. Boukerche, I.K. Sou, M. DeSouza, S.S. Yoo, J.P. Faurie: J. Vac. Sci. Technol. A **5**, 3119 (1987)
7.197 K.C. Woo, S. Rafol, J.P. Faurie: J. Vac. Sci. Technol. A **5**, 3093 (1987)
7.198 X. Chu, S. Sivananthan, J.P. Faurie: Appl. Phys. Lett. **50**, 597 (1987)
7.199 R.P. Ruth, Y. Marfaing, J.B. Mullin, J. Woods (eds.): Proc. 3rd Intl. Conf. II–VI Compounds, Monterey, 1987; J. Cryst. Growth **86**, nos. 1–4 (1988)
7.200 T.N. Casselman (ed.): Proc. 1985 U.S. Workshop on Phys. Chem. Mercury Cadmium Telluride; J. Vac. Sci. Technol. A **4**, no. 4 (1986)
7.201 H.F. Schaake (ed.): Proc. 1986 U.S. Workshop on Phys. Chem. Mercury Cadmium Telluride; J. Vac. Sci. Technol. A **5**, no. 5 (1987)
7.202 Proc. 1987 U.S.Workshop on Phys. Chem. Mercury Cadmium Telluride; J. Vac. Sci. Technol. A **6** (1988)
7.203 J.P. Faurie, A. Million: J. Cryst. Growth **54**, 582 (1981)
7.204 J.P. Faurie, A. Million: Appl. Phys. Lett. **41**, 264 (1982)
7.205 J.P. Faurie, A. Million, J. Piaguet: Appl. Phys. Lett. **41**, 713 (1982)
7.206 Y. Guldner, G. Bastard, J.P. Vieren, M. Voos, J.P. Faurie, A. Million: Phys. Rev. Lett. **51**, 907 (1983)
7.207 C. Fontaine, Y. Demay, J.P. Gailliard, A. Million, J. Piaguet: Thin Solid Films **130**, 327 (1985)
7.208 A. Million, L. DiCioccio, J.P. Gailliard, J. Piaguet: in [7.202]
7.209 S. Wood, J. Greggi, Jr., R.F.C. Farrow, W.J. Takei, F.A. Shirland, A.J. Noreika: J. Appl. Phys. **55**, 4225 (1984)
7.210 J.P. Faurie, S. Sivananthan, M. Boukerche, J. Reno: Appl. Phys. Lett. **45**, 1307 (1984)
7.211 L.A. Kolodziejski, R.L. Gunshor, N. Otsuka, X.C. Chang, S.K. Chang, A.V. Nurmikko: Appl. Phys. Lett. **47**, 882 (1985)
7.212 H.A. Mar, K.T. Chee, N. Salansky: Appl. Phys. Lett. **44**, 237 (1984)
7.213 J.M. Ballingall, M.L. Wroge, D.J. Leopold: Appl. Phys. Lett. **48**, 1273 (1985)
7.214 J. Humenberger: to be published in Z. Kristallogr., (1988)
7.215 J. Humenberger, H. Sitter: Proc. 7th Intl. Conf. Thin Films, New Delhi 1987, to be published in Thin Solid Films (1988)
7.216 S. Sivananthan, X. Chu, J. Reno, J.P. Faurie: J. Appl. Phys. **60**, 1359 (1986)
7.217 S. Sivananthan, X. Chu, J.P. Faurie: J. Vac. Sci. Technol. B **5**, 694 (1987)
7.218 J.M. Arias, S.H. Shin, E.R. Gertner: J. Cryst. Growth **86**, 362 (1988)
7.219 A. Kahn: Surf. Sci. **168**, 1 (1986)
7.220 K. Nishitani, R. Ohkata, T. Murotoni: J. Electron. Mater. **12**, 619 (1983)
7.221 R.N. Bicknell, R.Y. Yanka, N.C. Giles, J.F. Schetzina, T.J. Magee, C. Leung, H. Kawayashi: Appl. Phys. Lett. **44**, 313 (1984)
7.222 J.T. Cheung, T.J. Magee: J. Vac. Sci. Technol. A **1**, 1604 (1983)
7.223 P.P. Chow, D.K. Greenlaw, D. Johnson: J. Vac. Sci. Technol. A **1**, 562 (1983)

7.224 N. Otsuka, L.A. Kolodziejski, R.L. Gunshor, S. Datta, R.W. Bicknell, J.F. Schetzina: Appl. Phys. Lett. **46**, 860 (1985)
7.225 J.P. Faurie, C. Hsu, S. Sivananthan, X. Chu: Surf. Sci. **168**, 473 (1986)
7.226 R. Srinivasa, M.B. Panish, H. Temkin: Appl. Phys. Lett. **50**, 1441 (1987)
7.227 C. Hsu, S. Sivananthan, X. Chu, J.P. Faurie: Appl. Phys. Lett. **48**, 908 (1986)
7.228 N. Magnea, F. Dalbo, J.L. Pautrat, A. Million, L. DiCioccio, G. Feuillet: Mater. Res. Soc. Symp. Proc. **90**, 455 (1987)
7.229 C. Fontaine, J.P. Gailliard, S. Magli, A. Million, J. Piaguet: Appl. Phys. Lett. **50**, 903 (1987)
7.230 J.P. Faurie, M. Boukerche, J. Reno, S. Sivananthan, C. Hsu: J. Vac. Sci. Technol. A **3**, 55 (1985)
7.231 J. Reno, I.K. Sou, P.S. Wijewarnasuriya, P.J. Faurie: Appl. Phys. Lett. **47**, 1168 (1985)
7.232 S. Sivananthan, X. Chu, M. Boukerche, J.P. Faurie: Appl. Phys. Lett. **47**, 1291 (1985)
7.233 W.E. Spicer, J.A. Silberman, I. Lindau, A.B. Chen, A. Sher, J.A. Wilson: J. Vac. Sci. Technol. A **13**, 1735 (1983)
7.234 A. Sher, D. Eger, A. Zemel: Appl. Phys. Lett. **46**, 59 (1985)
7.235 J.P. Faurie, J. Reno, M. Boukerche: J. Cryst. Growth **72**, 111 (1985)
7.236 J.P. Faurie, M. Boukerche, S. Sivananthan, J. Reno, C. Hsu: Superlattices and Microstructures **1**, 237 (1985)
7.237 C.E. Jones, T.N. Casselman, J.P. Faurie, S. Perkowitz, J.N. Schulman: Appl. Phys. Lett. **47**, 140 (1985)
7.238 L. DiCioccio, A. Million, J.P. Gailliard, M. Dupny: Rev. Phys. Appl. **22**, 465 (1987)
7.239 D.J. Olego, J.P. Faurie: Phys. Rev. B **33**, 7357 (1986)
7.240 J.M. Berroir, Y. Guldner, J.P. Vieren, M. Voos, J.P. Faurie: Phys. Rev. B **34**, 891 (1986)
7.241 J. Reno, J.P. Faurie: Appl. Phys. Lett. **49**, 409 (1986)
7.242 J.P. Faurie, S. Sivananthan, X. Chu, P.A. Wijewarnasuiya: Appl. Phys. Lett. **48**, 785 (1986)
7.243 P.S. Wijewarnasuriya, I.K. Sou, Y.J. Kim, K.K. Mahavadi, S. Sivananthan, M. Boukerche, J.P. Faurie: Appl. Phys. Lett. **51**, 2025 (1987)
7.244 P.M. Raccah, J.W. Garland, Z. Zhang, A.H.M. Chu, J. Reno, J.K. Sou, M. Boukerche, J.P. Faurie: J. Vac. Sci. Technol. A **4**, 2077 (1986)
7.245 M. Boukerche, P.S. Wijewarnasuriya, J. Reno, I.K. Sou, J.P. Faurie: J. Vac. Sci. Technol. A **4**, 2072 (1986)
7.246 M. Boukerche, J. Reno, I.K. Sou, C. Hsu, J.P. Faurie: Appl. Phys. Lett. **48**, 1733 (1986)
7.247 J.M. Arias, S.H. Shin, J.G.Pasko, E.R. Gertner: Appl. Phys. Lett. **52**, 39 (1988)
7.248 J.P. Faurie, A. Million, R. Boch, J.L. Tissot: J. Vac. Sci. Technol. A **1**, 1593 (1983)

Chapter 8

8.1 M.B. Panish: Private communication
8.2 S. Shimizu, O. Tsukakoshi, S. Komiya, Y. Makita: Jpn. J. Appl. Phys. **24**, 1130 (1985)
8.3 S.Shimizu, O. Tsukakoshi, S. Komiya, Y. Makita: J. Vac. Sci. Technol. B **3**, 554 (1985)
8.4 S. Shimizu, Y. Makita: "Molecular and Ion-Beam Epitaxy of III–V Compound Semiconductors", in ISIAT '86, Proc, 10th Symp. on Ion Sources and Ion-Assisted Technology, ed. by T. Takagi, Tokyo 1986
8.5 S. Shimizu, S. Komiya: J. Cryst. Growth **81**, 243 (1987)
8.6 I. Yamada, H. Takaoka, H. Usui, T. Takagi: J. Vac. Sci. Technol. A **4**, 722 (1986)
8.7 T. Koyanagi, K. Matsubara, H. Takaoka, T. Takagi: J. Appl. Phys. **61**, 3020 (1987)
8.8 K.H. Müller: J. Appl. Phys. **61**, 2516 (1987)
8.9 T. Katsuyama, M.A. Tischler, N.H. Karam, N. El-Masry, S.M. Bedair: Appl. Phys. Lett. **51**, 529 (1987)
8.10 L.M. Fraas, P.S. McLeod, L.D. Partain, J.A. Cape: J. Vac. Sci. Technol. B **4**, 22 (1986)

8.11 L.M. Fraas, P.S. McLeod, R.E. Weiss, L.D. Partain, J.A. Cape: J. Appl. Phys. **62**, 299 (1987)
8.12 M.B. Panish, H. Temkin, S. Sumski: J. Vac. Sci. Technol. B**3**, 657 (1985)
8.13 M.B. Panish: J. Cryst. Growth **81**, 249 (1987)
8.14 H. Temkin, S.N.G. Chu, M.B. Panish, R.A. Logan: Appl. Phys. Lett. **50**, 956 (1987)
8.15 W.T. Tsang, E.F. Schubert: Appl. Phys. Lett. **49**, 220 (1986)
8.16 K. Tai, J. Hegarty, W.T. Tsang: Appl. Phys. Lett. **51**, 86 (1987)
8.17 H. Tanaka, Y. Kawamura, S. Nojima, K. Wakita, H. Asahi: J. Appl. Phys. **61**, 1713 (1987)
8.18 P.A. Claxton, J.S. Roberts, J.P.R. David, C.M. Sotomayor-Torres, M.S. Skolnick, P.R. Tapster, K.J. Nash: J. Cryst. Growth **81**, 288 (1987)
8.19 W.T. Tsang, J.C. Campbell, G.J. Qua: IEEE Trans. EDL-**8**, 294 (1987)
8.20 H. Temkin, G. Gershoni, M.B. Panish: Appl. Phys. Lett. **50**, 1176 (1987)
8.21 W.T. Tsang: J. Cryst. Growth **81**, 261 (1987)
8.22 W.T. Tsang: J. Appl. Phys. **58**, 1415 (1985)
8.23 W.T. Tsang, J.C. Campbell: Appl. Phys. Lett. **48**, 1416 (1986)
8.24 H. Temkin, M.B. Panish, P.M. Petroff, R.A. Hamm, J.M. Vandenberg, S. Sumski: Appl. Phys. Lett. **47**, 394 (1985)
8.25 D.A.B. Miller, D.S. Chemla, T.C. Damen, A.C. Gossard, W. Wiegmann, T.H. Wood, C.A. Burrus: Phys. Rev. B**32**, 1043 (1985)
8.26 W.D. Laidig, C.K. Peng, Y.F. Lin: J. Vac. Sci. Technol. B**2**, 182 (1984)
8.27 M. Quillec, L. Goldstein, G. LeRoux, J. Burgeat, J. Primot: J. Appl. Phys. **55**, 2904 (1984)
8.28 M.C. Tamargo, R. Hull, L.H. Greene, J.R. Hayes, A.Y. Cho: Appl. Phys. Lett. **46**, 569 (1985)
8.29 G.C. Osbourn: J. Vac. Sci. Technol. B**4**, 1423 (1986)
8.30 C. D'Anterroches, J.Y. Marzin, G. LeRoux, L. Goldstein: J. Cryst. Growth **81**, 121 (1987)
8.31 K. Hirose, T. Mizutani, K. Nishi: J. Cryst. Growth **81**, 130 (1987)
8.32 S. Hiyamizu, T. Fujii, S. Muto, T. Inata, Y. Nakata, Y. Sugiyama, S. Sasa: J. Cryst. Growth **81**, 349 (1987)
8.33 S.M. Liu, M.B.Das, C.K. Peng, J. Klem, T. Henderson, W. Kopp, H. Morkoç: J. Cryst. Growth **81**, 359 (1987)
8.34 F.Y. Juang, W.P. Hong, P.R. Berger, P.K. Bhattacharya, U. Das, J. Singh: J. Cryst. Growth **81**, 373 (1987)
8.35 P.M. Enquist, L.P. Ramberg, F.E. Najjar, W.J. Schaff, K.L. Kavanagh, G.W. Wicks, L.F. Eastman: J. Cryst. Growth **81**, 378 (1987)
8.36 H.T. Griem, K.H. Hsieh, I.J. D'Haenens, M.J. Delaney, J.A. Henige, W.W. Wicks, A.S. Brown: J. Cryst. Growth **81**, 383 (1987)
8.37 S. Söderström, T.G. Andersson, J. Westin: Superlattices and Microstructures **3**, 283 (1987)
8.38 S. Kaneda, S. Satou, T.Setoyama, S. Motoyama, M. Yokoyama, N. Ota: J. Cryst. Growth **76**, 440 (1986)
8.39 R.M. Park, H.A. Mar, N.M. Salansky: J. Vac. Sci. Technol. B**3**, 676 (1985)
8.40 R.M. Park, H.A. Mar, N.M. Salansky: Appl. Phys. Lett. **46**, 386 (1985)
8.41 H.A. Mar, R.M. Park: J. Appl. Phys. **60**, 1229 (1986)
8.42 T. Yao, Y. Okada, S. Matsui, K. Ishida, I. Fujimoto: J. Cryst. Growth **81**, 518 (1987)
8.43 T. Yao, T. Takeda: J. Cryst. Growth **81**, 43 (1987)
8.44 H. Cheng, S.K. Mohapatra, J.E. Potts, T.L. Smith: J. Cryst. Growth **81**, 512 (1987)
8.45 R.M. Park, H.A. Mar, N.M. Salansky: J. Vac. Sci. Technol. B**3**, 1637 (1985)
8.46 H. Fujiyasu, K. Mochizuki: J. Appl. Phys. **57**, 2960 (1985)
8.47 K. Mohammed, D.J. Olego, P. Newbury, D.A. Cammack, R. Dalby, H. Cornelissen: Appl. Phys. Lett. **50**, 1820 (1987)
8.48 M. Kobayashi, N. Mino, M. Konagai, K. Takahashi: J. Appl. Phys. **57**, 2905 (1985)
8.49 M. Kobayashi, N. Mino, H. Inuzuka, M. Konagai, K. Takahashi: J. Appl. Phys. **57**, 4706 (1985)
8.50 M. Kobayashi, N. Mino, H. Katagiri, R. Kimura, M. Konagai, K. Takahashi: Appl. Phys. Lett. **48** , 296 (1986)

8.51 M. Kobayashi, N. Mino, H. Katagiri, R. Kimura, M. Konagai, K. Takahashi: J. Appl. Phys. **60**, 773 (1986)
8.52 M. Kobayashi, R. Kimura, M. Konagai, K. Takahashi: J. Cryst. Growth **81**, 495 (1987)
8.53 M. Kobayashi, M. Konagai, K. Takahashi, K. Urabe: J. Appl. Phys. **61**, 1015 (1987)
8.54 A. Taike, N. Teraguchi, M. Konagai, K. Takahashi: Jpn. J. Appl. Phys. **26**, L989 (1987)
8.55 L.A. Kolodziejski, R.L. Gunshor, N. Otsuka, S. Datta, W.M. Becker, A.V. Nurmikko: IEEE J. QE-**22**, 1666 (1986)
8.56 Y. Hefetz, W.C. Goltsos, D. Lee, A.V. Nurmikko, L.A. Kolodziejski, R.L. Gunshor: Superlattices and Microstructures **2**, 455 (1986)
8.57 R.B. Bylsma, J. Kossut, W.M. Becker, L.A. Kolodziejski, R.L. Gunshor, R. Frohne: J. Appl. Phys. **61**, 3011 (1987)
8.58 R.L. Gunshor, L.A. Kolodziejski, N. Otsuka, B.P. Gu, D. Lee, Y. Hefetz, A.V. Nurmikko: Superlattices and Microstructures **3**, 5 (1987)
8.59 L.A. Kolodziejski, R.L. Gunshor, N. Otsuka, B.P. Gu, Y. Hefetz, A.V. Nurmikko: J. Cryst. Growth **81**, 491 (1987)
8.60 R.L. Gunshor, L.A. Kolodziejski, N. Otsuka, S. Datta: Surf. Sci. **174**, 522 (1986)
8.61 L.A. Kolodziejski, R.L. Gunshor, N. Otsuka, B.P. Gu, Y. Hefetz, A.V. Nurmikko: Appl. Phys. Lett. **48**, 1482 (1986)
8.62 D.L. Partin, R.F. Majkowski, D.E. Swets: J. Vac. Sci. Technol. B **3**, 576 (1985)
8.63 D.L. Partin, B.M. Clemens, D.E. Swets, C.M. Thrush: J. Vac. Sci. Technol. B **4**, 578 (1986)
8.64 P. Norton, M. Tacke: J. Cryst. Growth **81**, 405 (1987)
8.65 N. Koguchi, T. Kiyosawa, S. Takahashi: J. Cryst. Growth **81**, 400 (1987)
8.66 D.L. Partin: J. Electron. Mater. **13**, 493 (1984)
8.67 J. Fuchs, Z. Feit, H. Preier: Appl. Phys. Lett., to be published
8.68 L.J. Schowalter, R.W. Fathauer: J. Vac. Sci. Technol. A **4**, 1026 (1986)
8.69 C.W. Tu, S.A. Ajuria, H. Temkin: Appl. Phys. Lett. **49**, 920 (1986)
8.70 C. Webb, S.L. Weng, J.N. Eckstein, N. Missert, K. Char, D.G. Schlom, E.S. Hellman, M.R. Beasley, A. Kapitulnik, J.S. Harris, Jr.: Appl. Phys. Lett. **51**, 1191 (1987)
8.71 P.D.Dapkus: J. Cryst. Growth **68**, 345 (1984)
8.72 M.R. Leys, C. van der Opdorp, M.P.A. Viegers, H.J. Talen-van der Mheen: J. Cryst. Growth **68**, 431 (1984)
8.73 J.P. Duchemin, S. Hersee, M. Razeghi, M.A. Poisson: "Metal Organic Chemical Vapor Deposition" in *Molecular Beam Epitaxy and Heterostructures,* ed. by L.L. Chang, K. Ploog, NATO ASI Ser., Ser. E, no. 87 (Martinus Nijhoff, Dordrecht 1985)
8.74 N. Kondo, M. Kawashima: "The Enhancement of Surface Atom Migration by Ion Irradiation During MBE", in Inst. Phys. Conf. Ser. no. 79, Int. Symp. on GaAs and Rel. Comp., Karuizawa, Japan, 1985 (Adam Hilger, Bristol 1986) p. 97
8.75 K.H. Müller, J. Vac. Sci. Technol. A **4**, 184 (1986)
8.76 S.B. Ogale, A. Madhukar, M. Thomsen: Appl. Phys. Lett. **51**, 837 (1987)
8.77 B.A. Joyce: Rep. Prog. Phys. **48**, 1637 (1985)
8.78 S.V. Ghaisas, A. Madhukar: CRC Crit. Rev. Solid State Mater. Sci. **14**, 1 (1988)
8.79 T. Nishinaga, K.I. Cho: Jpn. J. Appl. Phys. **27**, L12 (1988)
8.80 K.H. Müller: Surf.Sci. **184**, L375 (1987)
8.81 J.W. Gadzuk, J.K. Norskov: J. Chem. Phys. **81**, 2828 (1984)
8.82 J.W. Gadzuk: Surf. Sci. **184**, 483 (1987)
8.83 A.C. Gossard: IEEE J. QE-**22**, 1649 (1986)
8.84 M. Suemitsu, T. Kaneko, N. Miyamoto: J. Vac. Sci. Technol. A **5**, 37 (1987)
8.85 M. Miyamoto, Y. Sumi, S. Komaki, K. Narushima, H. Ishimaru: J. Vac. Sci. Technol. A **4**, 2515 (1986)
8.86 J.R. Chen, K. Narushima, M. Miyamoto, H. Ishimaru: J. Vac. Sci. Technol. A **3**, 2200 (1985)
8.87 H.E. Elsayed-Ali, G.A. Morou: Appl. Phys. Lett. **52**, 103 (1988)
8.88 Workbook of MBE-V, Sapporo, Japan 1988, 5th Int. Confer. on MBE, to be published in Proceedings Vol., J. Cryst. Growth 1989

Subject Index

379

Printed by Publishers' Graphics LLC
MO20120905-364